Wireless Personal Area Networks

Wiley Series on Wireless Communications and Mobile Computing

Series Editors: Dr Xuemin (Sherman) Shen, *University of Waterloo, Canada*
Dr Yi Pan, *Georgia State University, USA*

The "Wiley Series on Wireless Communications and Mobile Computing" is a series of comprehensive, practical and timely books on wireless communication and network systems. The series focuses on topics ranging from wireless communication and coding theory to wireless applications and pervasive computing. The books offer engineers and other technical professionals, researchers, educators, and advanced students in these fields with invaluable insight into the latest developments and cutting-edge research.

Other titles in the series:

Perez-Fontan and Espiñeira: *Modeling the Wireless Propagation Channel: A Simulation Approach with Matlab*, April 2008, 978-0-470-72785-0

Takagi and Walke: *Spectrum Requirement Planning in Wireless Communications: Model and Methodology for IMT-Advanced,* April 2008, 978-0-470-98647-9

Myung: *Introduction to Single Carrier FDMA*, May 2008, 978-0-470-72449-1

Ippolito: *Satellite Communications Systems Engineering Handbook: Atmospheric Effects on Satellite Link Design,* May 2008, 978-0-470-72527-6

Stojmenovic: *Wireless Sensor and Actuator Networks: Algorithms and Protocols for Scalable Coordination and Data Communication*, December 2008, 978-0-470-17082-3

Qian, Muller and Chen: *Security in Wireless Networks and Systems*, December 2008, 978-0-470-51212-8

Wireless Personal Area Networks

Performance, Interconnections and Security with IEEE 802.15.4

Jelena Mišić and Vojislav B. Mišić

University of Manitoba, Canada

John Wiley & Sons, Ltd

Other Wiley Editorial Offices

John Wiley & Sons Inc., 111 River Street, Hoboken, NJ 07030, USA

Jossey-Bass, 989 Market Street, San Francisco, CA 94103-1741, USA

Wiley-VCH Verlag GmbH, Boschstr. 12, D-69469 Weinheim, Germany

John Wiley & Sons Australia Ltd, 42 McDougall Street, Milton, Queensland 4064, Australia

John Wiley & Sons (Asia) Pte Ltd, 2 Clementi Loop #02-01, Jin Xing Distripark, Singapore 129809

John Wiley & Sons Canada Ltd, 6045 Freemont Blvd, Mississauga, Ontario, L5R 4J3, Canada

Wiley also publishes its books in a variety of electronic formats. Some content that appears
in print may not be available in electronic books.

Library of Congress Cataloging-in-Publication Data

Mišić, Jelena
  Wireless personal area networks : performance, interconnections and
security with IEEE 802.15.4 / Jelena Mišić and Vojislav B. Mišić
     p. cm.
  Includes bibliographical references and index.
  ISBN 978-0-470-51847-2 (cloth)
  1. Personal communication service systems – Standards. 2. Wireless
LANs. 3. Bluetooth technology. I. Mišić, Vojislav B. II. Title.
  TK5103.485.M575 2007
  621.384 – dc22

                            2007033390

British Library Cataloguing in Publication Data

A catalogue record for this book is available from the British Library

ISBN   978-0-470-51847-2 (HB)

Typeset in 10/12pt Times by Laserwords Private Limited, Chennai, India

To Bratislav and Velibor

Contents

About the Series Editors

Xuemin (Sherman) Shen (M'97-SM'02) received his B.Sc degree in electrical engineering from Dalian Maritime University, China, in 1982, and the M.Sc. and Ph.D. degrees (both in electrical engineering) from Rutgers University, New Jersey, USA, in 1987 and 1990 respectively. He is a Professor and University Research Chair, and the Associate Chair for Graduate Studies, Department of Electrical and Computer Engineering, University of Waterloo, Canada. His research focuses on mobility and resource management in interconnected wireless/wired networks, UWB wireless communications systems, wireless security, and ad hoc and sensor networks. He is a co-author of three books, and has published more than 300 papers and book chapters on wireless communications and networks, control and filtering. Dr. Shen serves as a Founding Area Editor for *IEEE Transactions on Wireless Communications*; Editor-in-Chief for *Peer-to-Peer Networking and Application*; Associate Editor for *IEEE Transactions on Vehicular Technology*; *KICS/IEEE Journal of Communications and Networks*, *Computer Networks; ACM/Wireless Networks*; and *Wireless Communications and Mobile Computing* (Wiley), etc. He has also served as Guest Editor for *IEEE JSAC, IEEE Wireless Communications*, and *IEEE Communications Magazine*. Dr. Shen received the Excellent Graduate Supervision Award in 2006, and the Outstanding Performance Award in 2004 from the University of Waterloo, the Premier's Research Excellence Award (PREA) in 2003 from the Province of Ontario, Canada, and the Distinguished Performance Award in 2002 from the Faculty of Engineering, University of Waterloo. Dr. Shen is a registered Professional Engineer of Ontario, Canada.

Dr. Yi Pan is the Chair and a Professor in the Department of Computer Science at Georgia State University, USA. Dr. Pan received his B.Eng. and M.Eng. degrees in computer engineering from Tsinghua University, China, in 1982 and 1984, respectively, and his Ph.D. degree in computer science from the University of Pittsburgh, USA, in 1991. Dr. Pan's research interests include parallel and distributed computing, optical networks, wireless networks, and bioinformatics. Dr. Pan has published more than 100 journal papers with over 30 papers published in various IEEE journals. In addition, he has published over 130 papers in refereed conferences (including IPDPS, ICPP, ICDCS, INFOCOM, and GLOBECOM). He has also co-edited over 30 books. Dr. Pan has served as an editor-in-chief or an editorial board member for 15 journals including five IEEE Transactions and has organized many international conferences and workshops. Dr. Pan has delivered over 10 keynote speeches at many international conferences. He is an IEEE Distinguished Speaker (2000–2002), a Yamacraw Distinguished Speaker (2002), and a Shell Oil Colloquium Speaker (2002). He is listed in *Men of Achievement*, *Who's Who in America*, *Who's Who in American Education*, *Who's Who in Computational Science and Engineering*, and *Who's Who of Asian Americans*.

List of Figures

List of Tables

Preface

Wireless personal area networks and wireless sensor networks are rapidly gaining popularity, and the IEEE 802.15 Wireless Personal Area Working Group has defined no less than three different standards so as to cater to the requirements of different applications. One of them is the low data rate WPAN known as 802.15.4, which covers a broad range of applications that demand low power, low complexity scenarios typically encountered in home automation, sensor networks, logistics, and other similar applications. The initial standard, adopted in 2003, has enjoyed wide industry support and was even adopted by the ZigBee Alliance as the foundation for the ZigBee specification. In time, and partly because of the requirements of the ZigBee specification, a revised 802.15.4 standard was adopted in September 2006.

While industry support has been quite warm, researchers were slower to follow, and in-depth analyses of the operation and performance of 802.15.4-compliant networks were rather scarce. Reports on the operation of single-cluster 802.15.4 networks became more common only in 2006, while those pertaining to the operation of multi-cluster networks are still counted in single-digit numbers as of the time of this writing; security of 802.15.4 WPANs has also received little attention so far. The aim of this book is to fill this gap by providing sufficient insight into some of the most important aspects of wireless personal area networks with 802.15.4 – their performance, interconnections, and security – which has been our main research focus since 2004, in a single, coherent and informative volume. The book focuses on the MAC layer, where many variables exist that critically affect performance; it does not describe all the details of 802.15.4 technology (the official standard should be used to that effect), various application scenarios of 802.15.4 networks (other books deal with those topics), or the issues related to 802.15.4 communications at the physical layer (which are extensively covered by the research community). Furthermore, it relies on analytical techniques, rather than simulation, whenever possible, since we believe that rigorous mathematical techniques, in particular the tools of queueing theory, provide the best foundation for performance evaluation tasks.

The book is organized into four major parts. Part I consists of two chapters, one of which is devoted to the main tenets of wireless ad hoc networks, and wireless personal area networks and wireless sensor networks in particular, while the other presents a brief overview of the IEEE 802.15.4 standard and highlights some of its many features that will be useful in subsequent discussions.

Part II, most voluminous by far, models and analyzes the performance of single-cluster networks. Chapters 3 and 4 discuss the performance of a single-cluster network in cases with uplink and bidirectional traffic, respectively. Chapter 5 presents some shortcomings of the MAC layer, as defined by the current standard, that pose performance risks, and discusses small changes in the 802.15.4 specification that could easily alleviate those risks. Chapter 6 discusses activity management using both centralized and distributed algorithms, and shows that a simple and computationally inexpensive distributed activity management algorithm can improve the lifetime of the network. Finally, Chapter 7 discusses issues related to admission control.

Part III deals with performance-related aspects of multi-cluster networks utilizing hierarchical, tree-like topologies; Chapter 8 analyzes the impact of the number of child clusters and the bridge access mode on performance, while Chapter 9 introduces activity management, analyzes its impact, and shows the extension of the network lifetime it affords. Finally, Chapter 10 focuses on the performance of a slightly different multi-cluster topology in which ordinary nodes undertake the role of bridges (routers); advantages and shortcomings of this arrangement, as opposed to the hierarchical one used in Chapters 8 and 9, are presented and discussed.

Part IV introduces security issues in the context of both single- and multi-cluster networks. Chapter 11 presents security-related facilities provided by the most recent 802.15.4 standard, as well as a brief classification of possible attacks at the MAC and PHY layers. Chapter 12 analyzes the impact of the communication overhead caused by periodic key exchange/update on the performance of security-enabled networks.

Finally, Appendix A contains an introduction to the ZigBee standard, while Appendix B provides a brief refresher on the definitions and notation related to probability generating functions and Laplace-Stieltjes transforms thereof.

Parts II, III, and IV conclude with a very brief summary and overview of related work, both by us and by other researchers in the field, aided by an extensive bibliography at the end of the book. While we have done our best to make sure that none of the important contributions are left out, any claim as to exhaustiveness would obviously be an exaggeration, the more so because the problems addressed here are still an active research field and new results appear with increasing frequency.

Acknowledgments

Books like this cannot be written without the help, assistance, and encouragement of others. First and foremost, we are deeply indebted to Professor Xuemin (Sherman) Shen, of University of Waterloo, and Professor Yi Pan, of Georgia State University, who invited us to write this book and supported it most enthusiastically from its very start.

We express our gratitude to Ms. Shairmina Shafi for the simulation experiments on various aspects of single 802.15.4 clusters, done in the course of her MSc thesis work at the University of Manitoba. Some of these results, in particular those related to limitations of the MAC layer, activity management, and admission control, are presented in Chapters 5 through 7; others have helped confirm the analytical results presented in Chapters 3 and 4. We also thank Ms. C. J. Fung and Mr. R. Udayshankar, whose simulation experiments helped confirm the analytical results presented in several chapters of Part II, and Ms. J. Begum, who helped define the taxonomy of attacks presented in Chapter 11.

Those contributions notwithstanding, this book has been devised and written by us only, and we remain responsible for any errors that may have made it to its final version.

Last but not least, we would like to thank our sons Bratislav and Velibor who provide love, encouragement, and inspiration in our lives.

Jelena Mišić
Vojislav B. Mišić

Part I

WPANS and 802.15.4

1

Prologue: Wireless Personal Area Networks

1.1 Wireless Ad Hoc Networks

Wireless ad hoc networks are a category of wireless networks that utilize multi-hop relaying of packets yet are capable of operating without any infrastructure support (Perkins 2001; Ram Murthy and Manoj 2004; Toh 2002). Such networks are formed by a number of devices, possibly heterogeneous, with wireless communication capabilities that connect and disconnect at will. In addition, some or all of those devices may be mobile and are thus able to change their location frequently; ad hoc networks with mobile nodes are often referred to as mobile ad hoc networks, or MANETs. Even without mobility, nodes can join and/or leave an ad hoc network at will, and such networks need to possess self-organizing capability in terms of media access, routing, and other networking functions. Ad hoc networking includes such diverse applications as mobile, collaborative, and distributed computing; mobile access to the Internet; wireless mesh networks; military applications; emergency response networks; and others.

The design and deployment of those networks present a number of challenges which do not exist, or exist in rather different forms, in traditional wired networks:

- Self-organization, since individual nodes in an ad hoc networks must be able to attach to, and detach from, such networks at will, and without any fixed infrastructure. Protocols that can support and facilitate the tasks of topology construction, re-configuration, and maintenance, as well as routing, traffic monitoring and admission control, are needed.

- Scalability of the network refers to its ability to retain certain performance parameters regardless of large changes in the number of nodes deployed in that network. This is highly dependent on the amount of overhead at various layers (physical, medium access control, networking/routing, transport) of the network protocol stack.

Wireless Personal Area Networks   Jelena Mišić and Vojislav B. Mišić
© 2008 John Wiley & Sons, Ltd

- Delay is the critical parameter in certain types of applications, e.g., in military applications such as battlefield communications or detection and monitoring of troop movement, or in health care applications where patients with serious and urgent medical conditions must be continuously monitored for important health variables via ECG, EEG, or other probes. Low delays can be achieved by bandwidth reservation, scheduling, or through some kind of admission control; the last two mechanisms require the presence of a controller or coordinator to monitor and prevent network congestion.

- Throughput is the most important performance target in a number of collaborative, distributed computing applications and in mobile access to the Internet, which might include significant amounts of multimedia traffic. At the PHY (physical) layer level, throughput may be impaired by packet errors caused by noise and interference. At the MAC (Medium Access Control) level, throughput may be impaired by collisions, if a contention-based medium access mechanism is used, or by unfairness, if bandwidth reservation- or scheduling-based access mechanism is used. (Detailed descriptions of these mechanisms can be found below.) Cross-layer optimization that accounts for those effects – preferably, all of them – may be needed in order to achieve high throughput.

- Packet and data losses. Loss of information is not tolerated in ad hoc networks, and active measures to restore reliability of data transfers must be undertaken both at the MAC and at the upper layers.

- Fairness among different nodes, applications, and/or users is also of importance.

- Power management is important when nodes operate on battery power, although facilities to recharge the batteries may be readily available at home or in the office.

- Finally, all maintenance tasks in ad hoc networks should be automated, or (at worst) simple enough to be undertaken by non-specialist human operators such as owners of laptop computers and PDAs.

1.2   Design Goals for the MAC Protocol

The medium access control (MAC) protocol is that part of the overall network functionality that deals with problems of achieving efficient, fair, and dependable access to the medium shared by a number of different devices (Stallings 2002). The role of the MAC protocol is particularly important in wireless networks which differ from their wired counterparts in many aspects. The most important among those differences stem from the very nature of the wireless communication medium, where two devices need not be explicitly connected in order to be able to communicate–instead, it merely suffices that they are within the radio transmission range of each other.

For example, when two or more packets are simultaneously received, the receiver may encounter problems. At best, the unwanted packets are treated as noise which impairs the reception of the packet intended to be received but can be filtered out. At worst, the correct packet may be damaged beyond repair and the receiver may be unable to make any sense out of it; this condition is referred to as a collision. Collisions waste both network bandwidth and power resources of individual devices, transmitters and receivers alike, and active measures should be taken to reduce the likelihood of their occurrence.

Common approaches for collision minimization in wired networks include detection and avoidance. Collision detection is widely used in wired networks, where it involves the simple act of listening while transmitting. However, this is not feasible in wireless communication, where few devices are equipped with the required capability (Stallings 2002). Furthermore, packet collisions in wireless networks may occur in scenarios that cannot occur in wired ones, such as the so-called hidden and exposed terminal problems (Ram Murthy and Manoj 2004).

Since collision detection is not available, MAC protocols for wireless networks must rely on collision avoidance techniques, which include explicit scheduling, bandwidth reservation, and listening to the medium before attempting to transmit a packet. This last procedure is commonly known as clear channel assessment (IEEE 2003a, 2006; O'Hara and Petrick 1999), although other terms may be occasionally encountered as well.

Obviously, MAC protocols in wireless networks face both traditional challenges encountered in wired networks and new ones that stem from the use of the wireless communication medium. According to Ram Murthy and Manoj (2004), the most important features of MACs in ad hoc wireless networks can be summarized as follows:

- The operation of the protocol should be distributed, preferably without a dedicated central controller. If the use of such a controller cannot be avoided, the role should be only temporary, and devices with appropriate capabilities must be allowed to undertake it for a certain period of time.

- The protocol should be scalable to large networks.

- The available bandwidth must be utilized efficiently, including the minimization of packet collisions and minimization of the overhead needed to monitor and control network operation. In particular, the protocol should minimize the effects of hidden and exposed node problems.

- The protocol should ensure fair bandwidth allocation to all the nodes. Preferably, the fairness mechanism should take into account the current level of congestion in the network.

- The MAC protocol should incorporate power management policy, or policies, so as to minimize the power consumption of the node and of the entire network.

- The protocol should provide quality of service (QoS) support for real-time traffic wherever possible. Real-time, in this context, implies data traffic with prescribed performance bounds; these may include throughput, delay, delay jitter, and/or other performance indicators.

Two additional issues deserve to be mentioned. First issue is time synchronization among the nodes, which is required for the purpose of bandwidth reservation and allocation. Time synchronization is usually achieved by having one of the nodes periodically broadcast some sort of synchronization signal (the beacon) which is then used by other nodes. While the use of periodic beacon transmissions facilitates the process of placing the reservation requests and subsequent broadcasting of reservation allocations, it requires that some node is capable of, and willing to, act as the central controller – somewhat contrary to the distributed, self-organizing character of an ad hoc network. In particular, additional provisions must be made to replace the controller node when it departs from the network or experiences a failure; this is part of the self-healing property of ad hoc networks described above. Furthermore, the use of beacons consumes the bandwidth and affects the scalability of the MAC algorithm.

The second issue is related to the interference from neighbouring nodes. As this interference is harmful, steps have to be taken to reduce it, most often through appropriate multiplexing techniques. According to Stallings (2002), multiplexing techniques are available in the following domains:

- in the frequency domain (FDMA), wherein different frequency bands are allocated to different devices or subnetworks;

- in the code domain (CDMA), wherein different devices use different code sequences;

- in the time domain (TDMA), wherein different devices transmit at different times; and/or

- in the space domain, where the range and scope of transmissions are controlled through the use of transmitter power control and directional antennas, respectively.

Strictly speaking, all these techniques belong to the PHY layer; while the MAC layer is completely oblivious to the first two techniques, it can utilize the latter two (multiplexing in time and space domain), or even integrate them to a certain extent. (For example, time multiplexing is a close relative of scheduling.) This cross-layer integration and optimization allow the MAC protocol to better address the requirements outlined above. We note that such integration is not too common in ad hoc networks, where the MAC layer is more likely to cooperate with the network and, possibly, transport layers above it, than with the PHY layer below; however, MAC protocols exist that make use of it (Ram Murthy and Manoj 2004).

1.3   Classification of MAC Protocols for Ad Hoc Networks

Before we present some of the important MAC protocols for wireless ad hoc networks, we will give a brief overview of some among the possible criteria for classifying those protocols; the reader will thus be able to grasp main features of different MAC protocols and identify the important similarities as well as differences among them.

Mechanism for accessing the medium.   Probably the most intuitive among the classification criteria is the manner of accessing the medium, which comes in three main flavours:

- Contention-based protocols are those in which a potential sender node must compete with all others in order to gain access to the medium and transmit its data.

- Bandwidth reservation-based protocols have provisions for requesting and obtaining bandwidth (or time) allocations by individual senders.

- Finally, scheduling-based protocols, in which the transmissions of individual senders are scheduled according to some predefined policy which aims to achieve one or more of the objectives outlined above, such as the maximization of throughput, fairness, flow priority, or QoS support.

Note that the third option requires the presence of an entity which is responsible for implementing the aforementioned policy. In most cases, this requirement translates into the requirement for a permanent or temporary central controller. Note also that the policy to be pursued should be adaptive, depending on the traffic and/or other conditions in the network. The presence of a central controller is sometimes needed in protocols that use the second option as well.

Quite a few among the existing MAC protocols offer more than one of those mechanisms. This may be accomplished by slicing the available time into intervals of fixed or variable size, referred to as cycles or superframes (IEEE 2003a, 2006; O'Hara and Petrick 1999), and assigning certain portions of those intervals to different categories of access from the list above. For example, the IEEE 802.11 Point Coordinator Function (PCF) uses superframes in which the first part is reserved for (optional) contention-free access, while the second part is used for contention-based access (ANSI/IEEE 1999; O'Hara and Petrick 1999). A similar approach is adopted in the IEEE 802.15.4 protocol in its beacon enabled, slotted CSMA-CA mode (IEEE 2006), except that the contention access period precedes the contention-free period in the superframe. More details on the structure of the superframe are presented in the next chapter.

On the other hand, some MAC protocols offer optional features which modify the manner in which the protocol operates, and effectively introduce a different mechanism for medium access control. For example, the IEEE 802.11 Distributed Coordinator Function (DCF) utilizes pure contention-based access in its default form, but allows bandwidth reservation on a per-packet basis through the optional RTS/CTS handshake (ANSI/IEEE 1999).

Alternative classifications on the basis of medium access mechanism. An alternative classification criterion could be devised by assuming that contention-based access will always be present, and then using the presence or absence of the latter two access mechanisms as the basis for classification. This approach results in the common (and marginally more practical) classification into pure contention-based MACs, contention-based MACs with reservation mechanisms, and contention-based MACs with scheduling mechanisms (Ram Murthy and Manoj 2004). A variant of this approach distinguishes between contention- or random access-based protocols, scheduling or partitioning ones, and polling-based ones. Yet even these classifications are neither unambiguous, as the presence of optional features outlined above leads to the same protocol being attached to more than one category, nor comprehensive, as some of the existing protocols cannot be attached to any single category (Ram Murthy and Manoj 2004); on account of these shortcomings, it is listed as an alternative only.

Mechanism used for bandwidth reservation and its scope. These two criteria applies only to MAC protocols that employ some form of bandwidth reservation, and thus actually represent sub-classifications within the previous one based on the mechanism used to access the medium. With respect to the mechanism used for bandwidth reservation, we can distinguish between the protocols that use some kind of handshake, e.g., RTS/CTS, and those that use out-of-band signalling, most notably the Busy Tone approach which is an extension of the familiar concept from the traditional telephony systems.

With respect to the scope of bandwidth reservation, we can distinguish between the protocols which request bandwidth for a specified time (i.e., for a single packet or for a group of consecutive packets, commonly referred to as a burst) and those that request bandwidth allocation for an unspecified time. In both cases, time can be measured in absolute units or in data packets. In the former case, bandwidth allocation is valid for the transmission of a specified number of packets only, while in the latter, it has to be explicitly revoked by some central authority, or perhaps waived by the requester itself.

Another scheme based on the concept related to bandwidth reservation is the family of the so-called multi-channel MAC protocols. Namely, most communication technologies use only one channel out of several available in the given frequency band. Multi-channel MACs exploit this feature to employ channel hopping in order to improve bandwidth utilization and/or reduce congestion.

Presence and scope of synchronization. The presence or absence of time synchronization among the nodes in the network is another criterion that can be used to classify MAC protocols for wireless ad hoc networks. Synchronization, if present, may be required to extend to all the nodes in the network (global synchronization); alternatively, it may apply to just a handful of nodes which are physically close to one another (local synchronization). In the former case, a central controller may be needed to initiate and broadcast the necessary synchronization information.

Synchronization is most often required in protocols that use scheduling or bandwidth reservation, as basic synchronization intervals serve to apportion the available bandwidth to appropriate sender nodes. However, bandwidth reservation and allocation can be accomplished in an asynchronous manner, in particular when reservation is requested on a per-packet basis, while synchronous protocols can be used even with pure contention-based access. For example, the IEEE 802.15.4 protocol in its beacon enabled, slotted CSMA-CA mode without guaranteed time slots uses pure contention-based access, yet all transmissions must be synchronized to the beacon frames periodically sent by the network coordinator (IEEE 2006).

Synchronization is one of the most important factors that may affect scalability of the network. As the size of the network grows, synchronization becomes more difficult and more costly to establish and maintain. In particular, protocols which rely on global synchronization will suffer the most degradation; for example, it has been shown that the construction and maintenance of a globally optimal schedule in a multi-level Bluetooth network (a scatternet) is an NP-complete problem (Johansson et al. 2001).

Presence of a controller and its permanence. Another possible classification criterion is the presence and permanence of a central network controller or coordinator. While wireless ad hoc networks, by default, should be able to function without a permanent or dedicated central controller, quite a few protocols rely on certain monitoring and control functions that can only be provided by a local or global controller. This is the case with several of the MAC protocols that use bandwidth reservation, as well as with all of the MAC protocols which use scheduling. In fact, even some pure contention-based protocols rely on the presence of a controller for administrative tasks such as time synchronization and sometimes even node admission.

Again, the presence of a controller affects the scalability of the network, as the amount of work the controller has to do – most of which is administrative and control overhead – must grow with the number of nodes. Hierarchical decomposition or layering is often used to reduce this overhead, but it leads to additional problems regarding synchronization and delays.

Interdependence of the classification criteria. As can be seen, not all of the classification criteria outlined above are entirely independent of each other; rather, they exhibit a certain overlap or redundancy. Still, they are useful in the study of MAC protocols, as they tend to highlight different aspects of their design and operation.

1.4 Contention-Based MAC Protocols

We will now look at two contention-based MAC protocols: the basic CSMA protocol and the IEEE 802.11 DCF. They are interesting because the 802.15.4 protocol uses a variant of CSMA which is rather similar to those two. While many other protocols exist, contention-based, polling-based, and those that use bandwidth reservation, multiple channels, out-of-band signalling, and directional antennas, they are beyond the scope of the present work.

1.4.1 Basic CSMA

Many MAC protocols are derived from the basic Carrier Sense Multiple Access (CSMA) mechanism (Bertsekas and Gallager 1991). CSMA is a pure distributed protocol without centralized control, which operates as follows. The node that wants to transmit a packet first performs the clear channel assessment procedure, i.e., it listens to the medium, for a prescribed time. If the medium is found to be clear (or idle) during that time, the node can transmit its packet. Otherwise, i.e., if another transmission is in progress, the node backs off – i.e., waits for a certain time before undertaking the same procedure again.

Different MAC algorithms use different ways to calculate the time they need to listen to the channel during the clear channel assessment procedure and to calculate the time to wait (i.e., the duration of the backoff period) before the next transmission attempt.

It is possible that the transmissions from two or more nodes overlap in time, which results in a collision and loss of all packets involved. If lossless communication is desired, collisions must be detected so that the lost packets can be retransmitted. Since a collision can be detected only at the receiver side, some form of acknowledgment from the receiver may be needed; some MAC protocols provide this facility, while others leave it to some of the upper layers – most likely, the transport layer. The former approach is more efficient in terms of reaction time, whereas the latter allows for much simpler implementation of the MAC protocol used.

In the basic CSMA protocol, carrier sensing is performed only at the sending node. Therefore, the hidden terminal problem is still present. Moreover, the exposed terminal problem leads to deferred transmissions and thus reduces bandwidth utilization.

1.4.2   IEEE 802.11 MAC

The IEEE 802.11 protocol (O'Hara and Petrick 1999) is, strictly speaking, intended for wireless local area networks (LANs), rather than wireless ad hoc networks. However, it is interesting to examine it in some detail, mainly on account of its ubiquity, and because it uses most of the main concepts which are reused in many MAC protocols for ad hoc networks. The protocol covers the functional areas of access control, reliable data delivery, and security; in the following we will focus on the first two areas, as the last one (security) is beyond the scope of this chapter.

Reliable transfer is achieved through the use of special acknowledgment (ACK) packets or frames, sent by the destination node upon successfully receiving a data packet. Medium access is regulated in two ways, the first of which is a distributed contention-based mechanism known as Distributed Coordination Function (DCF), which does not require a centralized controller. The DCF, based on the CSMA protocol described above, operates as follows. The node that wants to transmit a packet first performs the clear channel assessment procedure, i.e., it listens to the medium, for a time equal to Interframe Space (IFS). If the medium is found to be clear (or idle) during that time, the node can transmit its packet immediately; otherwise, i.e., if another transmission is in progress, the node waits for another IFS period. If the medium remain idle during that period, the node backs off for a random interval and again senses the medium. During that time (referred to as the backoff window or contention window), if the medium becomes busy, the backoff counter is halted; it resumes when the medium becomes idle again. When the backoff counter expires and the medium is found to be idle, the node can transmit the packet.

A possible scenario in which this procedure is applied is shown in Figure 1.1. There are several points worth mentioning. First, the backoff interval is chosen as a random number from a predefined range. After each collision, the range is doubled in order to reduce the likelihood of a repeated collision. After each successful transmission, the range is reset to its initial value, which is typically small. This approach is known as binary exponential backoff, or BEB (Stallings 2002). In this manner, the protocol ensures a certain level of load smoothing in case of frequent collisions caused by heavy traffic.

Second, in order to enhance reliability and avoid the hidden/exposed terminal problems to a certain extent, the RTS/CTS handshake – well known from wired communications – may optionally be used. In this case, the node that wants to send a data packet first sends a Request To Send (RTS) packet to the designated receiver which, if ready, responds with a Clear To Send (CTS) packet. Both RTS and CTS packets contain information about

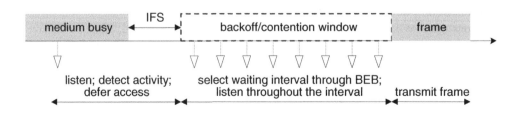

Figure 1.1 Basic access method in IEEE 802.11 DCF.

the duration of the forthcoming transmission, including the optional acknowledgment. Once the sender receives the CTS packet, it may begin actual data transmission, which may optionally be followed by an ACK packet. The RTS/CTS handshake constitutes a simple form of bandwidth reservation on a per-packet or per-group basis, as will be explained below.

Reliability of transmission is enhanced because the RTS and CTS packets are generally much shorter than data packets; if they collide, the time waste is not high – but the risk that subsequent data packets will experience a collision is substantially reduced. The hidden terminal problem is avoided because other nodes within the transmission range of the receiver, upon hearing the CTS packet, become aware of a forthcoming data transmission and defer their transmission for the time interval specified. On the other hand, a transmission from an exposed terminal may prevent the sender from initiating the RTS/CTS handshake. However, once the sender receives a proper CTS packet, it can assume that the receiver is not affected by the interfering transmission and can, thus, proceed with the data packet transmission.

Third, in order to ensure the proper functioning of the protocol, three different IFS intervals are used: a short IFS (SIFS), a medium duration Point Coordination Function IFS (PIFS), and a long duration one, referred to as Distributed Coordination Function IFS (DIFS). The existence of several IFS intervals of different duration actually serves to implement different priority levels for different types of access. The DIFS interval is used for ordinary asynchronous traffic, while the SIFS interval, being the shortest, is used in the following cases:

- When the receiver sends an ACK packet upon successful reception of a data packet; in this manner, ACK packets are safe from collisions since regular data packets wait longer.

- When the sender wants to send another data packet upon receiving an ACK packet for a previous one. In this manner, a burst of packets (commonly obtained by segmenting a longer packet from the upper layers) can be delivered quickly and with little risk from collision. However, such transmissions can result in unfairness, since there is limit on the duration of the burst that can be transmitted.

- When the node sends a CTS packet upon receiving a RTS packet from a prospective sender; again, the use of the SIFS interval minimizes the risk that the CTS packet will experience a collision.

The PIFS interval is used in an alternative access method known as the Point Coordination Function (PCF), which is implemented on top of DCF. The PCF requires the presence of a central point coordinator, hence the name. The point coordinator defines an interval known as superframe. In the first part of the superframe, the coordinator issues polls to all nodes configured for polling. The polls are sent using the regular CSMA algorithm outlined above. When a poll packet is sent, the polled node may respond using the SIFS interval. If the coordinator receives the response, it issues another poll but using the PIFS interval. The polling continues in round-robin fashion (i.e., one node at a time), until all the nodes are polled. Then, the point coordinator remains idle until the end of the superframe, which allows for DCF-style contention-based access by all other nodes. The duration of the superframe is fixed, but an ongoing transmission may force the coordinator to defer the beginning of a polling cycle; in this case the useful duration of the superframe will be reduced.

While the IEEE 802.11 DCF is able to deal with asynchronous traffic, the presence of synchronous traffic with specified (and reasonably stable) throughput over prolonged periods of time is well served by its PCF counterpart. Still, the PCF functionality is designated as an optional facility in the 802.11 standard (ANSI/IEEE 1999), and it is rarely used in practice.

1.5   New Kinds of Ad Hoc Networks

Recently, new families of wireless ad hoc networks for specialized applications have emerged, most notably sensor and personal area networks.

Wireless sensor networks, or WSNs, are aimed at monitoring environmental phenomena (e.g., temperature, humidity, light but also the presence of a specific object or movements of persons and objects) in a given physical space. Such networks find increasing use in areas as diverse as military applications, object surveillance, structural health monitoring, and agriculture and forestry, among others.

Wireless Personal Area Networks, or WPANs, are intended to provide advanced capabilities such as cable replacement, interconnection of various electronic devices, monitoring of physical parameters on the human body, and the like, all within a person's workspace. Different application areas for WPANs have widely differing requirements in terms of data rate, power consumption, and quality of service, such networks are typically classified into the following three classes:

- High data rate WPANs are needed for real-time and multimedia applications. Such applications are supported through the IEEE 802.15.3 standard (IEEE 2003a), with the maximum data rate of 55 Mbps (megabits per second).

- Medium data rate networks for cable replacement and consumer devices. This was the original use of WPANs, as envisioned in the IEEE 802.15.1 (Bluetooth) communications standard. The original Bluetooth specification (Bluetooth SIG 2003; IEEE 2002) allowed raw data rates of up to 1 Mbps, but recent improvements allow data rates of up to 3 Mps (Bluetooth SIG 2004; IEEE 2005).

- Finally, low data rate WPANs are intended for use in wireless sensor networks and other similar application scenarios. A typical example of a LR-WPAN is the 802.15.4 standard (IEEE 2003b, 2006), which allows data rates of up to 250 kbps (kilobits per second).

In this book, we will focus on the performance of WPANs that utilize the 802.15.4 standard in its various configurations.

1.6   Sensor Networks

Sensor networks are a class of wireless networks intended for monitoring environmental phenomena in a given physical space; such networks find increasing use in areas as diverse as military applications, object surveillance, structural health monitoring, and agriculture and forestry, among others. Monitoring may be continuous, with a prescribed data rate which may change over time; it may also be triggered by an explicit demand from a controlling node or a specific event in the environment. Environmental phenomena to be monitored include simple physical variables such as temperature, humidity, light, pressure, pH value, and the like; but other phenomena such as the presence or absence of a specific object (say, an inventory item with a RFID tag), or movements of persons and objects (e.g., cars) can be monitored as well. The spaces to be monitored include rooms, hallways, foyers, homes, backyards, streets, larger buildings and structures (e.g., bridges), but also open spaces such as fields or forests. Sensor nodes can be deployed in large numbers, from tens through hundreds to even thousands. Sensor networks are often expected to operate autonomously, with little or no human intervention, for prolonged periods of time. Sensor nodes are seldom mobile, and even when mobility is present, not all of the nodes are equipped with appropriate capabilities. Given such a diverse set of applications and requirements, it should come as no surprise that the constraints which guide the design and deployment of wireless sensor networks differ, sometimes substantially, from those that hold in wireless ad hoc networks (Achir and Ouvry 2004; Sohrabi et al. 2000). Let us now discuss those constraints in more detail.

Energy efficiency. Probably the most important difference is due to the fact that sensor nodes typically operate on limited battery power, which means that the maximization of network lifetime (and, consequently, minimization of power consumption) is a sine qua non for sensor networks. On the contrary, power consumption is seldom the critical requirement for ad hoc networks.

According to Jones et al. (2001), the constraint of minimal energy consumption translates into two distinct, yet closely related design requirements:

1. The communication efficiency has to be maximized through the design of simple yet flexible and effective communication protocols and functions.

2. Those protocols and functions have to be implemented by small chips with limited computational and memory resources.

Simultaneous achievement of these objectives necessitates some kind of cross-layer protocol optimization in which the MAC layer would use the information obtained from, and control the operation of the PHY layer. At the same time, optimal operation of the upper, network and transport layers requires the knowledge of appropriate information from both the PHY and MAC layers. Again, such tight integration is not too common in ad hoc networks.

An important consequence of the requirement for energy efficiency is the limited transmission range of most sensor node radio subsystems; few real devices have a transmission range of more than 100 meters (300 feet), and ranges of 10 meters (30 feet) and even less are not uncommon.

Protocol efficiency. Regarding communication protocols, the main sources of inefficiency are packet collisions, but also overly complex handshake protocols, receiving packets destined for other nodes, and idle listening to the medium (Ye et al. 2004). Actual power consumption of sensor nodes, often called motes, depends mostly on the radio subsystem and its operating mode. In most (but not all) cases, transmitting and receiving use about the same amount of energy, depending on the power level used for transmission. However, most savings can be made by putting the node to sleep, when power consumption drops by one to two orders of magnitude, depending on the hardware (Jung and Vaidya 2005; van Dam and Langendoen 2003).

Use of redundant sensors. Since nodes are small and cheap to produce and the network lifetime needs to be maximized, it is often feasible to deploy the sensors in a given physical space in much larger numbers than necessary to obtain the desired rate of information flow. If redundant sensors are used, they can be periodically sent to sleep in order to minimize their duty cycle, which extends the lifetime of individual sensors and of the entire network and reduces or eliminates the need for operator intervention, thus reducing the operational cost of the network (Akan and Akyildiz 2005). The use of redundant sensors has profound implications on the design of MAC protocols, as will be seen below.

Node specialization. Another important distinction is related to the role of individual nodes. An ad hoc network allows its nodes to choose the specific role, or roles, they would like to play – i.e., data source, destination, or intermediate router – at any given time. In most cases, a node is free to switch to a different role, or roles, whenever it finds appropriate or is instructed to do so by the specific application currently executing on it. On the contrary, nodes in a sensor network have specific roles that do not change often, or never change at all. Most of the nodes act as sensing nodes, some act as intermediaries which route the traffic and (possibly) perform some administrative duties, and a small number of nodes (sometimes only a single node) act as the network sink (or sinks) toward which all the sensed data ultimately flows (Akyildiz et al. 2002). A group of sensor nodes under the control of an intermediary is sometimes referred to as a sub-network or cluster, while the intermediary itself is known as cluster head. We note that the number of intermediate levels interposed between the sensing nodes and the network sink(s) depends on a number of variables such as the size of the network, the size of the physical space which the network has to monitor, the transmission range of individual nodes, and (to some extent) the actual MAC protocol used.

Traffic characteristics. The traffic in sensor networks is rather asymmetric, as the bulk of it flows from the sensing nodes toward the network sink (this is often referred to as the uplink direction). The traffic in the opposite direction is generally much smaller and consists of control information and, possibly, queries issued by the network sink on behalf of the corresponding sensing application (Intanagonwiwat et al. 2003). Furthermore, traffic patterns in sensor networks are rather different than in ad hoc networks. For example, temperature or humidity monitoring might require periodic or nearly periodic transmissions – in essence, synchronous traffic with low data rate – while object surveillance and other event-driven sensing applications exhibit low average traffic volume and random bursts with considerably higher peak rates.

Furthermore, data packets are often much smaller in sensor networks. Original data from sensing nodes typically consists of only a few data values reported by appropriate sensors. Intermediate nodes may choose to aggregate those values in order to improve energy efficiency and reduce bandwidth and energy consumption; data aggregation is more common in networks with a larger number of hierarchical levels. At the same time, the number of sensor nodes and their spatial density may be very large, depending on the size of the space to be monitored and the requirements of the sensing application.

Quality of Service requirements.   Delay considerations are of crucial importance in certain classes of applications, for example, in military applications such as battlefield communications and detection and monitoring of troop movement, or in health care applications where patients in special care units must be monitored for important health variables (via ECG or EEG) due to a serious and urgent medical condition. Maintaining prescribed delay bounds in a network of resource-constrained nodes with limited transmission range is a complex issue. Low delays can be achieved either by bandwidth reservation, as utilized in variations of the TDMA approach, or by some kind of admission control that will prevent network congestion, if the CSMA approach is used.

At the same time, the requirement for maximum throughput is relaxed due to the following. First, the exact value of the throughput requirement is usually prescribed by the sensing application, unlike general networks where the goal is to obtain as much throughput as possible. Second, energy efficiency dictates the use of protocols that incorporate power control, which will strive to keep the nodes inactive for as long as possible (Akan and Akyildiz 2005). In order to obtain the desired throughput, it suffices to adjust the mean number of active nodes.

Even packet losses can be catered to in this manner, since we don't care whether a given packet from a given node will reach the network sink – as long as the sink receives sufficient number of packets from other nodes. Any packet loss can be compensated for (in the long term) by varying the mean number of active nodes. In a certain sense, fairness is not needed at the node and packet level as long as it is maintained at the cluster level (Callaway, Jr. 2004). On the contrary, fairness at the node/packet level is important in ad hoc networks.

Differences from ad hoc networks.   The requirements outlined above lead to a number of important differences between sensor networks and ad hoc networks, most notably the following:

- Power efficiency and lifetime maximization are the foremost requirements for sensor networks.

- Self-organization is important in both ad hoc and sensor networks. In the former case, this is due to dynamicity and node mobility, which cause frequent topology changes and make self-organization more difficult; in the latter, this is mostly caused by sensor nodes exhausting their battery power (i.e., dying), although mobile sensors are used in some applications.

- Throughput maximization is often required in ad hoc networks but is not too common in sensor networks.

- Delay minimization is typically assigned much higher priority in sensor networks than in their ad hoc siblings.

- The use of redundant sensors allows for a certain level of fault tolerance; on the contrary, packet losses are intolerable in ad hoc networks.

- Scalability is an important issue due to the potentially large number of sensors; scalability is also important in ad hoc networks, but it is limited by the available bandwidth and the desired throughput.

- Nodes in ad hoc networks are often mobile, while most sensor networks have no mobile nodes.

In more than one sense, wireless ad hoc networks are a class of networks with flexible topology but without infrastructure, that should cater to all kinds of networking tasks. On other hand, sensor networks are highly specialized networks that perform a rather restricted set of tasks under severe computational and communication restrictions.

2

Operation of the IEEE 802.15.4 Network

From its inception, 802.15.4 technology was intended to be the key enabler for low complexity, ultra low power consumption, and low data rate wireless connectivity among inexpensive fixed, portable and moving devices (IEEE 2006). Thus it is an ideal candidate for the implementation of low data rate wireless personal area networks as well as wireless sensor networks. Let us now describe the basic characteristics of 802.15.4 communication technology in more detail.

2.1   Physical Layer Characteristics

We start with the physical (PHY) layer of the 802.15.4 protocol stack. IEEE 802.15.4 networks utilize three RF (radio frequency) bands: 868 to 868.6 MHz, 902 to 928 MHz and 2400 to 2483.5 MHz; these will be referred to as 868, 915, and 2450 MHz bands, respectively. The last band is commonly known as the Industrial, Scientific and Medical (ISM) band. Since it does not require licensing, it is used by a number of different communication technologies, including *b* and *g* variants of the 802.11 wireless LAN (also known as Wi-Fi) standard, various WPAN standards such as 802.15.1 (Bluetooth) and 802.15.3, but also other devices such as microwave ovens. While the possibility of unlicensed work is certainly attractive, the possibility that many devices using different communication technologies may be present means that the level of noise and interference might be rather high.

In the original standard (IEEE 2003b), frequency bands at 868 and 915 MHz utilized Direct Sequence Spread Spectrum (DSSS) with a comparatively low chip rate and binary phase shift keying (BPSK) modulation, which resulted in maximum attainable data rates of only 20 kbps and 40 kbps, respectively. In that case, each data bit represents one modulation symbol which is further spread with the chipping sequence. In the ISM (2450 MHz) band, Orthogonal Quadrature Phase Shift Keying (O-QPSK) modulation, in which four data bits comprise one modulation symbol which is further spread with the 32-bit spreading

Table 2.1 Frequency bands and data rates

PHY option	Frequency (MHz)	Type of modulation	Bit rate (kbps)	Symbol rate (ksymbols/s)
868/915	868-868.6	BPSK	20	20
	902-928	BPSK	40	40
868/915	868-868.6	ASK	250	12.5
(2006)	902-928	ASK	250	50
868/915	868-868.6	O-QPSK	100	50
(2006)	902-928	O-QPSK	250	62.5
2450	2400-2483.5	O-QPSK	250	62.5

Note: PHY specifications from the 2006 standard (IEEE 2006) are optional.

16 channels @2 MHz, located at 2405 + 5(k – 11) MHz, k = 11..26

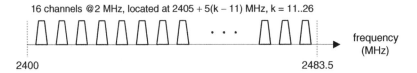

frequency (MHz)

2400   2483.5

Figure 2.1 Channel structure in the ISM band.

sequence, is used before spreading. As a result, the maximum raw data rate in this band is 250 kbps.

The actual types of spread spectrum and modulation techniques, together with the resulting data rates, are shown in Table 2.1.

In the revised standard (IEEE 2006), additional combinations of spread spectrum and modulation techniques were introduced to allow data rates of up to 250 kbps in the lower RF bands.

The original standard divided the available spectrum in the three bands into a total of 27 channels:

- channel $k = 0$, at the frequency of 868.3 MHz;

- channels $k = 1 .. 10$, at frequencies $906 + 2(k - 1)$ MHz; and

- channels $k = 11 .. 26$ in the ISM band, at frequencies $2405 + 5(k - 11)$ MHz.

Channel allocation in the ISM band is illustrated in Figure 2.1.

Each device must support at least one PHY option, and it must support all channels specified for the supported option. It is worth noting that, in some regions or countries, not all channels within a given PHY option may be allowed by regulations.

With additional spread spectrum and modulation choices introduced in the revised standard (IEEE 2006), the concept of channel pages was introduced. In this setup, a total of 32 channel pages are defined:

- channel page 0 contains the 27 channels from the original standard;

- channel page 1 contains 11 channels (0 to 10) in the 868/915 PHY that use ASK modulation;

- channel page 2 contains 11 channels (0 to 10) in the 868/915 PHY that use O-QPSK modulation;

- channel pages numbered 3 to 31, as well as the higher-numbered channels (11 to 26) in pages 1 and 2, are reserved for future use.

We will now mention a few other characteristics of the PHY layer specification which are relevant to our further discussions.

The minimum values for the long and short interframe spacing interval (LIFS and SIFS, respectively) are fixed at 40 and 12 symbol periods, respectively, in all PHY options listed above.

The PHY layer can handle protocol data units (i.e., packets) with a payload of up to 127 bytes or octets each.

The time required for the radio subsystem to switch from transmitting (TX) to receiving (RX) mode or vice versa must not exceed the value of *aTurnaroundTime*, the default value of which is 12 symbols.

The standard contains other common provisions such as the ability to adjust the transmitter power, the ability to measure the strength and/or quality of the received signal for each packet (through the so-called Link Quality Indicator, or LQI), and the ability to check for the activity on the medium. This last feature, known as Clear Channel Assessment or CCA, is used to guide the behavior of both slotted and unslotted versions of the CSMA-CA algorithm, as described in Sections 2.3 and 2.7.1, respectively. The CCA can operate in one of three modes:

- in Mode 1, the receiver measures the received energy and reports that the channel is busy if this energy exceeds a predefined threshold;

- in Mode 2, the receiver reports the channel is busy if it is able to detect a carrier signal that uses the same modulation and spreading characteristics as the PHY option currently employed by the device, regardless of its energy level;

- in Mode 3, the medium is deemed busy if a combination of a valid carrier signal and energy above the predefined level is detected; the combination operator may be AND or OR.

Regardless of the chosen CCA mode, the detection time is equal to 8 symbol periods.

In an IEEE 802.15.4-compliant WPAN, a controller device commonly referred to as the PAN coordinator builds a personal area network or cluster with other devices or nodes within a small physical space known as the personal operating space. Two topologies are supported: in the star topology network, shown in Figure 2.2(a), all communications, even those between the devices themselves, must go through the PAN coordinator. In the peer-to-peer topology, shown in Figure 2.2(b), ordinary devices can communicate with one another directly (as long as they are within physical range of each other) and/or with the coordinator – which must be present. Let us now investigate the operation of the cluster with the star topology in more detail.

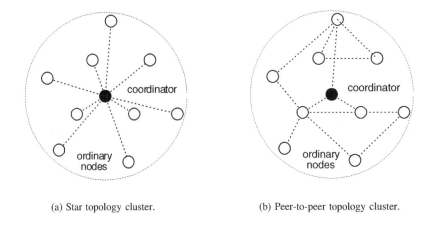

(a) Star topology cluster.   (b) Peer-to-peer topology cluster.

Figure 2.2 Topology of, and communications within, an 802.15.4 cluster. The enclosing circle denotes the transmission range of the cluster coordinator, while the dashed lines denote possible communication links.

In order to accommodate hardware devices with different levels of complexity, the standard distinguishes between two types of devices that can participate in 802.15.4 networks. A full-function device (FFD) is capable of acting as a PAN coordinator, a (cluster) coordinator, or an ordinary device. (The distinction between the two types of coordinators will be described in more detail in Section 2.8.) On the other hand, a reduced-function device (RFD) may function as an ordinary device but not as a coordinator of either type. As a result, an FFD can talk to any device in the network, be it an RFD or another FFD, whereas an RFD can only talk to an FFD; in fact, an RFD may only associate with a single FFD at a time. The use of RFDs is typically limited to simple application scenarios in which individual nodes possess basic communication capability but little computational capability.

2.2   Star Topology and Beacon Enabled Operation

The networks with the star topology use the so-called beacon enabled operating mode, in which the coordinator periodically emits a special frame or packet known as the *beacon frame*. The time between two successive beacon frames is known as the superframe or (more precisely) as the beacon interval. It is divided into an active portion and an optional inactive period. The structure of the superframe is shown in Figure 2.3(a).

All communications in the cluster take place during the active portion of the superframe. Individual nodes can send their data to the coordinator, or receive data from it; these two directions of communication are referred to as *uplink* and *downlink*, respectively.

The active portion of the superframe is divided into equally sized slots, each of which lasts for exactly $2^{SO} \cdot aBaseSlotDuration$ symbols; the *aBaseSlotDuration* contains exactly

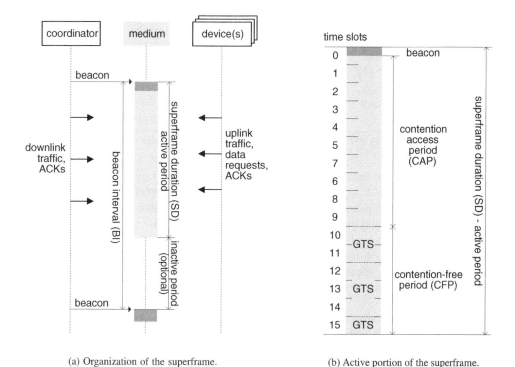

(a) Organization of the superframe.   (b) Active portion of the superframe.

Figure 2.3 Structure of the superframe in a cluster operating in beacon enabled mode.

three backoff periods. The duration of the backoff period is always equal to the time it takes to transmit 20 symbols; this time depends on the PHY option employed, as per Table 2.1.

The beacon frame is transmitted at the beginning of slot 0, and the contention access period (CAP) of the active portion starts immediately afterward. During the CAP, channel access is contention-based and all nodes, including the coordinator, must use the slotted CSMA-CA access mechanism (with a few exceptions described below). Furthermore, a device must complete all of its contention-based transactions within the CAP of the current superframe. The CAP is optionally followed by the contention-free period (CFP), in which an individual device may be granted exclusive access to the medium; this is explained in Section 2.6 below.

Figure 2.3(b) shows the structure of the active portion of the superframe.

During the inactive period (if present), individual nodes, as well as the coordinator, may reduce power consumption, e.g., by turning off the radio subsystem or by switching into low power mode. The inactive period may also be utilized to implement cluster interconnection, as explained in Chapter 8.

Table 2.2   Timing parameters in beacon enabled operating mode

Time period	MAC attribute	Duration (symbols)
Unit backoff period	*aUnitBackoffPeriod*	20
Basic superframe slot	*aBaseSlotDuration*	$3 \cdot aUnitBackoffPeriod = 60$
Superframe slot		$aBaseSlotDuration \cdot 2^{SO}$
Superframe duration	*SD*	$aBaseSuperframeDuration \cdot 2^{SO}$
Beacon interval	*BI*	$aBaseSuperframeDuration \cdot 2^{BO}$

Note: Values of both *BO* and *SO* must be less than 15 in the beacon enabled mode.

The duration of the beacon interval and the active portion of the superframe are controlled through two MAC layer attributes known as the beacon order, *BO*, and superframe order, *SO*, respectively, using the simple formulae presented in Table 2.2. Note that the values of these two attributes must satisfy the constraint $0 \leq SO \leq BO \leq 15$, but the formulae are valid only for values of 14 or below. Namely, when *BO* is set to 15, the coordinator does not transmit beacon frames unless specifically requested to do so, which means that the superframe, strictly speaking, does not exist; in that case, the value of superframe order *SO* is conventionally set to 15. This feature is used in the peer-to-peer topology described in Section 2.7 below.

In order to synchronize with the beacon, each node in a beacon enabled cluster must listen for the beacon for $aBaseSuperframeDuration \cdot (2^{BO} + 1)$ symbols. If a valid beacon frame is not received during that time, the procedure is repeated. If the number of missed beacons exceeds $aMaxLostBeacons = 4$, the MAC layer assumes that synchronization is lost and notifies the higher layers of the protocol stack.

All packet transmissions must be synchronized with backoff periods derived from the periodic beacon frames. Consequently, the so-called slotted carrier sense multiple access mechanism with collision avoidance (CSMA-CA) is used as the main medium access mechanism, as described below.

2.3   Slotted CSMA-CA Medium Access

Nodes in clusters that operate in beacon enabled mode must utilize the slotted CSMA-CA access mechanism, with a few exceptions. The flowchart shown in Figure 2.4 describes the slotted CSMA-CA algorithm which is executed when a packet is ready to be transmitted. The algorithm begins by setting the appropriate variables to their initial values:

1. Retry count *NB*, which refers to the number of times the algorithm was required to back off due to the unavailability of the medium during channel assessment, is set to zero.

2. Contention window *CW*, which refers to the number of backoff periods that need to be clear of channel activity before the packet transmission can begin, is set to 2.

3. Backoff exponent *BE* is used to determine the number of backoff periods a device should wait before attempting to assess the channel. If the device operates on battery power, in which case the attribute *macBattLifeExt* is set to true, *BE* is set to 2 or

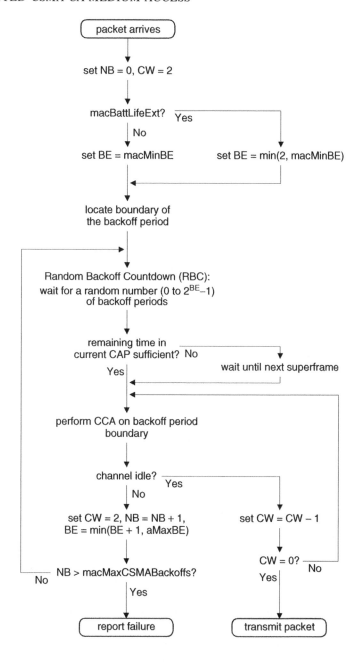

Figure 2.4 Operation of the slotted CSMA-CA algorithm.

to the constant *macMinBE*, whichever is less; otherwise, it is set to *macMinBE*, the default value of which is 3.

Then, the boundary of the next backoff period is located, and a random number in the range $0 .. 2^{BE} - 1$ is generated. The algorithm then counts down for this number of backoff periods; this period is referred to as the Random Backoff Countdown or RBC. During the RBC period, channel activity is not assessed and the backoff counter is not stopped if such activity takes place, unlike the similar CSMA mechanism utilized in 802.11 networks (O'Hara and Petrick 1999). For obvious reasons, the countdown will be suspended during the inactive portion of the beacon interval, and will resume immediately after the beacon frame of the next superframe.

Once the backoff count reaches zero, the algorithm first checks to see whether the remaining time within the CAP area of the current superframe is sufficient to accommodate the necessary number of CCA checks, the actual packet transmission, and subsequent acknowledgment. If this is the case, the algorithm proceeds to perform the CCA checks; otherwise, it pauses until the (active portion of the) next superframe. This feature poses an actual performance risk, as explained in Chapter 5.

CCA check is repeated on *CW* successive backoff period boundaries. It can use any of the three available modes, as explained on p. 19. If all CCA checks pass, the channel is deemed idle and the packet may be transmitted. Otherwise, if any of the CCAs detect activity on the channel, the node concludes that there is an ongoing transmission by another node and the current transmission attempt is immediately aborted. The CSMA-CA algorithm is then restarted; the number of retries, *NB*, and the backoff exponent, *BE*, are incremented by one, while the CCA count, *CW*, is reset to two. Note that the backoff exponent *BE* cannot exceed *macMaxBE*, the default value of which is 5.

However, if the number of unsuccessful backoff cycles *NB* exceeds the limit of *macMaxCSMABackoffs*, the default value of which is 5, the algorithm terminates with channel access failure status. Failure is reported to higher protocol layers, which can then decide whether to abort the packet in question or re-attempt to transmit it as a new packet.

Together, the limit on the number of retries and the manner in which the backoff exponent is incremented, impose a restriction on the range of allowable backoff countdown values. In non-battery-powered operation (when the variable *macBattLifeExt* is false), the random backoff countdown values will not exceed 7, 15, 31, 31, and 31, in successive retries. However, if the node is operating on battery power, the limits of the available range will be between zero and 3, 7, 15, 31, and 31, respectively. Presumably, smaller countdown values will lead to shorter countdowns and, by extension, to lower power consumption and longer battery lifetime.

Note that the backoff unit boundaries of every device should be aligned with the superframe slot boundaries defined by the beacon frame, i.e., the start of first backoff unit of each device is aligned with the beginning of the beacon frame. The MAC layer should also ensure that the PHY layer starts all of its transmissions on the boundary of a backoff unit.

2.4 Acknowledging Successful Transmissions

The use of acknowledgments is optional: they are sent by the receiver only at the sender's explicit request. If requested, the sender should wait for an acknowledgment for at most *macAckWaitDuration*, which amounts to 54 or 120 symbols, again depending on the PHY option utilized. If the acknowledgment packet is not received within *macAckWaitDuration* after the original data frame, the originator may safely assume that the frame has been lost and initiate re-transmission (Figure 2.5). Re-transmission must fit within the remaining time in the CAP portion of the current superframe; otherwise, it is deferred to the CAP portion of the next superframe. The standard allows up to *macMaxFrameRetries* = 3 repeated transmission attempts; if these are not successful, the MAC layer declares a transmission failure and notifies the higher layers of the network protocol stack.

Transmission of an acknowledgment frame must begin at the backoff period boundary between *aTurnaroundTime* and *aTurnaroundTime* + *aUnitBackoffPeriod* after the data frame, which amounts to a delay of 12 to 32 symbol periods, depending on the PHY option utilized. Since one backoff period takes 20 symbols, this time interval may include at most one backoff period at which the channel is assessed idle. However, a node that has finished its random countdown will need at least two CCAs before attempting transmission: while the first one may find the medium idle in between the data frame and the acknowledgment, the second one will coincide with the acknowledgment and cause the CSMA-CA algorithm to revert to the next iteration of the backoff countdown. Consequently, the acknowledgment packet cannot possibly collide with the data packet sent by another node.

2.5 Downlink Communication in Beacon Enabled Mode

As explained above, uplink transmissions in the star topology cluster operating in beacon enabled mode always use the CSMA-CA mechanism outlined above. A node initiates an uplink transmission whenever an application executing on it prepares a packet to be sent to the coordinator. Furthermore, both the original uplink transmission from a node to the coordinator and the subsequent acknowledgment must occur within the active portion of the same superframe.

Data transfers in the downlink direction, from the coordinator to a node, are more complicated. When a downlink packet is received by the MAC layer of the coordinator, it must first announce it to the destination node. The announcement is made through the beacon frame, in the form of a list of nodes that have pending downlink packets. When the destination node learns about a data packet to be received, it undertakes the so-called downlink data extraction procedure as follow. The node transmits a data request packet, which the coordinator must acknowledge by transmitting an appropriate acknowledgment packet. After receiving the acknowledgment, the destination node listens for the period of *aMaxFrameResponseTime*, during which the coordinator must send the data frame. An optional acknowledgment is sent upon successful reception of the downlink data packet. This message exchange is schematically depicted in Figure 2.6.

If the coordinator does not receive proper acknowledgment for a downlink packet, it will not attempt retransmission; instead, the destination node must explicitly request the data frame using a data request packet.

The standard allows the coordinator to send a data frame 'piggybacked' after the request acknowledgment packet, i.e., without using CSMA-CA. However, such transmission is contingent upon the following conditions:

- The coordinator must be able to commence the transmission of the data packet within the interval between *aTurnaroundTime* and *aTurnaroundTime* + *aUnitBackoffPeriod*.

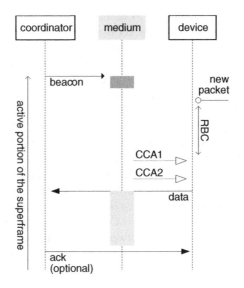

(a) Successful transmission.

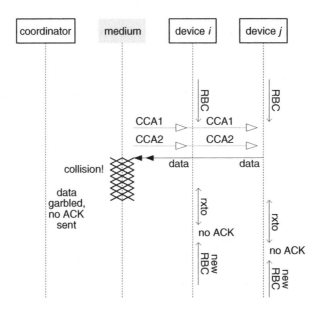

(b) Unsuccessful transmission.

Figure 2.5 Uplink packet transmission, beacon enabled mode.

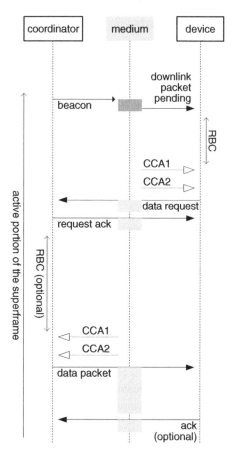

Figure 2.6 Downlink packet transmission, beacon enabled mode.

- The remaining time in the CAP of the current superframe must suffice to send the data frame and receive the acknowledgment, together with the appropriate inter-frame spacing.

If either of these conditions does not hold, the data frame must be sent using the CSMA-CA mechanism (IEEE 2006). The former condition depends on the capabilities of the coordinator hardware, but the latter depends on the actual traffic. Thus, piggybacking of downlink data frames onto the request acknowledgment packets cannot be guaranteed, and some downlink data will ultimately have to be sent using CSMA-CA.

It is worth noting that the node that does not have a pending downlink packet or a queued uplink packet at the time the beacon frame ends, may achieve further power savings by simply disabling its receiver until the next beacon frame.

2.6   Guaranteed Time Slots

In some applications, contention-based access may not offer adequate performance due to uncertainty caused by collisions and collision avoidance. When a node needs contention-free access, it can request a guaranteed time slot (GTS) of appropriate duration. The coordinator then decides whether to accept or reject the request. If the request is accepted, the node can use the bandwidth provided by the GTS for contention-free communication with the coordinator; other nodes are not allowed to transmit any data at that time. Interestingly enough, the node that is granted a GTS is still permitted to use the contention-based access during the CAP.

The coordinator must ensure that all GTSs are contained within the contiguous CFP period at the end of the active portion of the superframe, as shown in Figure 2.3(b). The coordinator must also make sure that the duration of the CAP period after all GTS allocations have been made remains sufficient for contention-based access of other nodes in the cluster. According to the standard, the CAP period must last for at least *aMinCAPLength* = 440 symbols, except temporarily for the purpose of a GTS announcement. Namely, GTS allocations are announced in the beacon frame, and the announcement is repeated for *aGTSDescPersistenceTime* = 4 superframes; this announcement is allowed to extend into the CAP period, thus reducing its effective duration.

The coordinator must be able to store the information needed to manage up to seven GTSs, including the device ID, direction, starting slot within the active portion of the superframe, and length of the GTS. An allocated GTS is reserved for the traffic between the requesting node and the coordinator in either uplink or downlink direction. Hence, a node that needs contention-free communication in both directions must request two such slots; this is the maximum number of such slots that can be allocated to any single node in the cluster. Acknowledgments may be sent within the same GTS that hosted the data frame to be acknowledged, provided the timing constraints are respected and the duration of the GTS is sufficient to accommodate it.

The receiving device, be it the coordinator (for an uplink/transmit GTS) or the node itself (for a downlink/receive GTS) must enable its receiver for the entire duration of the GTS in question. All other nodes need not listen and, thus, can disable their receivers during that time.

If a device with a GTS does not send or receive a data frame within $2n$ superframes, where $n = 2^{8-BO}$, for values of BO between 0 and 8 (both inclusive), and $n = 1$ otherwise, the coordinator may assume that the GTS is no longer used and may deallocate it. (In fact, the coordinator may deallocate a GTS at any time.) When a receive/downlink GTS is used, the coordinator may or may not request that successful transmissions are acknowledged; but in the latter case, it has no way of knowing whether the receiving node is still present in the cluster or not (as the data request packet need not be sent).

While the presence of one or more GTSs allows contention-free access to the nodes in question, it does shorten the CAP and, by extension, reduces the bandwidth available for contention-based access. In this book, we analyze the impact of GTS allocation on performance in the context of multi-cluster networks described in Chapter 8.

2.7 Peer-to-Peer Topology and Non-Beacon Enabled Operation

Networks with peer-to-peer topology use the so-called *non-beacon enabled operating mode* in which the coordinator emits the beacon frame only upon specific request from another device in the network. Any synchronization with the beacon is, therefore, performed with the sole purpose of eliminating the accumulated drift of the device clock, rather than to synchronize activities related to transmission and reception.

Non-beacon enabled mode is activated by setting the values of beacon order and superframe order, *BO* and *SO*, both to 15. In this case, the concepts of the superframe, its active and inactive portion, or contention-free access period, simply do not apply. While the backoff period of 20 symbols' duration is still used as the time unit, individual nodes are free to commence their transmissions at any time, as there is no periodic beacon frame to synchronize with. In this operating mode, individual devices or nodes are allowed to communicate with one another directly, rather than through the coordinator.

2.7.1 Unslotted CSMA-CA medium access mechanism

Medium access in the peer-to-peer topology uses a simpler, unslotted version of the CSMA-CA algorithm, which is described by the flowchart in Figure 2.7.

The meaning of parameters such as backoff retry count, *NB*, and backoff exponent, *BE*, is the same as in the slotted version of the algorithm described in Section 2.3. The main differences from that algorithm are as follows:

- While the countdown duration is determined in the same manner as in the slotted CSMA-CA algorithm, there is no synchronization to the backoff period boundary; the random backoff countdown begins immediately upon the arrival of the data packet from the upper layers of the protocol stack.

- Since there is no superframe, the node can perform the CCA check, followed by the packet transmission and subsequent acknowledgment (if requested), as soon as the random backoff countdown is finished.

- When the random backoff countdown reaches zero, only one CCA check is performed and, if successful, data packet transmission can begin immediately; neither of these activities need to be synchronized to the backoff period boundary.

2.7.2 Uplink and downlink communication

Furthermore, the procedures for uplink and downlink communication differ from the corresponding procedures in beacon enabled networks that use slotted CSMA-CA medium access:

- First, an uplink communication can be attempted as soon as the packet arrives, as shown in Figure 2.8(a).

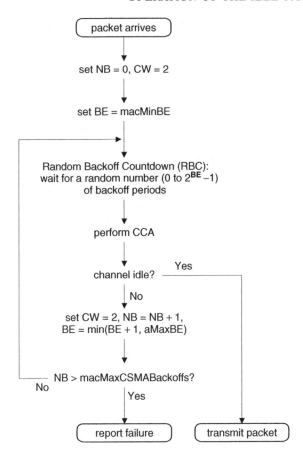

Figure 2.7 Operation of the unslotted CSMA-CA algorithm.

- Second, a node can find out about a pending downlink packet by simply sending a data request packet to the coordinator, which then sends the request acknowledgment and proceeds with sending the actual data packet using unslotted CSMA-CA, as shown in Figure 2.8(b).

It is worth noting that the standard is ambiguous as to the possibility for the coordinator to simply send the downlink data packet to the destination node as soon as it arrives, without waiting for the node to explicitly request it. The reason for this most likely lies in the anticipated application scenarios for 802.15.4 PANs, many of which include the requirement for ordinary nodes to spend prolonged periods of time sleeping. An unsolicited downlink transmission sent to an inactive node may easily exceed the retry count of *mac-MaxCSMABackoffs*, in which case the packet will be dropped. Storing the packet in the outgoing queue and waiting for the destination node to explicitly request it is a less risky solution. However, a packet instructing the node to leave the cluster (referred to as the 'disassociation command') may be sent directly to the device, as explained in Section 2.8.

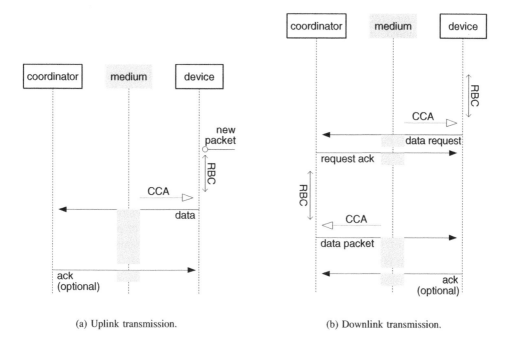

(a) Uplink transmission.

(b) Downlink transmission.

Figure 2.8 Uplink and downlink transmissions, non-beacon enabled mode.

The absence of superframe structure means that a node that wants to save energy by disabling its receiver for a prolonged period must notify all potential transmitters about it, or risk missing some packets transmitted to it during that time. It also means that all transmissions will be subject to contention, as the cluster operating in non-beacon enabled mode and using unslotted CSMA-CA cannot support contention-free access akin to GTS allocation.

In the chapters that follow, we will not consider the unslotted CSMA-CA access mechanism, as it is similar to the one used in IEEE 802.11 standard for which a number of excellent analyses exist in the literature.

2.8 Device Functionality and Cluster Formation

Let us now describe the process through which clusters can be started and maintained.

A device that wishes to start a new cluster or join an existing one must first scan the available channels to find out if there are other clusters or PANs in its personal operating space (i.e., within its transmission range). The process of scanning involves looking for beacon frames transmitted by the coordinators of any other clusters that are present in the personal operating space. This can be accomplished in several ways:

- In the active scan procedure, the device first selects one of the available channels, then sends a beacon request frame, and listens on the channel for a certain time. Only beacon frames received during this time interval are processed; all other frames are simply ignored. Active scanning capability is mandatory for FFDs but optional for RFDs.

- In the passive scan procedure, the device selects the channel and listens for a certain time. Again, only beacon frames received in this way are processed; all other frames are ignored. In this manner, the device can learn about other clusters operating in its personal operating space, and decide whether to start a new cluster or to join an existing one. Passive scanning capability is mandatory for both FFDs and RFDs.

- Scanning can also measure the peak energy in each channel, in which case all received frames are simply ignored. This approach is referred to as Energy Detection (ED) scanning, and it can provide the information about the least used channel, which can then be actively scanned to learn whether there are clusters operating there. Nonetheless, ED scanning is just a shortcut that cannot replace active or passive scanning in the process of looking for operational clusters within the personal operating space. ED scanning is mandatory for FFDs but optional for RFDs.

Once the scan is performed, the device can choose among the following courses of action.

Starting a new cluster. A device that decides to start a new cluster sets the cluster parameters as appropriate and begins broadcasting beacon frames that announce the new PAN to other prospective members. Parameters in question include the PAN identifier (a 16-bit integer set to a value that is currently unused), the operating mode (Sections 2.2 and 2.7) and, in the beacon enabled operating mode, appropriate superframe parameters (Section 2.2). Only a FFD can function as a coordinator.

Joining an existing cluster. A device that decides to join an existing cluster requests admission (or 'association', as it is referred to in the standard) from the cluster coordinator. Upon receiving such a request, the coordinator will first acknowledge it, and then decide whether to admit the requesting device or not. The admission decision is mostly based on the resources available in the cluster, i.e., the number of devices already admitted and their activity (traffic and sleep) patterns. The standard allows the coordinator to specifically disallow admission requests in cases where cluster resources are deemed insufficient to support additional members.

The decision to admit or reject an admission request must be reached within the time interval of *macResponseWaitTime* symbols, and a downlink packet with appropriate content is prepared by the coordinator. If positive, the decision contains the information necessary for the requesting node to participate in the cluster, in particular, the PHY channel page and actual channel, the cluster (PAN) identifier, and a device identifier. Both identifiers are two-byte integers; under certain circumstances, device identifiers are expressed as extended addresses of eight bytes.

It is worth noting that the admission response packet cannot be directly sent to the requesting node; instead, the appropriate procedure for downlink transmission must be initiated by that node. The procedure in question is described in Sections 2.5 and 2.7 for beacon enabled and non-beacon enabled operating modes, respectively. Successful reception of the admission response packet must be acknowledged by the destination node.

Synchronizing with the cluster. All devices that have joined a cluster are required to maintain their synchronization with it. If the cluster operates in beacon enabled mode, this is accomplished by acquiring and maintaining synchronization with periodic beacon frames. If the cluster operates in non-beacon enabled mode, this is accomplished by explicitly requesting the beacon frame from the coordinator, whenever appropriate.

Orphans and realignment. If several communication attempts fail within a certain time, the upper layers of the protocol stack may conclude that the device has been orphaned. The device may then reset the MAC layer and repeat the association procedure above. The device may also attempt to realign itself with the cluster, in which case it performs an orphan scan.

Orphan scanning, which both FFDs and RFDs must be capable of, is performed by switching to the appropriate RF channel and issuing an orphan notification command; the device then listens during the next *macResponseWaitTime* symbol intervals. If the coordinator receives an orphan notification command from a device previously associated with its cluster, it will send a coordinator realignment command which allows the orphaned device to re-establish its membership in the cluster. Orphan notification commands received from a device that was not the member of the cluster are ignored.

If an orphan scan fails to locate the appropriate cluster and its coordinator, the device is free to revert to active or passive scan in order to find another cluster to join, or perhaps start its own cluster.

Extending an existing PAN. The revised version of the 802.15.4 standard (IEEE 2006) introduced the possibility of constructing so-called multi-cluster networks in an iterative fashion. Namely, a member of an existing cluster may decide (or be instructed by the coordinator) to extend this cluster by forming a second cluster as its coordinator. In that case, the device broadcasts its own beacon frames periodically, but delayed with respect to the beacon frames sent by the original coordinator. (Note that the coordinator of the second cluster is still a member of the initial cluster.) This process can be repeated as many times as necessary, to form the so-called multi-cluster tree network. An example of a two-cluster tree is shown in Figure 2.9(a).

A notable characteristic of the multi-cluster tree is that all clusters use the same RF channel, which is accomplished in the following manner. The PAN coordinator (i.e., the coordinator of the first cluster) should set its timing parameters in such a way that the inactive period $BI - SD$, is much longer than the active period SD. (In reality, this means that $BO \geq SO + 1$.) The second coordinator then repeats the beacon frame received from the PAN coordinator, after a delay equal to a preset value of *StartTime* $\geq SD$. Note that the values of the beacon order and superframe order, BO and SO, are the same for all clusters in the tree. In this manner, the active intervals of different clusters are effectively interleaved within the superframe. An example of such timing arrangement in a two-cluster tree is shown in Figure 2.9(b).

The coordinator of the first cluster is conventionally designated as the PAN coordinator, while others are simply referred to as coordinators (IEEE 2006). Note that the ordering of clusters and coordinators is purely arbitrary. The standard allows any of the cluster coordinators to undertake the role of the coordinator of the entire PAN; the original PAN coordinator is then relegated to the role of the coordinator.

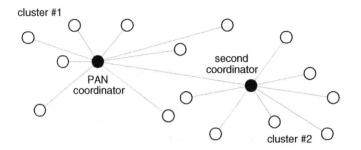

(a) Topology of a two-cluster tree.

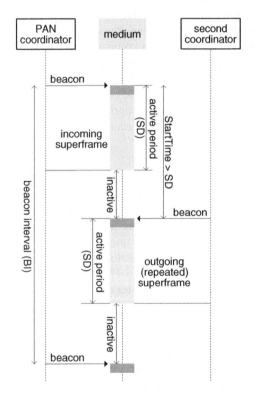

(b) Superframe timing in a two-cluster tree.

Figure 2.9 A two-cluster tree.

The facility described above may be used to build complex multi-cluster PANs in cases where all the nodes in the network are within the transmission range of the PAN coordinator. The presence of multiple coordinators, in this case, facilitates beacon discovery and synchronization.

The facility described above may also be used to build multi-cluster PANs and even span distances that exceed the transmission range of coordinator nodes. Interleaving of active periods allows devices with appropriate capabilities to participate in two or more clusters. The resulting tree may then function as a multi-hop network, which is particularly attractive in a number of PAN and sensor network scenarios.

Leaving the cluster.   A device may decide to leave (or 'disassociate from') the cluster it currently belongs to, which is accomplished by sending a disassociation notification to the coordinator. A device may also be asked to leave by the cluster coordinator, which prepares a disassociation command to be sent (in the form of a downlink packet) to the device. This command may be directly sent to the device, or a regular downlink procedure may be used. In the latter case, the coordinator announces the pending downlink packet in the beacon frame; the device must then explicitly request that this packet is transmitted. Either way, the packet that announces disassociation must be acknowledged by the receiving device.

2.9   Format of the PHY and MAC frames

Finally, we will briefly describe the format of protocol data units (PDUs or packets) at both PHY and MAC layers.

Each PHY layer packet consists of the following elements:

- it begins with a synchronization header, which allows the receiver to synchronize to the incoming transmission;

- it contains a PHY header, which informs the receiver about the size of the subsequent payload;

- it contains a payload – the MAC layer packet.

The PHY header contains a 7-bit field referred to as the Frame Length field. Its value can be between zero and *aMaxPHYPacketSize* (the value of which is 127), but some values are reserved. The only allowed values of Frame Length are 5, for acknowledgment packets, and 9 or above, for other kinds of MAC layer packets.

Each MAC layer packet consists of the header, which provides administrative and security information, a variable length payload, and a footer which contains the frame check sequence. The details of the MAC layer packet structure are shown in Table 2.3.

The Frame Control field announces the frame type, addressing fields, and other control information, as shown in Table 2.4.

Sequence Number field contains the beacon sequence number, in case of a beacon frame, or a data sequence number that is used to match an acknowledgment frame to the corresponding data or MAC command frame.

Destination and Source PAN Identifier fields, as well as the corresponding Address fields, are optional; their presence and format are guided through appropriate settings in

Table 2.3 MAC packet structure

Element	Field	Length (in bytes)
header	Frame Control	2
	Sequence Number	1
	Destination PAN Identifier	0 or 2
	Destination Address	0, 2, or 8
	Source PAN Identifier	0 or 2
	Source Address	0, 2, or 8
	Auxiliary Security Header	0, 5, 6, 10, or 14
payload	frame payload	variable
footer	Frame Check Sequence	2

Table 2.4 Structure of the Frame Control Field in the MAC packet header

Subfield	Bits	Allowed values and their meaning	
Frame Type	0-2	000	Beacon
		001	Data
		010	Acknowledgment
		011	MAC command
Security Enabled	3	1	frame is protected
Frame Pending	4	1	more data is pending
Acknowledgment Request	5	1	acknowledgment is requested
PAN ID Compression	6	1	destination and source PAN identifiers, equal – the latter can be omitted
Destination Addressing Mode	10-11	00	PAN ID and address not present
		10	16-bit short addresses used
		11	64-bit extended addresses used
Frame Version	12-13	00	frame compliant with 2003 standard
		01	frame compliant with 2006 standard
Source Addressing Mode	14-15	00	PAN ID and address not present
		10	16-bit short addresses used
		11	64-bit extended addresses used

Note: Values not shown are reserved for future use.

the Destination and Source Addressing Mode subfields. The short address value of $FFFF_{16}$ (where the subscript 16 indicates that the number is written using hexadecimal digits) indicate a broadcast transmission that should be received by all devices currently listening to the channel.

The Auxiliary Security Header field contains information related to security processing; more details on security provisions of the 802.15.4 standard are given in Chapter 11.

The MAC footer contains a two-byte ITU-T CRC value which is calculated using a standard polynomial of degree 16. It protects the MAC header and the payload.

Beacon frame format. The header for a beacon frame contains the following fields: Frame Control, Sequence Number, Addressing, and Auxiliary Security Header. The addressing fields do not contain any destination address fields, as the beacon is essentially broadcast to both current and prospective members of the cluster (PAN). The payload contains the superframe specification, the GTS information fields, the Pending Address information fields, and beacon payload.

The Superframe Specification field occupies two bytes (16 bits), and contains the current values of the following subfields:

- Beacon Order and Superframe Order, which determine the duration of the beacon interval and active portion of the superframe;

- the final superframe slot used by the CAP, the duration of which must not be shorter than *aMinCAPLength* symbols;

- Battery Life Extension, which is set to one if frames transmitted to the device emitting the beacon are required to start in or before *macBattLifeExtPeriods* full backoff periods after the IFS following the beacon;

- PAN Coordinator, which is set to one if the device emitting the beacon is the PAN coordinator (as explained in Section 2.8); and

- Association Permit, which is set to one if the coordinator is accepting association requests.

The GTS information, which is optional, consists of the following:

- the GTS specification field, which specifies the number of subsequent GTS descriptors (or, rather, the number of GTSs currently allocated in the CFP) and a GTS Permit subfield which is set to one if the coordinator is currently accepting GTS requests;

- the GTS directions field, actually a 7-bit mask that specifies the directions of currently allocated GTSs in the superframe (one for downlink-only, zero for uplink-only); and

- the GTS List field, which lists the GTS descriptors of currently allocated GTSs (of which there can be up to seven); each descriptor consists of the device short address, the starting slot and the duration of the appropriate GTS.

The Pending Address information, which is optional, list the addresses of devices that currently have messages pending with the coordinator. It consists of the following:

- the Pending Address Specification field, which contains the number of short addresses pending as well as the number of extended addresses pending; followed by

- the actual addresses, of which there may be at most seven, either short or extended ones (but the short ones must appear before the extended ones).

If the coordinator has more than seven pending packets at a given time, it may announce them in successive beacon frames, but no beacon frame can contain more than seven pending addresses.

Data frame format.   The data frame format generally conforms to the rules for the MAC packet outlined above.

Acknowledgment frame format.   The acknowledgment frame contains just the Frame Control Field, the Sequence Number equal to the Sequence Number of the data or MAC command frame being acknowledged, and the Frame Check Sequence.

MAC command frame format.   A number of MAC commands may be transmitted using this format, which generally conforms to the rules for the MAC frame outlined above but with the payload consisting of a Command Frame Identifier and the corresponding Command Payload. The available MAC commands include Association request and response, Disassociation notification, Data request (for extracting the pending downlink packet), Orphan notification, Beacon request (in non-beacon enabled networks), GTS request, Coordinator realignment, and PAN ID conflict notification. The interested reader can find more details on these in the standard (IEEE 2006).

Part II

Single-Cluster Networks

3

Cluster with Uplink Traffic

We begin our analysis by considering a single 802.15.4 cluster with star topology, which operates in beacon enabled, slotted CSMA-CA mode, with uplink traffic only. Where specific parameter values are necessary, we will assume that the cluster operates in the ISM band using the 2450 MHz PHY option as discussed in Section 2.1, although our analysis can easily be applied to clusters that employ any other of the allowed PHY options. The cluster consists of n devices and packet arrivals to each device follow the Poisson process with the mean arrival rate of λ and each node accepts new packets through a buffer with the finite size of L packets. When the buffer is empty, the device will not attempt any transmission; when the buffer is full, the device will reject new packets coming from the upper layers of the protocol stack.

The operation of the cluster is modeled using the theory of discrete time Markov chains and M/G/1/K queues. The model considers uplink transmissions only and includes the impact of different parameters such as the packet arrival rate, packet size, the duration of the inactive period between the beacons, the number of nodes (which are assumed to have identical characteristics), and buffer size at each node. After the initial description of this model, we will proceed to derive basic performance indicators such as the probability of accessing the medium, the probability that the medium is idle, the probability of transmission success, the queue length distribution in individual device, the probability distribution of the packet service time, and the probability distribution of access delay. In all derivations, we will use the backoff period as the time unit.

3.1 The System Model – Preliminaries

We begin by defining the probability generating functions (PGFs) for the variables to be used in our analyses. (A brief refresher of the probability generating functions and Laplace transforms can be found in Appendix B.) Since the cluster uses the slotted CSMA-CA algorithm (Section 2.3), all state changes may be considered to happen at the boundaries of backoff intervals. Therefore, all variables have discrete probability distributions and can be represented as weighted sums of multiples of backoff periods. For all variables, the corresponding PGFs will then be polynomials in z (Grimmett and Stirzaker 1992), e.g.,

$$V(z) = \sum_{z=z_{low}}^{z_{high}} p_k z^k \tag{3.1}$$

for a given variable V.

As for various parameters used to model the cluster, their values will be chosen to correspond to the 2450 MHz PHY option, as follows:

- The basic beacon length is 17 bytes (using short device addresses of four bytes each and including MAC and physical layer headers), and it does not contain GTS announcements or beacon payload. The PGF of the duration of the beacon frame which rounds this value to the closest number of backoff periods is denoted as $B_{ea}(z) = z^2$, and its mean is $\overline{B'_{ea}(1)}$. We will use this PGF to model all uplink transmissions. As downlink transmissions follow a more complex procedure explained in Section 2.5, they must be modeled with a different PGF; this model will be introduced in Chapter 4.

- The PGF of data packet length is denoted with $G_p(z) = z^k$. For simplicity, we assume that packets in 802.15.4 cluster have identical length dictated by the nature of the sensing application. For clusters operating in the ISM band, the packet size is $\overline{G_p} = G'_p(1) = k$ backoff periods or $10k$ bytes. The available range for $k \in (3 \ldots 13)$ is determined by the limitation (introduced by the standard), that the minimum MAC and PHY layer header size is 15 bytes, and that the maximum packet size, including both MAC and PHY layer headers, is 127 bytes.

- The time interval between the end of a packet and the arrival of the corresponding acknowledgment is between *aTurnaroundTime* and *aTurnaroundTime* + *aUnitBackoffPeriod*, which lasts between one and two backoff periods. For simplicity, we rounded it to the next higher integer – two; hence, the PGF of this interval is $t_{ack}(z) = z^2$.

- The duration of the acknowledgment is 11 bytes; again, we round it up to one backoff period. Hence, $G_a(z) = z$ stands for the PGF of the acknowledgment duration, and its mean value is $\overline{G_a} = 1$ backoff period.

3.1.1 Acknowledged vs. non-acknowledged transfer

As explained in Chapter 2, the sender may request that a successful transmission be acknowledged by the receiver. The lack of acknowledgment, then, is interpreted as transmission failure and leads to retransmission of the packet. (Acknowledgments are not used for transmissions during the CFP period of the superframe.) In our analysis, we consider both acknowledged and non-acknowledged transmission, as different applications may need to employ one or the other approach.

Non-acknowledged transfer. In the case of non-acknowledged transmission, the sending node does not request the acknowledgment of the packets which are successfully received by the cluster coordinator, and the coordinator does not acknowledge them. As a result, any packets that might be lost due to noise and interference (at the PHY layer), or collisions at the MAC layer, will not be re-transmitted.

Disregarding the random backoff countdown, the PGF for the transmission time of a data packet is $D_d(z) = z^2 G_p(z)$, where the z^2 term accounts for the two backoff periods needed to conduct the two CCA checks. The mean value of the transmission time is $\overline{D_d} = 2 + G'_p(1)$. The reader will recall from the description of the slotted CSMA-CA algorithm, presented in Section 2.3, that the CCA checks and actual packet transmission may be deferred to the next superframe if the remaining time within the CAP period of the current superframe does not suffice to complete the transmission. This feature is modeled as an additional delay, the probability of which can be approximated as $P_d = (2 + G'_p(1))/SD$, where SD denotes the duration of the active portion of the superframe. Exact calculation of the probability P_d will be presented in Chapter 4, as part of the model of downlink transmission.

Acknowledged transfer. The aim of acknowledged transfer is to achieve reliable packet transfer at the MAC layer. Acknowledgments must be explicitly requested by the sending node, which sets the Acknowledgment Request subfield in the Frame Control Field of the MAC packet header (Section 2.9). All successful packet transmissions which are received by the respective cluster coordinator and placed in its buffer are acknowledged.

The PGF for a single transmission attempt (ignoring the immediately preceding random backoff countdown) and the probability that the remaining time within the superframe will not suffice to complete the transmission, will have the same values as in the non-acknowledged case.

However, when acknowledgment is requested, the sending node will wait for it and repeat the transmission attempt if proper acknowledgment is not received within the prescribed time window. According to the standard (IEEE 2006) transmission may be re-attempted at most *macMaxFrameRetries* more times, the default value of which is three, after which the MAC layer announces a transmission failure to the higher layers and drops the packet. (This number should not be confused with the number of backoff attempts which is limited to *macMaxCSMABackoffs* = 5.) This approach may be denoted as *partially reliable transmission*. In that case, the final reliability of transmission is the responsibility of higher layers of the network protocol stack. However, it is also possible to employ *fully reliable transmission* approach, in which the node attempts transmission until it receives proper acknowledgment, regardless of the number of attempts. Both of these approaches will be considered in the following.

3.1.2 The queueing model

Clearly, the packet queue in the device buffer should be modeled as a $M/G/1/K$ queueing system. But before we discuss the queueing model, it helps to identify the possible states for a node in the cluster. From the discussion of the slotted CSMA-CA algorithm in Section 2.3, an ordinary node may be in one of the following states:

1. In the random backoff countdown mode.

2. In the contention window mode where it senses the channel to be idle but does not access it, $CW > 0$.

3. In the contention window mode where it senses the channel to be busy, but is allowed to backoff and wait, $NB < macMaxCSMABackoffs$.

4. In the (ready for) transmission mode, where it senses the channel to be idle and accesses it, $CW = 0$.

5. In the failure mode, where it senses the channel to be busy, and the index of the current backoff attempt is $NB > macMaxCSMABackoffs$.

6. In the delay mode, wherein the random backoff countdown is finished but the node is waiting for the beginning of the active portion of the next superframe to undertake the CCA checks and, if successful, the actual transmission.

In order to analyze the cluster behavior in this case, we will introduce the following variables:

1. $b(t)$ represents the value of the backoff time counter which, at the beginning of the backoff countdown, can take any value in the range $0 .. 2^{BE} - 1$. During the countdown, the counter decrements at the boundary of each backoff unit period. The value $b(t)$ will be frozen during the inactive portion of the beacon interval, and countdown will resume when the next superframe begins.

2. $n(t)$ represents the value of $NB \in 0 .. macMaxCSMABackoffs - 1$, at time t.

3. $c(t)$ represents the value of CW at time t; it may be 0, 1, or 2.

4. $d(t)$ represents the current value of the 'delay line counter', which is started if the transmission cannot be finished within the current superframe. The number of backoff periods necessary to complete the transmission within the current superframe is equal to $\overline{D_d}$.

We begin by considering the cluster in which the superframe has no inactive part, and then augment the model by including the modifications necessary to accommodate the more general case of the superframe with both active and inactive parts.

3.2   Superframe with an Active Period Only

The process $\{n(t), c(t), b(t), d(t)\}$ defines the state of the device at backoff unit boundaries. However, to reduce the notational complexity, we will show last tuple member $d(t)$ only within the delay line and omit it in other cases, where its value is zero. Also, the idle state with no packets to transmit will be denoted only as state x_0. The discrete-time Markov chain which depicts this process is presented in Figures 3.1 and 3.2. The T_r boxes model the time taken by the actual packet transmission.

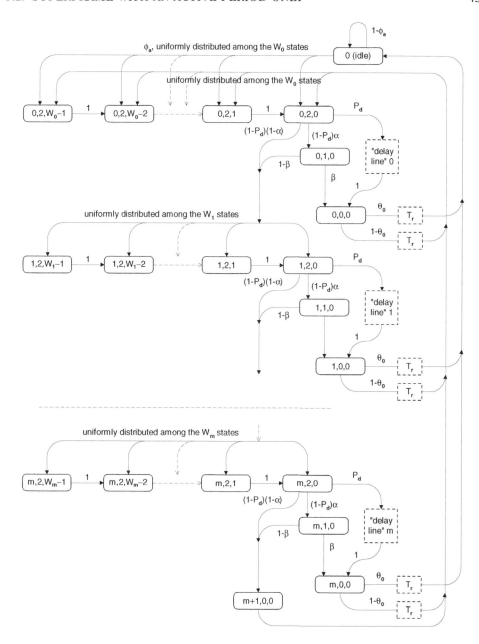

Figure 3.1 Markov chain model of a node executing the slotted CSMA-CA algorithm, in a cluster where the superframe has no inactive period. Adapted from J. Mišić, V. B. Mišić, and S. Shafi, 'Performance of IEEE 802.15.4 beacon enabled PAN with uplink transmissions in non-saturation mode – access delay for finite buffers,' *Proc. BROADNETS 2004*, pp. 416–425, © 2004 IEEE.

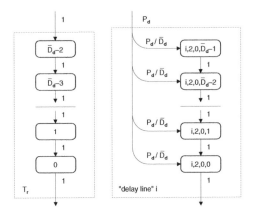

Figure 3.2 Delay lines for the Markov chain of Figure 3.1. Adapted from J. Mišić, V. B. Mišić, and S. Shafi, 'Performance of IEEE 802.15.4 beacon enabled PAN with uplink transmissions in non-saturation mode – access delay for finite buffers,' *Proc. BROADNETS 2004*, pp. 416–425, © 2004 IEEE.

For clarity of presentation we have tried to keep the notation simple. Thus, the probabilities $P\{n(t+1) = i, c(t+1) = j, b(t+1) = k-1, d(t+1) = l-1 \mid n(t) = i, c(t) = j, b(t) = k, d(t) = l\}$ are written simply as $P\{i, j, k-1, l-1 \mid 0, j, k, l\}$. Also, the maximum number of transmission attempts *macMaxFrameRetries* will be denoted with a, while the constant *macMaxCSMABackoffs*, representing the maximum value of the variable *NB*, is denoted with m.

For convenience, let W_0 stand for $2^{macMinBE}$, and let i represents the current value of *NB* during the execution of the algorithm $(i = 0 .. m)$. Then, the maximum value of the random waiting time (expressed in units of backoff periods) that corresponds to i will be $W_i = W_0 2^{min(i, 5 - macMinBE)}$ (note that the value of *BE* is limited to *macMaxBE = 5*).

The Markov chain from Figure 3.1 is general in the sense that the probability of leaving the transmission state θ_0 is generally labeled. Depending on the desired reliability option, it may take the following values:

- In the case of non-acknowledged transfer, this probability is equal to the probability that the device's buffer is empty after a packet transmission, i.e., $\theta_0 = \pi_0$ (this value will be derived later in the text).

- In the case of acknowledged transfer with full reliability, $\theta_0 = \pi_0 \gamma \delta$ where γ denotes the probability that the transmitted packet will not collide with any other transmission, and δ denotes the probability that the packet is not corrupted by noise. Given the bit error rate of the physical medium *BER*, the latter can be calculated as the probability that none of the bits in the packet nor in the acknowledgment is corrupted by noise, i.e., $\delta = (1 - BER)^{8(\overline{G}_p + \overline{G}_a)}$.

- In the case of acknowledged transfer with at most a attempts, the probability of switching to the idle state is $\theta_0 = (\gamma \delta / P_a) \pi_0$, where we make use of the probability that transfer is successfully completed within a attempts:

$$P_a = \sum_{i=0}^{a-1}(1 - \gamma\delta)^i\gamma\delta = 1 - (1 - \gamma\delta)^a. \tag{3.2}$$

The probability of leaving the idle state is equal to the probability of having at least one packet arrival during backoff period, i.e. $\phi_a = 1 - \exp(-\lambda) \approx \lambda$.

The non-null transition probabilities can be described with the following equations:

$$P\{0, 2, k \mid i, 0, 0\} = \frac{1 - \theta_0}{W_0}, \qquad \text{for } i = 0 \ldots m; k = 0 \ldots 2^{BE} - 1$$

$$P\{0 \mid i, 0, 0\} = \theta_0, \qquad \text{for } i = 0 \ldots m$$

$$P\{0, 2, k \mid 0\} = \frac{\phi_a}{W_0}, \qquad \text{for } i = 0 \ldots m; k = 0 \ldots 2^{BE} - 1$$

$$P\{i, 2, k - 1 \mid i, 2, k\} = 1, \qquad \text{for } i = 1 \ldots m; k = 1 \ldots 2^{BE} - 1$$

$$P\{i, 1, 0 \mid i, 2, 0\} = \alpha(1 - P_d), \qquad \text{for } i = 0 \ldots m$$

$$P\{i, 0, 0 \mid i, 1, 0\} = \beta, \qquad \text{for } i = 0 \ldots m$$

$$P\{i + 1, 2, k \mid i, 2, 0\} = \frac{(1 - \alpha)(1 - P_d)}{W_{i+1}}, \qquad \text{for } i = 0 \ldots m; k = 0 \ldots 2^{BE} - 1 \tag{3.3}$$

$$P\{i + 1, 2, k \mid i, 1, 0\} = \frac{1 - \beta}{W_{i+1}}, \qquad \text{for } i = 0 \ldots m; k = 0 \ldots 2^{BE} - 1$$

$$P\{i, 2, 0, l \mid i, 2, 0\} = \frac{P_d}{D_d}, \qquad \text{for } i = 0 \ldots m; l = 0 \ldots \overline{D_d} - 1$$

$$P\{i, 2, 0, l - 1 \mid i, 2, 0, l\} = 1, \qquad \text{for } i = 0 \ldots m; l = 0 \ldots \overline{D_d} - 1$$

$$P\{i, 0, 0 \mid i, 2, 0, 0\} = 1 \qquad \text{for } i = 0 \ldots m$$

$$P\{0, 2, k \mid m + 1, 0, 0\} = \frac{1}{W_0}, \qquad \text{for } k = 0 \ldots 2^{BE} - 1$$

The first transition probability in the set (3.3) represents the probability of choosing a random duration of the backoff period after a channel access that did not leave the node packet buffer empty. Since the range to choose from is 0 to $2^{BE} - 1$, this probability is equal to $\frac{1 - \theta_0}{W_0}$. Note that the random backoff period always precedes packet transmission, regardless of whether the packet to be transmitted is 'brand new', or it is simply an earlier packet that could not have been transmitted in the previous attempt due to a collision.

The second probability corresponds to the case when the node buffer becomes empty after the channel access and the node enters the idle state. The third expression corresponds to the probability of moving from the idle state to the one backoff state. The fourth equation shows the probability that the backoff time is decremented after each *aUnitBackoffPeriod*.

The fifth equation determines the probability α that the channel is sensed idle at the first CCA check, after the backoff counter reaches zero. The sixth equation determines the probability β that the channel is sensed to be idle at the second CCA check, i.e., after it was already sensed idle in the previous backoff period. Note that, when the backoff counter reaches zero and CCA senses a busy channel, the ongoing packet transmission may have started one or more backoff periods earlier, as shown in Figure 3.3(a). However, when the first CCA senses that the medium is idle but the second one finds it busy, that packet transmission must have started in that same backoff period, as shown in Figure 3.3(b). As a result, the probabilities that the medium is not idle at the first and second CCA α and β, respectively, differ.

The seventh and eighth equations describe the probabilities that the device, upon sensing the channel to be busy, chooses another random backoff in the range $0 .. W_{i+1} - 1$.

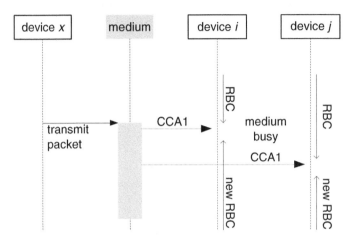

(a) When the first CCA fails, the ongoing transmission may have started in that backoff period (at the CCA1 executed by device i), or in an earlier one (at the CCA1 executed by device j).

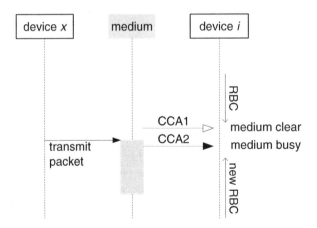

(b) When the second CCA fails, the ongoing transmission must have started in that same backoff period.

Figure 3.3 Pertaining to the difference between success probabilities of the first (α) and second (β) CCA.
Note: RBC denotes random backoff countdown.

The next three equations describe the delay line which is entered if there is insufficient time to complete the CCA checks, packet transmission, and optional acknowledgment in the current superframe.

Finally, the last equation describes the event when m backoff attempts were unsuccessful and the backoff window size has to be reset to W_0.

Let the stationary distribution of the chain be $x_{i,j,k,l} = \lim\limits_{t\to\infty} P\{n(t) = i, c(t) = j, b(t) = k, d(t) = l\}$, for $i = 0 .. m$; $j = 0, 1, 2$; $k = 0 .. 2^{BE} - 1$; $l = 0 .. \overline{D_d} - 1$. For brevity, we will omit l whenever it is zero, and introduce the auxiliary variables C_1, C_2, and C_3:

$$
\begin{aligned}
x_{0,1,0} &= x_{0,2,0}(1 - P_d)\alpha &&= x_{0,2,0}C_1 \\
x_{1,2,0} &= x_{0,2,0}(1 - P_d)(1 - \alpha\beta) &&= x_{0,2,0}C_2 \\
x_{0,0,0} &= x_{0,2,0}((1 - P_d)\alpha\beta + P_d) &&= x_{0,2,0}C_3
\end{aligned}
\tag{3.4}
$$

which simplify the following relations:

$$
\begin{aligned}
x_0 &= x_{0,0,0}\frac{\theta_0\left(1 - C_2^{m+1}\right)}{\phi_a(1 - C_2)} \\
x_{i,0,0} &= x_{0,0,0}C_2^i, &&\text{for } i = 0 .. m \\
x_{i,2,k} &= x_{0,0,0}\frac{W_i - k}{W_i} \cdot \frac{C_2^i}{C_3}, &&\text{for } i = 1 .. m; k = 0 .. W_i - 1 \\
x_{i,1,0} &= x_{0,0,0}\frac{C_1 C_2^i}{C_3} &&\text{for } i = 0 .. m \\
x_{0,2,k} &= x_{0,0,0}\frac{W_0 - k}{W_0 C_3} &&\text{for } k = 1 .. W_0 - 1 \\
x_{m+1,0,0} &= x_{0,0,0}\frac{C_2^{m+1}}{C_3} \\
\sum_{l=0}^{\overline{D_d}-1} x_{i,2,0,l} &= \frac{x_{0,0,0}C_2^i P_d(\overline{D_d} - 1)}{2C_3}
\end{aligned}
\tag{3.5}
$$

Since the sum of all probabilities in the Markov chain must be equal to one, we obtain:

$$
\begin{aligned}
x_0 &+ \left(\overline{D_d} - 2\right)\sum_{i=0}^{m}\sum_{k=0}^{W_i-1} x_{i,2,k} + \sum_{i=0}^{m} x_{i,0,0} \\
&+ \sum_{i=0}^{m} x_{i,1,0} + x_{m+1,0,0} + \sum_{i=0}^{m}\sum_{l=0}^{\overline{D_d}-1} x_{i,2,0,l} = 1
\end{aligned}
\tag{3.6}
$$

from which the value for $x_{0,0,0}$ can be obtained as

$$
x_{0,0,0} = \cfrac{1}{\displaystyle\sum_{i=0}^{m} \frac{C_2^i(W_i + 1)}{2C_3} + \frac{1 - C_2^{m+1}}{1 - C_2}\left(\overline{D_d} - 2 + \frac{C_1}{C_3} + \frac{\theta_0}{\phi_a} + \frac{P_d(\overline{D_d} - 1)}{2C_3}\right) + \frac{C_2^{m+1}}{C_3}}
\tag{3.7}
$$

The total probability to access the medium is

$$\tau = \sum_{i=0}^{m} x_{i,0,0} = x_{0,0,0} \frac{1 - C_2^{m+1}}{1 - C_2} \tag{3.8}$$

However, we distinguish between the probability τ_1 to access the medium when the transmission is deferred to the next superframe due to lack of space, and the probability τ_2 to access the medium in the current superframe:

$$\begin{aligned} \tau_1 &= \frac{P_d}{C_3} \tau \\ \tau_2 &= \left(1 - \frac{P_d}{C_3}\right) \tau \end{aligned} \tag{3.9}$$

3.2.1 Medium behavior: success probabilities

Let us now calculate the probabilities that the medium is found idle at the first and second CCA. We assume that q and $n - 1 - q$, which denote the probabilities of non-delayed and delayed packet transmissions, respectively, follow a binomial distribution with probability $P_q = \binom{n-1}{q}(1 - P_d)^q P_d^{n-1-q}$.

In order to calculate the probability α that the medium is idle at the first CCA, we have to find the mean number of busy backoff periods within the superframe. This number will be divided into the total number of backoff periods in the superframe wherein the first CCA can occur. Note that the first CCA will not take place if the remaining time in the superframe is insufficient to complete the transaction, which amounts to $SM = SD - \overline{D_d} + 1$ backoff periods. Then, the probability that one or more packet transmissions will take place at the beginning of the superframe is

$$n_1 = 1 - (1 - \tau_1)^{n-1-q}(1 - \tau_2)^q \tag{3.10}$$

while the number of backoff periods that are found to be busy due to these transmissions is $n_1(G'_p(1) + G'_a(1))$, for acknowledged transfer, and $n_1 G'_p(1)$, for the non-acknowledged one.

The occupancy of the medium after the first transmission time can be found by dividing the superframe into chunks of $\overline{D_d}$ backoff periods and calculating the probability of transmission within each chunk. As the total arrival rate of non-deferred packets is $q\tau_2$, the probability that the number of transmission attempts during the period $\overline{D_d}$ will be non-zero is $n_2 = q\tau_2\overline{D_d}$. The total number of backoff periods in which the first CCA may occur is SM. Then, the probability that the medium is idle at the first CCA, in the case of acknowledged transfer, is

$$\alpha = \sum_{q=0}^{n-1} P_q \left(1 - \left(\frac{n_1 \overline{D_d}}{SM} + \frac{n_2(SD - 2\overline{D_d} + 1)}{SM}\right) \frac{(G'_p(1) + G'_a(1))}{\overline{D_d}}\right) \tag{3.11}$$

while the corresponding probability for non-acknowledged transfer is

$$\alpha = \sum_{q=0}^{n-1} P_q \left(1 - \left(\frac{n_1 \overline{D_d}}{SM} + \frac{n_2(SD - 2\overline{D_d} + 1)}{SM}\right) \frac{G'_p(1)}{\overline{D_d}}\right) \tag{3.12}$$

It is easy to see that the two expressions differ only insofar as the term $G'_a(1)$ is absent from the latter one.

The probability β that the medium is idle on the second CCA for a given node equals the probability that neither the coordinator nor any of the remaining $n - 1$ nodes have started a transmission in that backoff period, as shown in Figure 3.3(b). The second CCA can be performed in any backoff period in the time interval from the second backoff period in the superframe to the period in which there is no more time for packet transmission, which amounts to SM. Then, the probability β is

$$\beta = \sum_{q=0}^{n-1} P_q \left(\frac{1}{SM} + \frac{(1 - \tau)^{n-1}}{SM} + \frac{SD - \overline{D_d} - 2}{SM} (1 - \tau_2)^q \right) \tag{3.13}$$

where the appropriate values of $\overline{D_d}$ is to be used for non-acknowledged and acknowledged transfer, respectively.

Finally, the probability γ that a packet will not collide with other packet(s) that have succeeded in their first and second CCAs can be calculated as the probability that there are no accesses to the medium by the other nodes or the coordinator during the period of one complete packet transmission time. (Note that a collision can happen in SM consecutive backoff periods starting from the third backoff period in the superframe.)

$$\gamma = \sum_{q=0}^{n-1} P_q \left(\frac{1}{SM} (1 - \tau)^{\overline{D_d}(n-1)} + \frac{SD - \overline{D_d}}{SM} (1 - \tau_2)^{\overline{D_d}q} \right) \tag{3.14}$$

where appropriate values of $\overline{D_d}$ are to be used for non-acknowledged and acknowledged transfer respectively.

3.3 Superframe with Both Active and Inactive Periods

The presence of the inactive period in the superframe, in the case where $SO < BO$, necessitates a slightly different Markov chain model, as shown in Figure 3.4. The difference pertains to the behavior of the node during the inactive period; however, the T_r and delay line blocks are the same as in Figure 3.2. Namely, when a packet arrives to an idle node during inactive period, its backoff countdown will start immediately after the beacon; this is shown in the topmost delay line in the graph. When a packet arrives to the idle node during the active superframe part, the backoff countdown will start immediately.

Probabilities that a new packet arrives to the node in the idle state in the active and inactive part of the superframe are denoted as ϕ_a and ϕ_i, respectively. θ_0 is the probability that the node buffer is empty after a successful packet transmission. As in the previous case, these probabilities take different values depending whether the transmission is non-acknowledged or acknowledged.

We note that the first backoff phase in the Markov chain actually has two parts. The part which is connected to the idle state with the probability ϕ_i represents the situation when a new packet arrives to the empty buffer during the inactive portion of the superframe. In that case, the first backoff countdown will start immediately after the beacon and the value of backoff counter will be in the range $0 .. W_0 - 1$. Those states will be denoted as

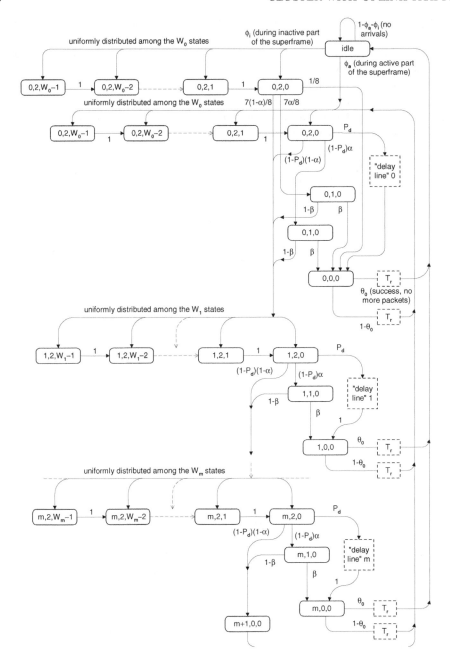

Figure 3.4 Markov chain model of a node executing the slotted CSMA-CA algorithm, in a cluster where the superframe has both active and inactive periods. Adapted from J. Mišić, J. Fung, and V. B. Mišić, 'Interconnecting 802.15.4 clusters in master-slave mode: queueing theoretic analysis,' *Proc. I-SPAN 2005*, pp. 378–385, © 2005 IEEE.

$x_{i,c,k}^s$. On the other hand, if the packet arrives to a node during the active portion of the superframe, the backoff countdown will start at a random position within the superframe, and those states will be denoted as $x_{i,c,k}$. As these two cases affect the behavior of the medium in different ways, they have to be separately modeled.

From the Markov chain, the probability to access the medium is

$$\tau = \sum_{i=0}^{m} x_{i,0,0} \qquad (3.15)$$

Then, the probability of the node switching into the idle state is $\tau\theta_0$. After setting up the balance equation for the idle state, we obtain the probability of being in the idle state as Prob(idle) $= P_z = \tau\theta_0/(\phi_a + \phi_i)$.

If we further consider the output from the idle state and set up the balance equations for the first backoff phase started after the inactive part of the superframe, we obtain

$$x_{0,2,W_0-1}^s = P_z\phi_i/W_0$$
$$x_{0,2,W_0-k}^s = kP_z\phi_i/W_0, \quad \text{for } 1 < k < W_0 - 1 \qquad (3.16)$$
$$x_{0,2,0}^s = P_z\phi_i$$

The first backoff phase in the active portion of the superframe may start in the following cases: after the packet arrival during the idle state, after a packet transmission (regardless of the transmission success), or after the last unsuccessful backoff phase. For clarity, let us represent the state after last unsuccessful backoff phase as $x_{m+1,0,0}$, although there is no physical meaning associated with it. The state probabilities of the first backoff phase started in the active superframe part are represented as $x_{0,2,k}$, $0 \le k < W_0$. The input probability for that set of states is equal to

$$\begin{aligned} U_a &= P_z\phi_a + \tau(1-\theta_0) + x_{m+1,0,0} \\ &= \tau\left(1 - \theta_0\frac{\phi_i}{\phi_i + \phi_a}\right) + x_{m+1,0,0} \end{aligned} \qquad (3.17)$$

By setting the balance equations, we obtain:

$$x_{0,2,W_0-1}^s = U_a/W_0$$
$$x_{0,2,W_0-k}^s = kU_a/W_0, \quad \text{for } 1 < k < W_0 - 1 \qquad (3.18)$$
$$x_{0,2,0}^s = U_a.$$

A similar approach can be utilized to derive $x_{i,2,k}$ for backoff attempts $i = 2, 3, \dots m$.

Using the transition probabilities indicated in Figures 3.4 and 3.2, we can derive the relationships between the state probabilities and solve the Markov chain. For brevity, we will omit index d whenever it is zero, and introduce the auxiliary variables C_1, C_1^s C_2, C_2^s, C_3 and C_3^s, defined via following equations:

$$\begin{aligned} x_{0,1,0} &= x_{0,2,0}(1-P_d)\alpha + x_{0,2,0}^s\left(\frac{7}{8}\alpha\right) \\ &= x_{0,2,0}C_1 + x_{0,2,0}^s C_1^s \\ x_{1,2,0} &= x_{0,2,0}(1-P_d)(1-\alpha\beta) + x_{0,2,0}^s\left(\frac{7}{8}(1-\alpha\beta)\right) \\ &= x_{0,2,0}C_2 + x_{0,2,0}^s C_2^s \end{aligned} \qquad (3.19)$$

$$
\begin{aligned}
&= \tau \left(\theta_0 \frac{\phi_i}{\phi_i + \phi_a} C_2^s + \left(1 - \theta_0 \frac{\phi_i}{\phi_i + \phi_a} \right) C_2 \right) + C_2 x_{m+1,0,0} \\
x_{0,0,0} &= x_{0,2,0} \left((1 - P_d)\alpha\beta + P_d \right) + x_{0,2,0}^s \left(\frac{7}{8}\alpha\beta + \frac{1}{8} \right) \\
&= x_{0,2,0} C_3 + x_{0,2,0}^s C_3^s
\end{aligned}
\tag{3.19}
$$

From the expressions that describe the Markov chain we obtain

$$
\begin{aligned}
x_{i,1,0} &= C_1 C_2^{(i-1)} x_{1,2,0}, & i &= 1..m \\
x_{i,2,0} &= C_2^{(i-1)} x_{1,2,0}, & i &= 1..m \\
x_{i,0,0} &= C_3 C_2^{(i-1)} x_{1,2,0}, & i &= 1..m+1
\end{aligned}
\tag{3.20}
$$

$$
\sum_{d=0}^{\overline{D_d}-1} x_{i,2,0,d} = x_{i,2,0} C_2^i (\overline{D_d} - 1)/2
$$

Of course, the sum of all probabilities in the Markov chain must be equal to one:

$$
\begin{aligned}
U &+ \sum_{k=0}^{W_0-1} x_{0,2,k}^s + x_{0,1,0}^s + \sum_{i=0}^{m} \sum_{k=0}^{W_i-1} x_{i,2,k} + \sum_{i=0}^{m} x_{i,0,0}(\overline{D_d} - 2) \\
&+ \sum_{i=0}^{m} x_{i,1,0} + x_{m+1,0,0} + \sum_{i=0}^{m} \sum_{d=0}^{\overline{D_d}-1} x_{i,2,0,d} = 1
\end{aligned}
\tag{3.21}
$$

which has to be solved for τ; note that this is just the mean value during the active portion of the superframe. However, access to the medium is prohibited in the first two and last $\overline{D_d} - 1$ backoff periods. Therefore, it becomes necessary to identify the parts of the superframe where some types of access can occur and scale the access probabilities accordingly.

Considering the transmissions of packets which arrive at an idle node during the inactive portion of the superframe, we find that access is possible during the interval starting from the third backoff period after the beacon, until the $W_0 + 2$-th backoff period of the superframe. Solving the Markov chain gives us $x_{0,2,0}^s$ as the probability that this access is possible throughout the entire active portion of the superframe, and we need to scale that value to the time interval where access can actually occur. Since the initial value for the countdown is chosen at random between 0 and $W_0 - 1$, the probability of access in the third backoff period after the beacon is

$$
\tau_{3,1} = x_{0,2,0}^s \frac{1}{W_0} \cdot \frac{SM - 2}{W_0},
\tag{3.22}
$$

where the term $SM - 2$ corresponds to the total number of backoff periods where any access can happen, while the denominator W_0 is the number of backoff periods where this access can really occur. The probability that the access will occur in some other (fourth to $W_0 + 2$-th) backoff period after the beacon is

$$
\tau_{3,2} = x_{0,2,0}^s \frac{W_0 - 1}{W_0} \cdot \alpha\beta \cdot \frac{SM - 2}{W_0 - 1}
\tag{3.23}
$$

The reason for separating $\tau_{3,1}$ from $\tau_{3,2}$ is that the former overlaps with the transmissions deferred from the previous superframe due to insufficient time. The probability of accessing the medium in this case is $SM - 2$ times higher than the average value over the whole superframe – since it can happen only in the third backoff period after the beacon. Therefore, the probabilities of deferred and non-deferred access are

$$\begin{aligned}
\tau_1 &= (SM - 2)\frac{P_d}{C_3}\left(\tau - x_{0,2,0}C_3^s\right) \\
\tau_2 &= \left(1 - \frac{P_d}{C_3}\right)\left(\tau - x_{0,2,0}C_3^s\right).
\end{aligned} \tag{3.24}$$

3.3.1 Acknowledged vs. non-acknowledged transfer

In order to proceed, we need the conditional probability that the packet arrives at an idle node during the inactive portion of the superframe, which is $P_{sync} = 1 - 2^{SO-BO}$.

Non-acknowledged transfer. The idle state of the Markov chain is reached when the buffer is empty after transmission, regardless of whether the packet suffered a collision or not. In that case, $\theta_0 = \pi_0$, as will be derived in Section 3.5 . Since the packet arrival rate at an individual node, λ, is assumed to be small, the probability of zero Poisson arrivals during unit backoff period can be approximated with the corresponding Taylor series, i.e., $exp(-\lambda) \approx 1 - \lambda$. Therefore, the probability of non-zero packet arrivals, i.e., the probability to leave the idle state, is simply λ, which further gives $\phi_i = P_{sync}\lambda$ and $\phi_a = (1 - P_{sync})\lambda$.

Acknowledged transfer with full reliability. In this case, the idle state is reached only if the node buffer is empty after the transmission, the transmission was successful, and the packet was accepted by the coordinator, hence $\theta_0 = \gamma\delta\pi_0$. Since the packet arrival rate to an individual node is λ, then $\phi_i = P_{sync}\lambda$ and $\phi_a = (1 - P_{sync})\lambda$.

Acknowledged transfer with partial reliability. As was the case for the presence of active period only, the probability that transfer is successfully completed in a attempts is equal to $P_a = 1 - (1 - \gamma\delta)^a$. Then, the probability of joining idle state is equal to $\theta_0 = \pi_0\gamma\delta/P_a$.

3.3.2 Medium behavior and success probabilities

In the presence of both active and inactive superframe parts, medium behavior changes slightly compared to the case where the superframe has no inactive portion. At any moment, q stations out of $n - 1$ are not delayed due to insufficient space that remains in the superframe, while $n - 1 - q$ are delayed to the start of next superframe. The values of q and $n - 1 - q$ follow a binomial distribution with the probability

$$P_q = \binom{n-1}{q}(1 - P_d)^q P_d^{n-1-q}. \tag{3.25}$$

Within q nodes which are not delayed, exactly r nodes have received the packet during the inactive part of the superframe with the probability

$$P_r = \binom{q}{r}(P_0 P_{sync}\lambda(1 - P^B))^r(1 - P_0 P_{sync}\lambda(1 - P^B))^{q-r}, \tag{3.26}$$

where P_0 denotes the probability that the node buffer is empty at arbitrary time (it will be derived in Section 3.5.2), while P^B denotes the probability that an incoming packet will be dropped because the input buffer is already filled to capacity. Note also that distinct values of $\overline{D_d}$ apply to non-acknowledged and acknowledged transfer, respectively.

Probability of success of the first CCA. As before, the probability that the medium is idle at the first CCA is obtained by dividing the mean number of busy backoff periods within the superframe into the total number of backoff periods in the superframe in which the first CCA can occur. As before, the first CCA will not take place if the remaining time in the superframe is insufficient to complete the transaction, which amounts to $SM = SD - \overline{D_d} + 1$ backoff periods. Then, the probability that any packet transmission will take place at the beginning of the superframe is

$$n_1 = 1 - (1 - \tau_1)^{n-1-q}(1 - \tau_2)^{q-r}(1 - \tau_{3,1})^r \tag{3.27}$$

and the number of busy backoff periods due to these transmissions is $n_1(G'_p(1) + G'_a(1))$, for acknowledged transfer, and $n_1 G'_p(1)$, for the non-acknowledged one.

The probability of a transmission attempt after the third backoff period within the period of $\overline{D_d}$ backoff periods is

$$n_2 = 1 - ((1 - \tau_2)^{q-r}(1 - \tau_{3,2})^r)^{\overline{D_d}} \tag{3.28}$$

The probability of a transmission in the remaining part of the superframe within $\overline{D_d}$ backoff periods is

$$n_3 = 1 - ((1 - \tau_2)^q)^{\overline{D_d}} \tag{3.29}$$

The occupancy of the medium after the first transmission time can be found by dividing the superframe into chunks of $\overline{D_d}$ backoff periods and calculating the probability of transmission within each chunk. The total number of backoff periods in which the first CCA may occur is $SM = SD - \overline{D_d} + 1$. The probability that the medium is idle at the first CCA for acknowledged transfer is

$$\alpha = \sum_{q=0}^{n-1}\sum_{r=0}^{q} P_q P_r \left(1 - \frac{n_1\overline{D_d} + n_2\overline{D_d} + n_3(SD - 3\overline{D_d} + 1)}{SM} \cdot \frac{G'_p(1) + G'_a(1)}{\overline{D_d}}\right) \tag{3.30}$$

The same expression holds in the case of non-acknowledged transfer, except that the term $G'_a(1)$ is not present.

Probability of success of the second CCA. The second CCA can be performed in any backoff period during the interval from the second backoff period in the superframe, up to the last SM backoff periods (in which there is no more time for packet transmission). The probability that the medium is idle at the second CCA is equal to the probability that neither of the remaining $n - 1$ nodes nor the cluster coordinator have started a transmission in that backoff period:

$$\begin{aligned} \beta = {} & \sum_{q=0}^{n-1}\sum_{r=0}^{q} P_q P_r \left(\frac{1}{SM} + \frac{(1 - \tau_1)^{n-1-q}(1 - \tau_2)^{q-r}(1 - \tau_{3,1})^r}{SM}\right.\\ & \left. {} + \frac{\overline{D_d} - 1}{SM}(1 - \tau_2)^{q-r}(1 - \tau_{3,2})^r + \frac{SM - \overline{D_d}}{SM}(1 - \tau_2)^q\right) \end{aligned} \tag{3.31}$$

Probability of successful transmission. Probability of successful transmission γ is the probability that a packet will not collide with other packet, or packets, that succeeded in their first and second CCAs. Note that a collision can happen in *SM* consecutive backoff periods, starting from the third backoff period in the superframe. This probability can be calculated as the probability that there are no accesses to the medium by the other nodes or the coordinator during the period of one complete packet transmission time.

$$
\gamma = \sum_{q=0}^{n-1} P_q \sum_{r=0}^{q} P_r \left(\frac{((1 - \tau_1)^{n-1-q}(1 - \tau_2)^{q-r}(1 - \tau_{3,1})^r)^{\overline{D_d}}}{SM} \right.
$$
$$
\left. + \frac{\overline{D_d} - 1}{SM}((1 - \tau_2)^{q-r}(1 - \tau_{3,2})^r)^{\overline{D_d}} + \frac{SM - \overline{D_d}}{SM}((1 - \tau_2)^q)^{\overline{D_d}} \right)
$$

(3.32)

3.4 Probability Distribution of the Packet Service Time

We can now derive the probability distribution of the packet service time at the MAC layer. The packet service time is, in fact, the service time for the packet queue at the network node. The time needed to transmit a packet from the head of the queue includes the time from the moment when the CSMA-CA algorithm has started (i.e., from the start of the backoff countdown procedure) to the moment when the receipt of packet has been acknowledged by the destination node. Again, we will separately model the cases where the superframe has an active period only, and where the superframe has both an active and an inactive portion. If the inactive period is not present, the PGF for the backoff period is simply $B_{off} = z$. In the presence of inactive period, we need to model the effect of freezing the backoff counter during the inactive period of the superframe. To that end, we calculate the probability that a given backoff period is the last one within the active portion of the superframe as $P_{last} = 1/SD$, and the PGF for the effective duration of the backoff period, including the duration of the beacon frame, as

$$
B_{off}(z) = (1 - P_{last})z + P_{last}z^{BI-SD+1}B_{ea}(z)
$$

(3.33)

The PGF for the duration of i-th backoff attempt is

$$
B_i(z) = \sum_{k=0}^{W_i-1} \frac{1}{W_i} B_{off}^k(z) = \frac{B_{off}^{W_i}(z) - 1}{W_i(B_{off}(z) - 1)}
$$

(3.34)

The transmission procedure will not start unless it can be finished within the current superframe. The number of backoff periods wasted due to the insufficient space in the current superframe can be described by the PGF of

$$
B_p(z) = \frac{1}{D_d} \sum_{k=0}^{D_d-1} z^k
$$

(3.35)

Then, in case of acknowledged transmission, the PGF of the data packet transmission time for deferred and non-deferred transmissions, respectively, is

$$
\begin{aligned}
T_{d1}(z) &= B_p(z)z^{BI-SD}B_{ea}(z)G_p(z)t_{ack}(z)G_a(z) \\
T_{d2}(z) &= G_p(z)t_{ack}(z)G_a(z)
\end{aligned}
$$

(3.36)

For the case of non acknowledged transmission, the corresponding PGFs are

$$
\begin{aligned}
T_{d1}(z) &= B_p(z)z^{BI-SD}B_{ea}(z)G_p(z) \\
T_{d2}(z) &= G_p(z)
\end{aligned}
\tag{3.37}
$$

3.4.1 PGF for one transmission attempt including backoffs

Let us denote the probability that a backoff attempt will be unsuccessful as $R_u = 1 - P_d - (1 - P_d)\alpha\beta$. (The subscript u indicates that the attempt is made to transmit a packet in the uplink direction.) The function that describes the time needed for the backoff countdown and the transmission attempt itself can be presented as

$$
\begin{aligned}
\mathcal{A}(z) &= \sum_{i=0}^{m}\prod_{j=0}^{i} B_j(z)R_u z^{2(i+1)}\left(P_d T_{d1}(z) + (1 - P_d)\alpha\beta T_{d2}(z)\right) \\
&+ R_u^{m+1}\prod_{j=0}^{m} B_j(z)z^{2(m+1)}\mathcal{A}(z)
\end{aligned}
\tag{3.38}
$$

where R_u^{m+1} denotes the probability that a total of $m + 1$ successive backoff attempts with non-decreasing backoff windows were not successful and, hence, the sequence of backoff windows has to be repeated starting from the smallest backoff window. If we substitute $z = 1$ into Equation (3.38), we obtain $\mathcal{A}(1) = 1$ which is a necessary condition for a function to be a proper PGF. From Equation (3.38), we obtain

$$
\mathcal{A}(z) = \frac{\displaystyle\sum_{i=0}^{m}\prod_{j=0}^{i}\left(B_j(z)R_u\right)z^{2(i+1)}\left(P_d T_{d1}(z) + (1 - P_d)\alpha\beta T_{d2}(z)\right)}{1 - R_u^{m+1}\displaystyle\prod_{j=0}^{m} B_j(z)z^{2(m+1)}}
\tag{3.39}
$$

Non-acknowledged transmission. The 'pure' transmission time from Equation (3.37) has to be substituted in Equation (3.39). Furthermore, the appropriate value of $\overline{D_d} = 2 + G'_p(1)$ has to be substituted in equations which determine α, β, γ, and δ, and these variables in turn have to be substituted in Equation (3.39). When all these items are taken into account, the PGF for the packet service time, in case of non-acknowledged transfer, has the value of

$$
T_t(z) = \mathcal{A}(z)
\tag{3.40}
$$

Acknowledged transmission with partial reliability. In this case we have to take into account that an unsuccessful transmission will be retried at most a times. The probability that the transmission is completed within a attempts, was previously computed as

$$
\begin{aligned}
P_a &= \sum_{i=0}^{a-1}(1 - \gamma\delta)^i\gamma\delta \\
&= 1 - (1 - \gamma\delta)^a
\end{aligned}
\tag{3.41}
$$

where γ is the probability of no collisions in the cluster, while δ is the probability that the packet or the acknowledgment will not be corrupted by the noise.

Then, the PGF for the packet service time, in the cluster where the superframe contains only the active period, becomes

$$
\begin{aligned}
T_t(z) &= \gamma \delta A(z) + \gamma \delta (1 - \gamma \delta) A(z)^2 + \ldots + \gamma \delta (1 - \gamma \delta)^{(a-1)} A(z)^a \\
&\quad + \left[1 - (1 - \gamma \delta)^a \right] z^0 \\
&= \frac{1 - A(z)^a (1 - \gamma \delta)^a}{1 - A(z)(1 - \gamma \delta)} + \left[1 - (1 - \gamma \delta)^a \right] z^0
\end{aligned} \tag{3.42}
$$

Acknowledged transmission with full reliability. In order to derive this PGF we take the limiting value of Expression (3.42) when $a \to \infty$, and the PGF for the packet service time becomes

$$
\begin{aligned}
T_t(z) &= \gamma \delta A(z) \sum_{a=0}^{\infty} (1 - \gamma \delta)^a A(z)^a \\
&= \frac{\gamma \delta A(z)}{1 - (1 - \gamma \delta) A(z)}
\end{aligned} \tag{3.43}
$$

However, in clusters where the superframe has both active and inactive periods, the packet service time has an additional component. This component is the initial waiting time from the packet arrival to an empty buffer during the inactive portion of the superframe up to the following beacon frame, after which the node will commence the random backoff countdown. Since the packet arrival process is totally oblivious to the superframe timing, new packets can arrive at any time during the inactive part of the superframe with the same probability. Consequently, this waiting time has a uniform probability distribution. The PGF of this synchronization time is

$$
T_{sync}(z) = \pi_0 P_{sync} \frac{z^{BI-SD} - 1}{(BI - SD)(z - 1)} + (1 - \pi_0 P_{sync}) z^0 \tag{3.44}
$$

and it should multiply $T_t(z)$.

3.5 Probability Distribution of the Queue Length

3.5.1 Queue length at the time of packet departures

Let us first note that the Laplace-Stieltjes Transform (LST) of the packet service time $T_t^*(s)$ can be obtained by substituting the variable z with e^{-s} in the expression for $T_t(z)$. Also, let the PDF of the packet service time be denoted with $T_{dt}(x) = \text{Prob}[T_{dt} \leq x]$, while the corresponding pdf is denoted with $t_{dt}(x)$. Then, the probability of exactly k packet arrivals to the device queue during the packet service time is

$$
a_k = \int_0^{\infty} \frac{(\lambda x)^k}{k!} e^{-\lambda x} t_{dt}(x) dx \tag{3.45}
$$

We also note (Takagi 1991) that the PGF for the number of packet arrivals to the device queue during the packet service time is

$$
A(z) = \sum_{k=0}^{\infty} a_k z^k = \int_0^{\infty} e^{-x\lambda(1-z)} dt_t(x) = T_t^*(\lambda - z\lambda) \tag{3.46}
$$

When the LST of the packet service time is known, the probability a_k can be obtained as

$$a_k = \frac{1}{k!} \frac{d^k A(z)}{dz^k}\bigg|_{z=0} \tag{3.47}$$

Let π_k denote the steady state probability that there are k packets in the device buffer immediately after the packet departure. Then, the steady state equations for state transitions are given by

$$\pi_k = \pi_0 a_k + \sum_{j=1}^{k+1} \pi_j a_{k-j+1}, \qquad \text{for } 0 \le k \le L - 2$$

$$\pi_{L-1} = \pi_0 \sum_{k=L-1}^{\infty} a_k + \sum_{j=1}^{L-1} \pi_j \sum_{k=L-j}^{\infty} a_k \tag{3.48}$$

The probability distribution of the device queue length at the time of packet departure can be found by solving the system in a recursive manner. If we introduce the substitution $\pi'_k = \pi_k/\pi_0$ (Takagi 1993), we obtain

$$\pi'_0 = 1$$

$$\pi'_{k+1} = \frac{1}{a_0}\left(\pi'_k - \sum_{j=1}^{k} \pi'_j a_{k-j+1} - a_k\right), \qquad \text{for } 0 \le k \le L - 2 \tag{3.49}$$

which provides the third equation that describes the devices with finite queues, i.e.,

$$\pi_0 = \frac{1}{\displaystyle\sum_{k=0}^{L-1} \pi'_k} \tag{3.50}$$

Note that the probability π_0 is the function of packet service time indirectly, through the probabilities a_k.

3.5.2 Queue length at arbitrary time

Of course, new packets can arrive to the device queue at any time. If the buffer size is L, the buffer occupancy at arbitrary time can have values from 0 to L, while the buffer occupancy immediately after the packet departure time can have values from 0 to $L - 1$ only. Obviously, the probability distribution of the queue length at arbitrary time is different from the probability distribution of the queue length immediately upon packet departure. However, both probability distributions must satisfy the system of Equations (3.48), and the individual mass probabilities are related as $P_k = c\pi_k$, for $0 \le k \le L - 1$. The proportionality constant may be easily found to be $c = 1 - P^B$, where the blocking probability P^B is equal to the probability that the input buffer contains exactly L packets, P_L, since

$$\sum_{k=0}^{L-1} P_k = 1 - P^B. \tag{3.51}$$

Thus, $P_k = (1 - P^B)\pi_k$, where $0 \le k \le L - 1$.

If we define the offered load as $\rho = \lambda T_t'(1)$, the equality $(1 - P^B)\rho = 1 - (1 - P^B)\pi_0$ leads to

$$P^B = 1 - \frac{1}{\pi_0 + \rho}, \tag{3.52}$$

and subsequently

$$P_0 = 1 - \lambda T_t'(1)\frac{1}{\pi_0 + \lambda T_t'(1)} \tag{3.53}$$

An important characteristic of such systems is that the probability π_0 that the queue is empty immediately upon packet departure differs from the probability P_0 that the queue is empty at an arbitrary time.

To summarize:

- if the superframe has no inactive portion, cluster behavior is described by the system consisting of Equations (3.8), (3.11), (3.13), (3.14), (3.50), and (3.53);

- if the superframe has both active and inactive periods, cluster behavior is described by Equations (3.21), (3.30), (3.31), (3.32), (3.50) and (3.53).

Each of the systems can be solved for the unknowns τ, α, β, γ, π_0, and P_0. Once these solutions are found, we can find the probability distribution of the queue length at arbitrary time by following these steps:

1. Calculate $T_t(z)$ from Equations (3.40), (3.42), or (3.43), depending on the acknowledgment mode.

2. Calculate the PGF for the number of packet arrivals during one packet service time $A(z)$ from Equation (3.46), and consequently calculate a_k, $k = 0 .. L$.

3. Go back to system of equations (3.49) and calculate π_k, $k = 1 .. L - 1$.

4. Calculate $P^B = 1 - 1/(\pi_0 + \rho)$.

5. Finally, find $P_k = (1 - P^B)\pi_k$.

3.6 Access Delay

We define access delay as the time interval from the packet arrival at the device queue until the successful packet departure from that queue. A successful packet departure is defined as the packet transmission which is successfully acknowledged. To calculate the access delay, we need to derive the joint probability distribution of the number of packets in the device queue and the remaining service time for the packet which is currently being serviced. These distributions may be obtained using the previously derived probability distribution of the queue size at packet departure time. To facilitate derivation we will introduce the following variables:

- The current queue length will be denoted with Q.

- The elapsed packet service time T_{t_-} is the time from the previous packet departure until some arbitrary time before the departure of the current packet. The probability density function of the elapsed packet service time is $(1 - T_{dt}(x))/T_t'(1)$.

- The remaining packet service time T_{t+} is the time from the arbitrary time between two successive packet departures up to the first packet departure.

- The number of packet arrivals at the device queue during the elapsed packet service time is $A(T_{t-})$.

$T_{dt}(x) = \text{Prob}[T_{dt} \leq x]$ will denote the probability distribution of the packet service time, while $t_{dt}(x)$ will denote the corresponding probability density function (pdf). According to the results of renewal theory (Kleinrock 1972), the probability density functions of the elapsed packet service time and of the remaining packet service time, respectively, are

$$
\begin{aligned}
t_{dt-}(x) &= \frac{1 - T_{dt}(x)}{T_t'(1)}, \\
t_{dt+}(x) &= \frac{t_{dt}(x)}{1 - T_{dt}(x)}.
\end{aligned}
\tag{3.54}
$$

The joint probability distribution of queue size and remaining packet service time is defined as

$$
\Pi_k^*(s) = \int_0^\infty e^{-sy} \text{Prob}[Q = k, \, y < T_{t+} < y + dy], \, 1 \leq k \leq L
\tag{3.55}
$$

where L denotes the size of the device buffer.

By using the probability distribution of queue size at packet departures and the number of packet arrivals during the elapsed service time T_{t-}, we obtain the expressions for system state at arbitrary time between departures as

$$
\begin{aligned}
\Pi_k^*(s) = \ & \lambda T_t'(1)(1 - P^B)\pi_0 E[e^{-sT_{t+}} | A(T_{t-}) = k - 1]\text{Prob}[A(T_{t-}) = k - 1] \\
& + \lambda T_t'(1)(1 - P^B) \sum_{j=1}^k \pi_j E[e^{-sT_{t+}} | A(T_{t-}) = k - j]\text{Prob}[A(T_{t-}) = k - j],
\end{aligned}
$$

$$
\text{for } 1 \leq k \leq L - 1
$$

$$
\begin{aligned}
\Pi_L^*(s) = \ & \lambda T_t'(1)(1 - P^B)\pi_0 \sum_{k=L-1}^\infty E[e^{-sT_{t+}} | A(T_{t-}) = k]\text{Prob}[A(T_{t-}) = k] \\
& + \lambda T_t'(1)(1 - P^B) \sum_{j=1}^{L-1} \pi_j \sum_{k=L-j}^\infty E[e^{-sT_{t+}} | A(T_{t-}) = k]\text{Prob}[A(T_{t-}) = k]
\end{aligned}
\tag{3.56}
$$

Equations (3.56) can be simplified using the substitution

$$
\begin{aligned}
\psi_k^*(s) &= E\left[e^{-sT_{t+}} | A(T_{t-}) = k\right] \text{Prob}[A(T_{t-}) = k] \\
&= \int_0^\infty \frac{(\lambda x)^k}{k!} e^{-\lambda x} t_{dt-}(x) dx \int_0^\infty e^{-sy} t_{dt+}(x + y) dy \\
&= \int_0^\infty \frac{(\lambda x)^k}{k!} e^{-\lambda x} \frac{1 - T_{dt}(x)}{T_t'(1)} dx \int_0^\infty e^{-sy} \frac{t_{dt}(x + y)}{1 - T_{dt}(x)} dy
\end{aligned}
\tag{3.57}
$$

Expression (3.57), for any $k = 0, 1, 2, \ldots$, can be further simplified to

$$
\begin{aligned}
\psi_k^*(s) &= \frac{1}{T_t'(1)} \int_0^\infty \frac{1}{i!} \frac{(\lambda x)^k}{k!} e^{-\lambda x} dx \int_0^\infty e^{-sy} t_{dt}(x+y) dy \\
&= \frac{1}{T_t'(1)} \int_0^\infty \frac{1}{k!} \frac{(\lambda x)^k}{k!} e^{(s-\lambda)x} dx \int_x^\infty e^{-su} t_{dt}(u) du \\
&= \frac{1}{T_t'(1)} \int_0^\infty e^{-su} t_{dt}(u) du \int_0^u \frac{1}{k!} \frac{(\lambda x)^k}{k!} e^{(s-\lambda)x} dx \\
&= \frac{\lambda}{T_t'(1)} \left(T_t^*(s) \left(\frac{\lambda}{\lambda - s} \right)^{k+1} - \sum_{l=0}^k a_l \left(\frac{\lambda}{\lambda - s} \right)^{k+1-l} \right)
\end{aligned}
\tag{3.58}
$$

where a_l was defined in Equation (3.45). Equations (3.56) can then be rewritten as:

$$
\Pi_k^*(s) = \lambda T_t'(1)(1 - P^B) \left(\pi_0 \psi_{k-1}^*(s) + \sum_{j=1}^k \pi_j \psi_{k-j}^*(s) \right), \quad \text{for } 1 \le k \le L - 1
$$

$$
\Pi_L^*(s) = \lambda T_t'(1)(1 - P^B) \left(\pi_0 \sum_{k=L-1}^\infty \psi_k^*(s) + \sum_{j=1}^{L-1} \pi_j \sum_{k=L-j}^\infty \psi_k^*(s) \right)
\tag{3.59}
$$

By substituting expressions (3.58) and (3.48) in Equation (3.59) we obtain:

$$
\begin{aligned}
\Pi_k^*(s) &= (1 - P^B) \left(T_t^*(s) \left(\pi_0 \left(\frac{\lambda}{\lambda - s} \right)^k + \sum_{j=1}^k \pi_j \left(\frac{\lambda}{\lambda - s} \right)^{k-j+1} \right) \right. \\
&\qquad \left. - \sum_{j=0}^{k-1} \pi_j \left(\frac{\lambda}{\lambda - s} \right)^{k-j} \right) \\
\Pi_L^*(s) &= \frac{\lambda}{s}(1 - P^B) \left(T_t^*(s) \left(\pi_0 \left(\frac{\lambda}{\lambda - s} \right)^{L-1} + \sum_{j=1}^{L-1} \pi_j \left(\frac{\lambda}{\lambda - s} \right)^{L-j} \right) \right. \\
&\qquad \left. - \sum_{j=0}^{L-1} \pi_j \left(\frac{\lambda}{\lambda - s} \right)^{L-1-j} \right)
\end{aligned}
\tag{3.60}
$$

Finally, we are able to determine the LST of the access delay. Since the access delay is the time elapsed from the packet arrival until the packet departure, the probability distribution of the queue size at packet arrival times is the same as the probability distribution of the queue size at arbitrary time. This is the consequence of the PASTA property (Poisson Arrivals See Time Averages). Access delay is equal to the sum of the remaining processing time of the packet currently in service (backoff, transmission, acknowledgment and potential retries), service times of all the packets queued before the arrival of the 'target' packet, and the service time of the target packet. Therefore, the LST of the access delay has the form

$$
D^*(s) = \frac{T_t^*(s)}{1 - P^B} \left(P_0 + \sum_{k=1}^{L-1} \Pi_k^*(s)(T_t^*(s))^{k-1} \right)
\tag{3.61}
$$

If we substitute P_0 from Equation (3.53), we obtain

$$D^*(s) = \frac{\pi_0 s T_t^*(s)\left(1 - \left(\frac{\lambda T_t^*(s)}{\lambda - s}\right)^L\right)}{s - \lambda + \lambda T_t^*(s)} + (T_t^*(s))^L \sum_{j=0}^{L-1} \pi_j \left(\frac{\lambda}{\lambda - s}\right)^{L-j} \tag{3.62}$$

The k-th moment of the access delay can be determined from the corresponding LST as

$$\overline{D^{(k)}} = (-1)^k \frac{d^k D^*(s)}{ds^k}\bigg|_{s=0} \tag{3.63}$$

In particular, the first and second moment of the access delay have the form

$$\begin{aligned}
\overline{D} &= \frac{1}{\lambda}\sum_{k=1}^{L-1} k\pi_k + \frac{L}{\lambda}\left(\pi_0 + \lambda T_t'(1) - 1\right) \\
\overline{D^{(2)}} &= (L-1)\left((L-2)\overline{T_t}^2 + \overline{T_t^{(2)}} - \frac{L\overline{T_t}\,p i_0}{\lambda} - \frac{2\overline{T_t}}{\lambda}\sum_{i=1}^{L-1} i\pi_{L-i}\right) \\
&\quad + \frac{1}{\lambda^2}\sum_{i=1}^{L-1} i(i+1)\pi_{L-i}
\end{aligned} \tag{3.64}$$

3.6.1 Throughput considerations

The output process from the device queue has the LST of

$$O^*(s) = (1 - \pi_0)T_t^*(s) + \pi_0 \frac{\lambda}{s+\lambda}T_t^*(s) \tag{3.65}$$

with the first two moments of

$$\begin{aligned}
\overline{O} &= T_t'(1) + \frac{\pi_0}{\lambda} \\
\overline{O^{(2)}} &= \overline{T_t^{(2)}} + \frac{2\pi_0(1 + T_t'(1)\lambda)}{\lambda^2}
\end{aligned} \tag{3.66}$$

where $\overline{T_t^{(2)}}$ represents the second moment of the packet service time.

We can also calculate an equivalent measure which is the number of packets (per second) correctly received by the cluster coordinator. In sensor network applications, this value is often referred to as the event sensing reliability and denoted with R (Sankarasubramaniam et al. 2003).

For non-acknowledged transfer, the event sensing reliability can be calculated as:

$$R = n\lambda(1 - P^B)\gamma\delta/t_{boff} \tag{3.67}$$

where t_{boff} denotes the duration of the basic backoff period; when the 802.15.4 cluster uses the 2.4 GHz PHY option, this period equals 32 μs.

For acknowledged, fully reliable transfer, event sensing reliability is equal to:

$$R = n\lambda(1 - P^B)/t_{boff} \tag{3.68}$$

For acknowledged, partially reliable transfer, event sensing reliability is equal to:

$$R = n\lambda(1 - P^B)P_a/t_{boff} \tag{3.69}$$

3.7 Performance Results

We will now demonstrate the results obtained through the queueing theoretic analysis presented above. For simplicity, we will assume that the cluster operates in the ISM band at 2.4 GHz, which results in raw data rate of 250 kbps (note, however, that the performance when some of the other PHY options is used can easily be obtained through appropriate scaling). In that case, one modulation symbol corresponds to four data bits, *aUnitBackoffPeriod* has 10 bytes, while *aBaseSlotDuration* has 30 bytes; as *aNumSuperframeSlots* is 16, the *aBaseSuperframeDuration* is exactly 480 bytes. The parameters of the slotted CSMA-CA MAC algorithm were set to their default values, i.e., the minimum value of the backoff exponent, *macMinBE*, is set to three; the maximum value of backoff exponent, *aMaxBE*, is set to five; and the maximum number of backoff attempts, *macMaxCSMABackoffs*, is set to five.

The packet size has been fixed at $G'_p(1) = 3$ backoff periods, while the device buffer had a fixed size of $L = 2$ packets. The packet size includes the PHY layer header of six bytes and the MAC layer header of nine bytes (including the Frame Check Sequence fields). Such a short MAC header implies that the destination addressing mode subfield (bits 10-11) within frame control field is set to 0, and source addressing mode field (bits 14-15) is set to short address mode. This means that packet is directed to the coordinator with the PAN identifier as specified in the source PAN identifier field. Finally, the bit error rate at the PHY layer was set to $BER = 10^{-4}$.

3.7.1 Superframe with active period only

In this case, the active part of the superframe is equal to the beacon interval, which is accomplished by setting to zero the values for both the superframe order *SO* and beacon interval *BO*. Other parameters were set to the values listed in Table 3.1. We will consider acknowledged, fully reliable transfer since it imposes the highest load to the cluster, and our intention is to explore limits of linear network operation. To that end, we have solved the system formed by Equations (3.7), (3.11), (3.13), (3.14), and (3.50), together with auxiliary equations which we substituted in the main ones.

Figure 3.5 shows various success probabilities in the cluster where the superframe has active period only. The diagrams show the probability α that the medium is idle on the first CCA, the probability β that the medium is idle on the second CCA, the probability τ to access the medium within the basic backoff period of $t_{boff} = 32$ μs, and overall probability

Table 3.1 Parameters used in performance analysis (no in-active period)

number of nodes, n	5–50
packet arrival rate per node, λ	30–300 packets per minute
buffer size at node, L	2 packets
raw data rate	250kbps
superframe duration, SD	480 bytes
transfer type	acknowledged, fully reliable
bit error rate, BER	10^{-4}

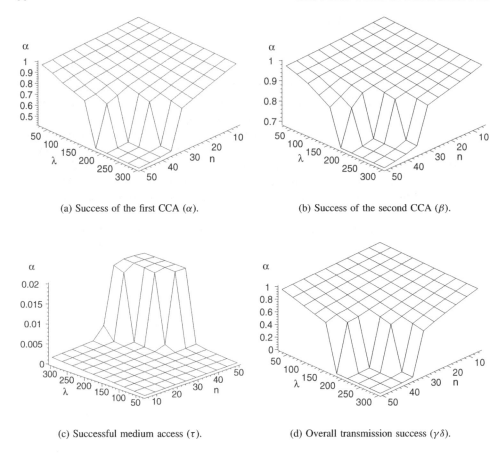

(a) Success of the first CCA (α).

(b) Success of the second CCA (β).

(c) Successful medium access (τ).

(d) Overall transmission success ($\gamma\delta$).

Figure 3.5 Success probabilities in the cluster where the superframe has no inactive period.

of successful transmission, $\gamma\delta$. Independent variables, in this case, are packet arrival rate per node per minute, denoted with λ, and number of nodes in the cluster, n.

Figure 3.6 shows the buffer blocking probability, the total traffic that reaches the coordinator (i.e., event sensing reliability), mean packet service time, and mean packet access delay. Event sensing reliability is expressed in packets per second, while packet delays are expressed in backoff periods. Two well-defined operating regions can be discerned in virtually all diagrams. In the first region, the probability of access can be seen to be rather close to the packet arrival rate scaled by the backoff period, i.e., $\tau \approx \frac{\lambda t_{boff}}{60}$. This means that most packet transmissions are successful (i.e., most packets will be transmitted only once) and that packet re-transmissions occur so seldom that the process of servicing the packets in the buffer is not affected. This is corroborated by the negligible packet blocking probability in that region, as shown in Figure 3.6(a), as well as by the small packet service times, Figure 3.6(c). In this region the values of success probabilities α, β, and $\gamma\delta$ decrease in an almost linear fashion, with their highest values being close to one.

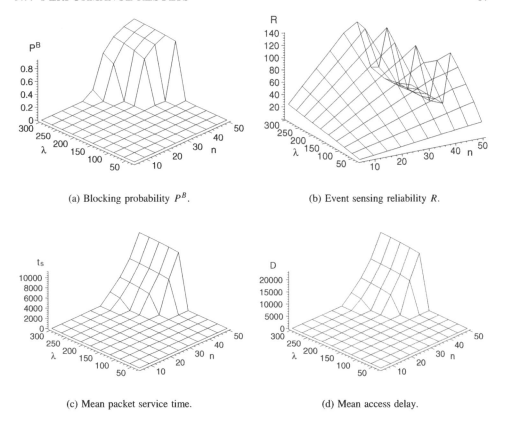

(a) Blocking probability P^B.

(b) Event sensing reliability R.

(c) Mean packet service time.

(d) Mean access delay.

Figure 3.6 Performance indicators in the cluster where the superframe has no inactive period.

Values of the access delay are initially small, which indicates that most packets do not wait too long before they are transmitted. However, when the arrival rate and/or cluster size increase, an increasing proportion of newly arriving packets will find that the node buffer already has a packet waiting to be serviced. As a result, the value of the access delay is approximately twice that of the individual packet service time, since it includes the service time of the packet which is already in the buffer. Finally, total traffic to reach the coordinator linearly increases with the increase of packet arrival rate per node and number of nodes.

However, once a certain threshold of λ and n is exceeded, the cluster switches over to a different operating region in a rather abrupt manner, as can be seen from the diagrams. In this region, the probabilities of success on first and second CCA flatten at their lower bounds which depend on the packet size, while the probability of successful transmission reaches its lower bound of zero. Packet service time and access delay increase to large values, while aggregate packet rate to reach the coordinator sharply drops toward zero. Large values of access probability which are not caused by a comparative increase in packet arrival rates indicate that nodes access the medium frequently and manage to transmit their packets – only to collide with other packet(s).

This region will be referred to as the 'saturation region', following the pioneering work in the analysis of 802.11 MAC (Bianchi 2000). However, saturation throughput in 802.11 networks is well above zero due to two facts: first, nodes listen to the medium during backoff countdown, and freeze their backoff counters when they detect activity on the medium; second, the sizes of backoff windows are much larger in 802.11. On the contrary, the packet throughput in an 802.15.4 cluster that reaches saturation drops to zero, which means that the cluster effectively ceases to operate. This is due to the reduced range of backoff windows and to the fact that backoff countdowns are allowed to proceed regardless of medium activity.

Saturation effects would be somewhat reduced (i.e., the onset of the saturation regime would simply be pushed toward higher packet arrival rates and/or network sizes) if the cluster were to operate with longer active periods. However, the differences would be quantitative, rather than qualitative, since the underlying mechanism remains the same; and the improvement that can be achieved in this manner would be small because the range of backoff countdown values is rather small – zero to 31 at most, as discussed earlier in Section 2.3.

3.7.2 Superframe with both active and inactive periods

In this case, we have set the superframe order and beacon order to $SO = 0$ and $BO = 1$, respectively, which result in active and inactive periods of equal duration of 480 bytes each; the remaining parameters were set to the values listed in Table 3.2. In order to find the limits of non-saturated operating regime, we consider only the acknowledged, fully reliable transfer. In this case, the system of equations that needs to be solved consists of Equations (3.21), (3.30), (3.31), (3.32), (3.50) together with the auxiliary related equations that define appropriate substitutions.

Figure 3.7 shows various success probabilities for this case, again with packet arrival rate per node per minute, λ, and number of nodes in the cluster, n, as independent variables. Figure 3.8 shows buffer blocking probability, event sensing reliability, mean packet service time, and mean packet access delay. As can be seen, maximum achievable event sensing

Table 3.2 Parameters used in performance analysis (inactive period present)

number of nodes	5–50
packet arrival rate per node	30–300 packets per minute
buffer size at each node	2 packets
raw data rate	250kbps
superframe size	960 bytes
transfer type	acknowledged, fully reliable
bit error rate BER	10^{-4}

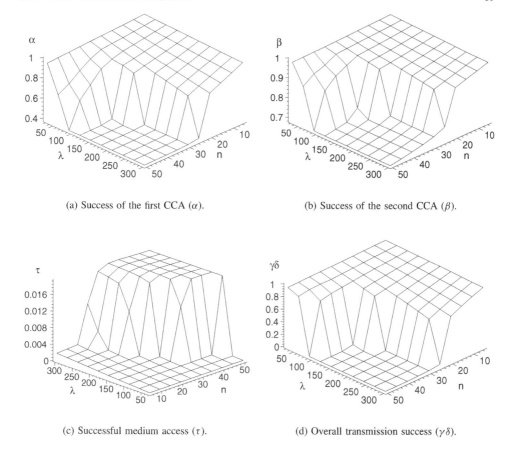

(a) Success of the first CCA (α).

(b) Success of the second CCA (β).

(c) Successful medium access (τ).

(d) Overall transmission success ($\gamma\delta$).

Figure 3.7 Success probabilities in the cluster where the superframe has both an active and an inactive period.

reliability is significantly less than half of the corresponding throughput in the cluster in which the superframe has no inactive period. This is due to the smaller available bandwidth (which is, at best, 50% of the one available in the previous case), but also to the effects of packet blocking (which is higher) and increased collision rate due to the use of fully acknowledged transfers.

The mean access delay is again about twice as high as the packet service time, as most newly arrived packets will find that the node buffer already has a packet waiting to be serviced (the node buffer size is $L = 2$ packets).

Overall, the linear (i.e., non-saturated) region of operation is much smaller that the corresponding limits obtained for the case where the superframe has no inactive period. Of course, the performance worsens with the increase of the ratio of duration of inactive over active period.

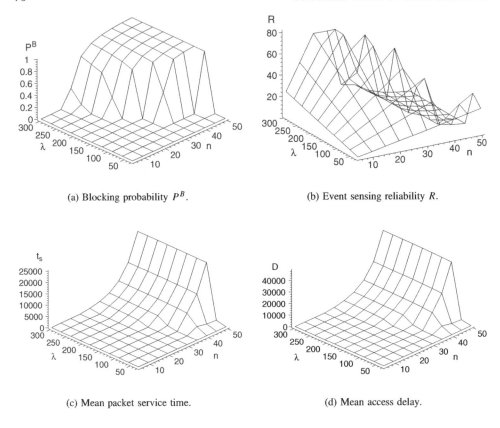

(a) Blocking probability P^B.   (b) Event sensing reliability R.

(c) Mean packet service time.   (d) Mean access delay.

Figure 3.8 Performance indicators in the cluster where the superframe has both an active and an inactive period.

It should be noted that the main justification for employing the superframe with an inactive period is not performance, but rather power efficiency. However, our results indicate that applications in which power consumption is the main concern should reduce their throughput requirements accordingly; they should employ fewer nodes and/or lower values of packet arrival rate per node.

4

Cluster with Uplink
and Downlink Traffic

In this chapter, we consider a star-topology 802.15.4 cluster with the coordinator and n ordinary nodes. The cluster operates in beacon enabled mode with both downlink and uplink traffic. As in the previous chapter, our analysis is based on the theory of discrete-time Markov chains and the theory of M/G/1 queues. Together, they allow us to derive the probability distributions of packet service time, access delay, queue lengths, and throughput in either direction. Furthermore, we derive the stability limits for individual queues and employ them to analyze the conditions that cause the network to enter the saturation region. Unlike the previous chapter, however, we consider only the acknowledged, fully reliable transmission, but not the non-acknowledged one; this allows us to closely explore the limits of linear, non-saturated network operation.

4.1 The System Model

4.1.1 Operational states

When both uplink and downlink traffic are present in a cluster, the analysis is more complex than in the case of a cluster that has uplink traffic only. The following operational states of the cluster coordinator may easily be identified from the description of uplink and downlink traffic in Section 2.5:

1. The coordinator may be transmitting the beacon frame.

2. The coordinator may be listening to the ordinary nodes and receiving uplink data packets.

3. The coordinator may also be listening to the ordinary nodes and receiving data request packets sent by nodes that have learned about pending downlink packets, which is immediately followed by the transmission of request acknowledgment packets.

Wireless Personal Area Networks   Jelena Mišić and Vojislav B. Mišić
© 2008 John Wiley & Sons, Ltd

4. Upon receiving and acknowledging a data request packet, the coordinator will undertake the transmission of the pending downlink data packet; as soon as the downlink transmission is finished (and, optionally, acknowledged), the coordinator switches back to the listening mode.

Note that the first two states are present in a cluster with uplink traffic only; the latter two occur only in relationship with downlink traffic.

Likewise, an ordinary (i.e., non-coordinator) node in the cluster can be in one of the following states:

1. The node may be transmitting an uplink data packet.

2. The node may be transmitting a data request packet to the coordinator.

3. The node may be in an uplink request synchronization state, which is a virtual state that lasts from the moment of the arrival of a new downlink packet at the coordinator (or the failure of the previous downlink reception) to the beginning of the CSMA-CA procedure for the uplink data request. (This state is analyzed in detail in Section 4.1.7.)

4. Upon successful transmission of a data request packet to the cluster coordinator, the node may be waiting for the downlink packet transmission.

5. The node may also be in an idle state, without any downlink or uplink transmission pending or in progress.

In this model, uplink transmissions follow the procedure outlined in Chapter 3. However, the procedure for downlink transmissions differs: each of these must be preceded by a successful transmission of the data request packet to, and the subsequent acknowledgment from, the coordinator. Those packets may experience collisions, or they may arrive while the coordinator is executing backoff countdown and thus are ignored. Thus, the behavior of the coordinator corresponds to the $M/G/1/1$ model. Upon receipt of a request, the coordinator will acknowledge it; the absence of acknowledgment means that the node must repeat the request transmission procedure.

If a downlink transmission was successful but the downlink queue towards the node is not empty, the coordinator will announce the presence of another pending packet in the next beacon frame and the node will start a new downlink transmission cycle. If the downlink queue was empty but the uplink queue contained a packet, the node will initiate the uplink transmission cycle.

After a successful uplink or downlink transmission, the node will enter the idle state if both of its data queues are empty. The node will leave the idle state upon the arrival of a packet to either queue during the current backoff period. If the packet has arrived to the uplink data queue, the node will enter the uplink transmission state; if the packet has arrived to the downlink queue, the node will enter the uplink data request synchronization state. In case of simultaneous packet arrivals to both the uplink and the downlink queues, the downlink transmission is given non-preemptive priority over the uplink one (IEEE 2006); in other words, the node will undertake the extraction of the pending downlink data packet before the transmission of an uplink data packet.

Both uplink and downlink packet arrivals to the node follow a Poisson process with the average arrival rate of λ_u and λ_d, respectively. The latter assumption needs to be explained in more detail. Namely, the queueing system is implemented in a distributed fashion, since the downlink data queue is physically hosted at the cluster coordinator, rather than at the node itself. Since this queue is actually fed by packets from all other uplink queues, exact analysis of its operation is rather involved. However, we may assume that downlink destinations for each uplink node are uniformly distributed. In this case, the arrival process for each downlink queue is a sum of a large number of independent arrivals at very low rates. This allows us to assume that the downlink queues operate independently of the uplink queues. By extension, the arrivals to the downlink queue may be assumed to follow a Poisson process. Note, however, that the corresponding announcements in the beacon (from which the target node finds out about those packets) do not follow the Poisson distribution.

Whenever necessary, different high-level states of a node will be distinguished through appropriate indices: ud for uplink data transmissions, ur for data request transmissions, and dd for the downlink data transmissions. Using this notation, the following performance descriptors can be identified:

- In the uplink data transmission state, we will need the uplink data packet service time, which has the PGF of $T_{ud}(z)$ and mean value of $\overline{T_{ud}}$. The offered load for the uplink data queue of the device i is denoted with $\rho_{ud} = \lambda_u \overline{T_{ud}}$.

- In the uplink request transmission state, the PGF for the uplink request packet service time is given with $T_{ur}(z)$ and average value $\overline{T_{ur}}$. The offered load for the uplink data queue of the device i is denoted with ρ_{ur}.

- In the downlink data reception state, the PGF for the downlink transmission is denoted as $T_{dd}(z)$ with average value of $\overline{T_{dd}}$. The total offered load for the downlink data queue at the coordinator is denoted with $\rho_{dtot} = n\lambda_d \overline{T_{dd}}$, where we assume that the cluster consists of n identical nodes. The offered load towards one node is $\lambda_d \overline{T_{dd}}$.

4.1.2 Markov chain model for a node

The states defined above and their interconnection are schematically shown in Figure 4.1. Packet transmissions in all three high level states follow the same CSMA-CA algorithm depicted in Figure 4.2, which is why it is represented as a single reusable component.

The packet queues in the device data buffer and the data request buffer are modeled as a $M/G/1$ queueing system, in which the packet request queue has non-preemptive priority over the data queue at the device. (It might be argued that the $M/G/1/K$ system would be more accurate, but the increase in complexity would be unjustifiably high.)

Note that the uplink data transmission, the uplink request transmission, and the downlink data transmission all use the same CSMA-CA algorithm from Section 2.3. This algorithm can be modeled with the discrete-time Markov chain block presented in Figure 4.2. As before, the case when the two CCAs, packet transmission, and (optional) acknowledgment are deferred to the next superframe is modeled as a 'delay line' shown in Figure 4.3. (The probability $P_{d,i}$ of this event happening after the i-th backoff attempt is derived in Section 5.1) below. We assume that this Markov chain, together with the higher level

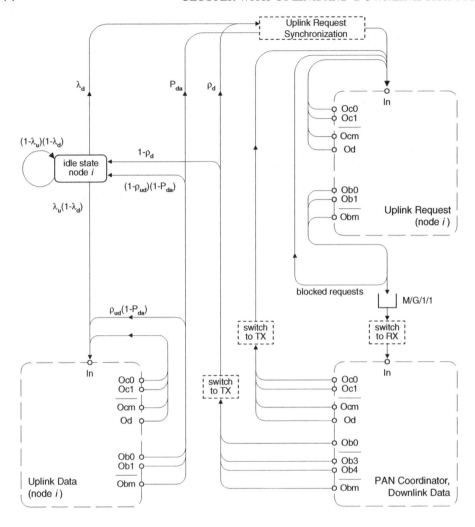

Figure 4.1 Complete Markov chain model of a node. Adapted from J. Mišić, S. Shafi, and V. B. Mišić, 'Performance of a beacon enabled IEEE 802.15.4 cluster with downlink and uplink traffic,' *IEEE Trans. Parallel Dist. Syst.*, **17**(4): 361–377, © 2006 IEEE.

structure into which it is incorporated, has a stationary distribution (Bianchi 2000). The process $\{i, c, k, d\}$ then defines the state of the device at backoff unit boundaries. In this notation, i is the index of the backoff attempt, c is the index of the CCA, k is the current value of the backoff counter, and d denotes the index of the state within the delay line mentioned above. (Although all of these variables are functions of time, we have omitted the time dependency $\cdot(t)$ for brevity.) In order to reduce notational complexity, the index

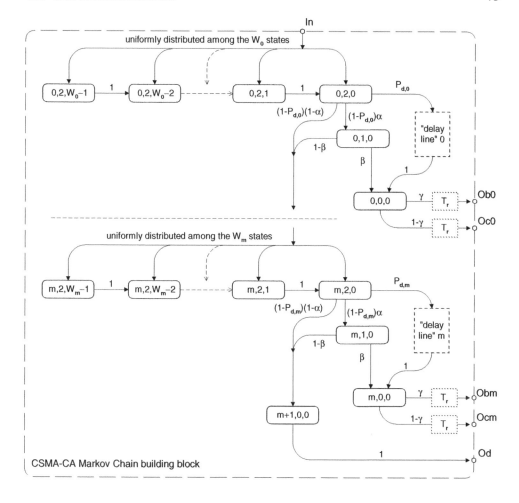

Figure 4.2 General Markov chain model of the slotted CSMA-CA algorithm representing a non-idle state of the node. Adapted from J. Mišić, S. Shafi, and V. B. Mišić, 'Performance of a beacon enabled IEEE 802.15.4 cluster with downlink and uplink traffic,' *IEEE Trans. Parallel Dist. Syst.*, **17**(4): 361–376, © 2006 IEEE.

d is shown only within the delay line; it is omitted in other cases where its value is undefined.

A number of relevant variables, most of which have already been defined in Chapter 3, are needed to analyze this Markov chain:

- the PGF for the duration of a beacon frame is $B_{ea}(z) = \sum_{k=2}^{5} b_k z^k$;

- the PGF for the data packet length is $G_p(z) = z^k$;

- the PGF for the duration of an uplink data request packet is $G_r(z) = z^2$;

- the PGF for the interval between the end of a packet and subsequent acknowledgment be $t_{ack}(z) = z^2$;

- the PGF for the duration of an acknowledgment packet is $G_a(z) = z$;

- the PGF for the total transmission time of an uplink data request packet is $D_r(z) = z^2 G_r(z) t_{ack}(z) G_a(z)$, while its mean value is $\overline{D_r} = 2 + G'_r(1) + t'_{ack}(1) + G'_a(z)$;

- finally, the PGF for the total transmission time of a data packet (regardless of the transmission direction) is $D_d(z) = z^2 G_p(z) t_{ack}(z) G_a(z)$, and its mean value is $\overline{D_d} = 2 + G'_p(1) + t'_{ack}(1) + G'_a(z)$.

For simplicity, we assume that *BO* has a constant value of zero, in which case the superframe duration is *SD = aBaseSuperframeDuration*. Different superframe durations, as well as the presence of inactive periods, can be accommodated with ease, in the manner outlined in Chapter 3. The beacon interval and the duration of the superframe are determined by the energy management policy of the network.

For convenience, let W_0 stand for $2^{macMinBE}$. Then, the maximum value of the random waiting time, expressed in units of backoff periods, that corresponds to i will be $W_i = W_0 2^m$, where $m = min(i, 5 - macMinBE)$. Note that the value of *BE* cannot exceed *macMaxBE* = 5.

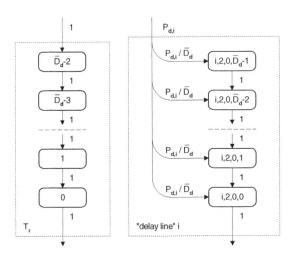

Figure 4.3 Delay lines for Figure 4.2. Adapted from J. Mišić, S. Shafi, and V. B. Mišić, 'Performance of a beacon enabled IEEE 802.15.4 cluster with downlink and uplink traffic,' *IEEE Trans. Parallel Dist. Syst.*, **17**(4): 361–376, © 2006 IEEE.

The non-null transition probabilities can be described by the following equations:

$$
\begin{aligned}
P\{0, 2, k\} &= \text{Prob}(In)/W_0, & k &= 0 .. 2^{BE} - 1 \\
P\{i, 2, k - 1 \mid i, 2, k\} &= 1, & i &= 0 .. m; k = 1 .. 2^{BE} - 1 \\
P\{i, 1, 0 \mid i, 2, 0\} &= \alpha(1 - P_{d,i}), & i &= 0 .. m \\
P\{i, 0, 0 \mid i, 1, 0\} &= \beta, & i &= 0 .. m \\
P\{i + 1, 2, k \mid i, 2, 0\} &= \frac{(1 - \alpha)(1 - P_{d,i})}{W_{i+1}}, & i &= 0 .. m; k = 0 .. 2^{BE} - 1 \\
P\{i + 1, 2, k \mid i, 1, 0\} &= (1 - \beta)/W_{i+1}, & i &= 0 .. m; k = 0 .. 2^{BE} - 1 \\
P\{i, 2, 0, l \mid i, 2, 0\} &= P_{d,i}/\overline{D_d}, & i &= 0 .. m; l = 0 .. \overline{D_d} - 1 \\
P\{i, 2, 0, l - 1 \mid i, 2, 0, l\} &= 1, & i &= 0 .. m; l = 0 .. \overline{D_d} - 1 \\
P\{i, 0, 0 \mid i, 2, 0, 0\} &= 1 & i &= 0 .. m \\
\text{Prob}(Ob_i) &= \gamma P\{i, 0, 0\} & i &= 0 .. m \\
\text{Prob}(Oc_i) &= (1 - \gamma)P\{i, 0, 0\} & i &= 0 .. m
\end{aligned}
\tag{4.1}
$$

The first equation in the set (4.1) shows the connection between two states of the node. The variable $\text{Prob}(In)$ denotes the probability of the transition into the block through the In port at the top Figure 4.2; likewise, $\text{Prob}(Ob_i)$ and $\text{Prob}(Oc_i)$ denote the probabilities that the CSMA-CA Markov chain block will be exited through the outgoing ports Ob_i and Oc_i, for $i = 0 .. m$.

The second equation shows the probability that the backoff time is decremented after each $aUnitBackoffPeriod$.

The third equation determines the probability α that the channel is sensed to be idle in the first CCA, i.e., when the backoff counter reaches zero, while the fourth equation determines the probability β that the channel is sensed to be idle in the second CCA, i.e., after it was already sensed idle for one backoff period. It is worth noting that both of these probabilities may take two values, depending on the transmission direction:

- When an ordinary node undertakes an uplink transmission, it competes for medium access against the coordinator and $n - 1$ other nodes.

- The coordinator that undertakes a downlink transmission has to compete against $n - 1$ ordinary nodes only, since the destination node is waiting for the downlink packet.

The corresponding probabilities are denoted with α_u and β_u, for the uplink transmission, and with α_d and β_d, for the downlink transmission; in general, $\alpha_u \neq \alpha_d$ and $\beta_u \neq \beta_d$.

The fifth and sixth equations describe the probabilities that the device, upon sensing the channel to be busy, will choose another random backoff in the range $0 .. W_{i+1} - 1$.

The seventh, eighth, and ninth equations describe the delay line which is entered if the remaining time in the CAP of the current superframe is insufficient for all the actions necessary to complete the transmission (Section 2.3).

The last two equations express the probabilities of transitions to other system blocks through appropriate output ports shown in Figure 4.2. The success probability γ denotes the probability that a transmission will not experience a collision; it has to be evaluated for uplink and downlink directions separately.

For simplicity, we assume that the bit error rate is equal to zero, $BER = 0$, in which case the probability that the packet is not corrupted by noise and interference is $\delta = 1$.

4.1.3   Integration of transmission blocks

Let the stationary distribution of the overall Markov chain be given with the probability of idle state x^z and probabilities of three active states $x^{ud}_{i,c,k,d}, x^{ur}_{i,c,k,d}, x^{dd}_{i,c,k,d}$ for $i = 0 \ldots m$; $c = 0, 1, 2; k = 0 \ldots 2^{BE} - 1$; and $d = 0 \ldots \overline{D_d} - 1$. (For brevity, the index d will be omitted whenever it is zero.) We will also introduce auxiliary variables $C_{1,i}, C_{2,i}$, and $C_{3,i}$ (where i denotes the index of the backoff attempt), through the following equations:

$$
\begin{aligned}
x^{ud}_{0,1,0} &= x^{ud}_{0,2,0}(1 - P_{d,0})\alpha_u = x^{ud}_{0,2,0}C_{1,0} \\
x^{ud}_{1,2,0} &= x^{ud}_{0,2,0}(1 - P_{d,0})(1 - \alpha_u\beta_u) = x^{ud}_{0,2,0}C_{2,0} \\
x^{ud}_{0,0,0} &= x^{ud}_{0,2,0}((1 - P_{d,0})\alpha_u\beta_u + P_{d,0}) = x^{ud}_{0,2,0}C_{3,0}.
\end{aligned}
\tag{4.2}
$$

The same relations hold for states of uplink request transmission and downlink data reception. Also, the following relations hold within each of the non-idle states:

$$
x^{ud}_{i,0,0} = x^{ud}_{0,0,0}\frac{C_{3,i}}{C_{3,0}}\prod_{j=0}^{i-1}C_{2,j}, \qquad\qquad i = 0 \ldots m
$$

$$
\sum_{i=0}^{m}x^{ud}_{i,0,0} = \sum_{i=0}^{m}\frac{C_{3,i}}{C_{3,0}}\left(\prod_{j=0}^{i-1}C_{2,j}\right)x^{ud}_{0,0,0}
$$

$$
= C_4 x^{ud}_{0,0,0}, \qquad\qquad i = 0 \ldots m
$$

$$
x^{ud}_{i,2,k} = x^{ud}_{0,0,0}\frac{W_i - k}{W_i}\cdot\frac{\prod_{j=0}^{i-1}C_{2,j}}{C_{3,0}}, \qquad i = 1 \ldots m; k = 0 \ldots W_i - 1
$$

$$
x^{ud}_{i,1,0} = x^{ud}_{0,0,0}\frac{C_{1,i}}{C_{3,0}}\prod_{j=0}^{i-1}C_{2,j} \qquad\qquad i = 0 \ldots m
\tag{4.3}
$$

$$
x^{ud}_{0,2,k} = x^{ud}_{0,0,0}\frac{W_0 - k}{W_0\,C_{3,0}} \qquad\qquad k = 1 \ldots W_0 - 1
$$

$$
x^{ud}_{m+1,0,0} = \frac{x^{ud}_{0,0,0}}{C_{3,0}}\prod_{j=0}^{m}C_{2,j}
$$

$$
= C_9 x^{ud}_{0,0,0}
$$

$$
\sum_{d=0}^{D_d-1}x^{ud}_{i,2,0,d} = \frac{x^{ud}_{0,0,0}}{2C_{3,0}}\prod_{j=0}^{i-1}C_{2,j}P_{d,i}(\overline{D_d} - 1).
$$

For brevity, we list only the equations relevant to the uplink data transmission state; analogous equations hold for the other two states. Note that we have introduced substitutions

$$
C_4 = \sum_{i=0}^{m}\frac{C_{3,i}}{C_{3,0}}\prod_{j=0}^{i-1}C_{2,j}
$$

$$
C_9 = \frac{\prod_{j=0}^{m}C_{2,j}}{C_{3,0}}
\tag{4.4}
$$

4.1.4 Transmission of uplink data packets

Additional equations are needed to represent the interconnection of the high-level states in Figure 4.1. Let us denote the probability of one or more downlink packet arrivals during the uplink transmission as $P_{da} = \lambda_d \overline{T_{ud}}$. Then, the following holds:

$$x_{0,2,0}^{ud} = (1 - \gamma_u) \sum_{i=0}^{m} x_{i,0,0}^{ud} + x_{m+1,0,0}^{ud} + x^z \lambda_u (1 - \lambda_d)$$

$$+ \rho_{ud} \gamma_u (1 - P_{da}) \sum_{i=0}^{m} x_{i,0,0}^{ud}. \tag{4.5}$$

Using Equations (4.2) and (4.3), we obtain

$$x_{0,0,0}^{ud} = \frac{x^z \lambda_u (1 - \lambda_d)}{1/C_{3,0} - (1 - \gamma_u)C_4 - C_9 - C_4 \gamma_u \rho_{ud}(1 - P_{da})}. \tag{4.6}$$

If we introduce the substitution $C_5 = \frac{1}{C_3} - (1 - \gamma_u)C_4 - C_9 - C_4 \gamma_u \rho_{ud}(1 - P_{da})$, the last equation becomes

$$x_{0,0,0}^{ud} = \frac{x^z \lambda_u (1 - \lambda_d)}{C_5}. \tag{4.7}$$

The sum of probabilities of all sub-states of the uplink data transmission state is

$$
\begin{aligned}
\Sigma^{ud} &= \sum_{i=0}^{m} \sum_{k=0}^{W_i-1} x_{i,2,k} + (\overline{D_d} - 2) \sum_{i=0}^{m} x_{i,0,0} + \sum_{i=0}^{m} x_{i,1,0} \\
&\quad + x_{m+1,0,0} + \sum_{i=0}^{m} \sum_{d=0}^{\overline{D_d}-1} x_{i,2,0,d} \\
&= x_{0,0,0}^{ud} \left(\sum_{i=0}^{m} \frac{W_i + 1}{2C_{3,0}} \cdot \prod_{j=0}^{i-1} C_{2,j} + C_4 \left(\overline{D_d} - 2 \right) \right. \\
&\quad \left. + \sum_{i=0}^{m} \frac{P_{d,i}(\overline{D_d} - 1)}{2C_{3,0}} \cdot \prod_{j=0}^{i-1} C_{2,j} + \sum_{i=0}^{m} \frac{C_{1,i}}{C_{3,0}} \cdot \prod_{j=0}^{i-1} C_{2,j} + C_9 \right) \\
&= C_6 x_{0,0,0}^{ud}
\end{aligned}
\tag{4.8}
$$

For clarity, state probabilities x and coefficients C are listed without the superscript ud, even though all of them apply to the uplink data transmission state.

4.1.5 Waiting for the downlink data

The state of waiting for a downlink packet is modeled according to the actions of the coordinator when it receives the valid request from node i. Before starting the backoff countdown procedure, the coordinator will switch its radio interface to transmission. During the backoff countdown, no further data requests are received, and they will have to be retransmitted. Therefore we can model the coordinator as a $M/G/1/1$ system, i.e., it keeps only the packet which is currently being serviced. Let us denote the probability that an uplink request is blocked by the coordinator as P^B. According to the theory of $M/G/1/1$ queues, if the offered load to the queue is ρ_x, the blocking probability is

$$P^B = 1 - \frac{1}{1 + \rho_x} \tag{4.9}$$

The exact expression for the blocking probability is derived in Section 4.3.2.

The outputs of the uplink data request sub-chain and the input of the downlink sub-chain are related through

$$x_{0,2,0}^{dd} = (1 - P^B)\gamma_u \sum_{i=0}^{m} x_{i,0,0}^{ur}. \tag{4.10}$$

Using Equations (4.2) and (4.3), but applied to the uplink data request state, we obtain

$$x_{0,0,0}^{dd} = (1 - P^B)\gamma_u C_{3,0}^{dd} C_4^{ur} x_{0,0,0}^{ur}, \tag{4.11}$$

where the superscript dd means that α_d, and β_d are used instead of their uplink counterparts. The sum of probabilities of sub-states in the downlink data transmission state is

$$\Sigma^{dd} = C_6^{dd} x_{0,0,0}^{dd}, \tag{4.12}$$

where C_6^{dd} is obtained from Equation (4.8) by replacing the coefficients C_{ud} pertaining to the uplink data transmission high-level state with their downlink data counterparts.

If the duration of the downlink packet service time exceeds *aMaxFrameResponseTime* = 61 backoff periods, the MAC layer will time-out and decide that the packet is not received. The data extraction procedure is then repeated, starting with the data request packet sent to the coordinator. We model this situation through a time-out at the node after reaching the $m + 1$-th backoff iteration; when this time-out occurs, the packet request is repeated. We use the maximum number of backoff iterations (five) allowed by the standard; given the maximum backoff window sizes in each iteration, the time-out condition corresponds to the sum of average window sizes for five backoff iterations.

By the same token, the data extraction procedure is repeated if the downlink data packet is lost due to a collision.

4.1.6   Uplink data request state

An uplink data request has to precede any downlink transmission state, regardless of whether it is a new packet transmission or a re-transmission of a previous packet because of collision or time-out. For simplicity, we assume that the time limit is not exceeded if the downlink transmission is completed with five or fewer backoff iterations. The input to the uplink request state may be described by

$$\begin{aligned}
x_{0,2,0}^{ur} &= x^z \lambda_d + (1 - \gamma_u) \sum_{i=0}^{m} x_{i,0,0}^{ur} + \gamma_u P^B \sum_{i=0}^{m} x_{i,0,0}^{ur} + x_{m+1,0,0}^{ur} \\
&\quad + \gamma_u P_{da} \sum_{i=0}^{m} x_{i,0,0}^{ud} + (1 - \gamma_d) \sum_{i=0}^{m} x_{i,0,0}^{dd} + C_9^{ur} x_{0,0,0}^{dd} + \gamma_d \rho_d \sum_{i=0}^{m} x_{i,0,0}^{dd}
\end{aligned} \tag{4.13}$$

The superscript r indicates that the appropriate coefficients have been derived with the probability $P_{r,i}$ of the request packet being delayed after the i-th backoff attempt. As this is an uplink transmission, α_u and β_u are used.

The last expression can be simplified through the use of Equations (4.2) and (4.3), when applied to the uplink request state, to read

$$
\begin{aligned}
x_{0,0,0}^{ur} &\left(\frac{1}{C_{3,0}^{ur}} - (1 - \gamma_u)C_4^{ur} - C_9^{ur} - \gamma_u C_4^{ur} P^B \right) \\
&= x^z \left(\lambda_d + P_{da} \frac{\gamma_u C_4^{ud}}{C_5^{ud}} \lambda_u (1 - \lambda_d) \right) + x_{0,0,0}^{dd} \left(C_4^{dd}(1 - \gamma_d) + C_9^{dd} + C_4^{dd}\gamma_d\rho_d \right)
\end{aligned}
\tag{4.14}
$$

By introducing substitutions

$$
\begin{aligned}
C_7^{ur} &= \frac{1}{C_{3,0}^{ur}} - (1 - \gamma_u)C_4^{ur} - C_9^{ur} - \gamma_u C_4^{ur} P^B \\
C_8^{dd} &= C_4^{dd}(1 - \gamma_d) + C_9^{dd} + \gamma_d\rho_d C_4^{dd},
\end{aligned}
\tag{4.15}
$$

Equation (4.14) becomes

$$
x_{0,0,0}^{ur} = \frac{x^z \left(\lambda_d + P_{da} \frac{\gamma_u C_4^{ud}}{C_5^{ud}} \lambda_u (1 - \lambda_d) \right)}{C_7^r - (1 - P^B)\gamma_u C_4^{dd} C_{3,0}^{dd} C_8^{dd}}
\tag{4.16}
$$

The sum of all the probabilities in the uplink request state is

$$
\Sigma^{ur} = C_6^{ur} x_{0,0,0}^{ur},
\tag{4.17}
$$

where C_6^{ur} has the same form as C_6^{ud} but with $\overline{D_d}$ replaced by $\overline{D_r}$.

4.1.7 Uplink request synchronization

Newly arrived downlink packets have to be announced in the beacon. According to the renewal theory, the probability distribution of the time between the packet arrival to the downlink queue at the coordinator and the beginning of the first subsequent beacon has the PGF of

$$
S_1(z) = \frac{1 - z^{SD}}{SD(1 - z)},
\tag{4.18}
$$

with the mean value of $SD/2$. However, according to the standard, at most seven stations can be advertised in the beacon which causes the synchronization time to depend on the number of pending downlink packets. We assume that coordinator that has more than seven downlink packets will announce their presence to the corresponding destination nodes in several beacon frames, in a round-robin fashion.

Let us consider the destination node and $n - 1$ other nodes that generate background traffic, and let us also assume that the coordinator uses the round-robin policy in advertising the destination nodes. The probability that the number of nodes that have a pending downlink packet is between $7(l - 1) + 1$ and $7l - 1$, where $l = 1 \ldots \lceil (n - 1)/7 \rceil$, and that the packet for the destination node advertised in the l-th coming beacon, is

$$s_l = \sum_{k=7(l-1)}^{\min(7l-1,n-1)} \binom{n-1}{k} \rho_{ur}^k (1 - \rho_{ur})^{n-1-k} \tag{4.19}$$

Then, the PGF for the beacon synchronization time after the idle or downlink reception state becomes

$$S(z) = \sum_{l=1}^{L} s_l S_1(z) z^{(l-1)SD} \tag{4.20}$$

However, if the destination node was in the uplink data transmission mode when the downlink packet arrives, the beacon synchronization and uplink transmission will run in parallel for some time. Therefore, we may assume that the node will enter the beacon synchronization state at a random time after the uplink transmission. According to the renewal theory, the duration of the beacon synchronization after the uplink transmission has the PGF of

$$S_u(z) = \frac{1 - S(z)}{S'(1)(1 - z)} \tag{4.21}$$

The Markov chain representation of the uplink request synchronization state is shown in Figure 4.4.

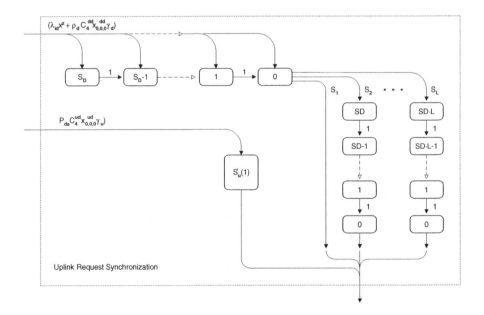

Figure 4.4 Markov chain model for the uplink request synchronization. Adapted from J. Mišić, S. Shafi, and V. B. Mišić, 'Performance of a beacon enabled IEEE 802.15.4 cluster with downlink and uplink traffic,' *IEEE Trans. Parallel Dist. Syst.*, **17**(4): 361–376, © 2006 IEEE.

The mean number of backoff periods inserted by this synchronization time is $S'(1)$, while the sum of probabilities of all states in the beacon delay line is

$$\Sigma^s = (\lambda_d x^z + \rho_d C_4^{dd} x_{0,0,0}^{dd} \gamma_d) S'(1) + P_{da} C_4^{ud} x_{0,0,0}^{ud} \gamma_u S_u'(1). \tag{4.22}$$

At this point, we can also calculate the probability distribution of the beacon frame length through its PGF, defined in Section 4.1:

$$B_{ea}(z) = \sum_{k=2}^{5} b_k z^k. \tag{4.23}$$

If we assume that each advertised address takes four bytes and that the overall header takes 17 bytes, and then round the beacon size to the next higher integer number of backoff periods, the unknown coefficients can be obtained as

$$b_k = \sum_{l=l_{lo}}^{l_{hi}} \binom{n}{l} \rho_{ur}^l (1 - \rho_{ur})^{n-l}, \tag{4.24}$$

where k takes values from the range $2 \ldots 5$, and values of l_{lo}, l_{hi} can easily be computed for each k.

4.1.8 Normalization condition

Finally, we need the normalization condition, which states that all the state probabilities for a single node should add up to one:

$$x^z + \Sigma^{ud} + \Sigma^{ur} + \Sigma^{dd} + \Sigma^s = 1. \tag{4.25}$$

From Equations (4.7), (4.11), (4.16), (4.22) and (4.25), we can express all the components of (4.25) as functions of the variable x^z, which depends on other performance parameters yet to be found.

The probability to access the medium with an uplink data request from node i is

$$\tau_{ur} = \sum_{i=0}^{m} x_{i,0,0}^{ur} = C_4^{ur} x_{0,0,0}^{ur}. \tag{4.26}$$

By the same token, the probability to access the medium by an uplink data packet is

$$\tau_{ud} = C_4^{ud} x_{0,0,0}^{ud}, \tag{4.27}$$

while the probability to access the medium with a downlink packet from the coordinator toward node i is

$$\tau_{dd} = C_4^{dd} x_{0,0,0}^{dd}. \tag{4.28}$$

The total probability of uplink access for one node is $\tau_u = \tau_{ur} + \tau_{ud}$. The total probability of downlink access by the coordinator is $\tau_{dtot} = \tau_{dd}(1 + (n-1)\rho_{ur})$. In the last expression, the first term corresponds to the destination node while the second corresponds to the background traffic with the remaining $n - 1$ nodes.

The reader may remember that some transmissions have to be deferred to the next superframe due to insufficient time remaining in the current CAP. The probabilities of deferred and non-deferred transmissions, denoted with τ_{ur1} and τ_{ur2}, respectively, are

$$\tau_{ur1} = \sum_{i=0}^{m} x_{i,2,0}^{ur} P_{r,i} = \frac{x_{0,0,0}^{ur}}{C_{3,0}^{ur}} \sum_{i=0}^{m} P_{r,i} \prod_{j=0}^{i-1} C_{2,j}^{r}, \tag{4.29}$$

$$\tau_{ur2} = \tau_{ur} - \tau_{ur1},$$

and the corresponding total uplink access probabilities are

$$\tau_{u1} = \frac{x_{0,0,0}^{ur}}{C_{3,0}^{ur}} \sum_{i=0}^{m} P_{r,i} \prod_{j=0}^{i-1} C_{2,j}^{ur} + \frac{x_{0,0,0}^{ud}}{C_{3,0}^{ud}} \sum_{i=0}^{m} P_{d,i} \prod_{j=0}^{i-1} C_{2,j}^{ud}, \tag{4.30}$$

$$\tau_{u2} = \tau_{u} - \tau_{u1}.$$

Similar equations hold for τ_{ud} and τ_{dd} as well.

4.2   Modeling the Behavior of the Medium

We saw that the coordinator and ordinary nodes will have different views of the activity on the medium: an ordinary node competes for medium access against the coordinator and the remaining $n-1$ nodes, while the coordinator competes against $n-1$ nodes only, since the node which is destination for the current downlink packet does not transmit. Furthermore, the distribution of packet transmissions within the superframe is uneven because of the transmissions deferred from one superframe to another. On account of these observations, we assume that the number of non-delayed and delayed uplink packet transmissions, denoted with q and $n-1-q$, respectively, follow a binomial distribution with the probability $P_q = \binom{n-1}{q}(1 - P_d)^q P_d^{n-1-q}$.

The probability α that the medium is idle on the first CCA.   In order to calculate this probability, we have to find the mean number of busy backoff periods within the superframe and divide it into the total number of backoff periods in the superframe wherein the first CCA can occur. The first CCA will not take place if the remaining time in the superframe is insufficient to complete the transaction, which amounts to $SD - \overline{D_d} + 1$ backoff periods. The probability that one or more packet transmissions will take place at the beginning of the superframe is

$$\begin{aligned} n_{1u} &= 1 - (1 - \tau_{u1})^{(n-1-q)}(1 - \tau_{dtot})(1 - \tau_{u2})^q \\ n_{1d} &= 1 - (1 - \tau_{u1})^{(n-1-q)}(1 - \tau_{u2})^q \end{aligned} \tag{4.31}$$

for an ordinary node and the coordinator, respectively; the number of busy backoff periods due to these transmissions is $n_{1u}(G_p'(1) + G_a'(1))$ and $n_{1d}(G_p'(1) + G_a'(1))$, respectively.

In order to find the occupancy of the medium after the first packet transmission time, we need to calculate the mean packet lengths. We begin by noting that the probability for a given packet to be a data packet is

$$K_u = \frac{(n-1)\tau_{ud} + \tau_{dtot}}{(n-1)\tau_{ud} + \tau_{dtot} + (n-1)\tau_{ur}} \tag{4.32}$$

$$K_d = \frac{\tau_{ud}}{\tau_{ud} + \tau_{ur}}$$

The mean packet size may be obtained as

$$\overline{D_{lu}} = K_u G'_p(1) + (1 - K_u) G'_r(1)$$
$$\overline{D_{ld}} = K_d G'_p(1) + (1 - K_d) G'_r(1)$$

(4.33)

while the mean transmission time is

$$\overline{D_{du}} = 2 + \overline{D_{lu}} + t'_{ack}(1) + G'_a(1)$$
$$\overline{D_{dd}} = 2 + \overline{D_{ld}} + t'_{ack}(1) + G'_a(1)$$

(4.34)

The subscripts u and d correspond to variables as 'viewed' by an individual node and the coordinator, respectively.

As the total arrival rate of non-deferred packets is $q\tau_{u2}$, the probability of a non-zero number of transmission attempts during the period $\overline{D_{du}}$ or $\overline{D_{dl}}$, is $n_{2u} = \overline{D_{du}}q(\tau_{u2} + \tau_{dtot2})$ and $n_{2d} = \overline{D_{dd}}q\tau_{u2}$, for the uplink and downlink, respectively. Then, the total number of backoff periods wherein the first CCA can be executed is $SD - \overline{D_{du}} + 1$ for uplink transmissions, and $SD - \overline{D_{dd}} + 1$ for the downlink ones. The probability that the medium is idle at the first CCA, for an ordinary node and the coordinator, respectively, is

$$\alpha_u = \sum_{q=0}^{n-1} P_q \left(1 - \left(\frac{n_{1u}\overline{D_{du}}}{SD - \overline{D_{du}} + 1} + \frac{n_{2u}(SD - 2\overline{D_{du}} + 1)}{SD - \overline{D_{du}} + 1} \right) \frac{(\overline{D_{lu}} + G'_a(1))}{\overline{D_{du}}} \right)$$

$$\alpha_d = \sum_{q=0}^{n-1} P_q \left(1 - \left(\frac{n_{1d}\overline{D_{dd}}}{SD - \overline{D_{dd}} + 1} + \frac{n_{2d}(SD - 2\overline{D_{dd}} + 1)}{SD - \overline{D_{dd}} + 1} \right) \frac{(\overline{D_{ld}} + G'_a(1))}{\overline{D_{dd}}} \right)$$

(4.35)

The average value of the probability α is

$$\alpha = \frac{n\tau_u \alpha_u + \tau_{dtot}\alpha_d}{n\tau_u + \tau_{dtot}}.$$

(4.36)

The probability β that the medium is idle on the second CCA. The second CCA can be performed in the interval from the second backoff period in the superframe to the period in which there is no more time for packet transmission, which amounts to $SD - \overline{D_{du}} + 1$ for uplink transmissions and $SD - \overline{D_{dd}} + 1$ for the downlink ones. It will succeed if none of the other $n - 1$ nodes nor the coordinator has begun transmission in this backoff period, as explained in relationship to Figure 3.3, on p. 48. The probability of success β (which, like α, depends on whether it is seen by an ordinary node or by the coordinator) may be obtained as

$$\beta_u = \sum_{q=0}^{n-1} P_q \left(\frac{1}{SD - \overline{D_{du}} + 1} + \frac{(1 - \tau_u)^{n-1}(1 - \tau_{dtot})}{SD - \overline{D_{du}} + 1} \right.$$
$$\left. + \frac{SD - \overline{D_{du}} - 2}{SD - \overline{D_{du}} + 1}(1 - \tau_{u2})^q (1 - \tau_{dtot2}) \right)$$

$$\beta_d = \sum_{q=0}^{n-1} P_q \left(\frac{1}{SD - \overline{D_{dd}} + 1} + \frac{1}{SD - \overline{D_{dd}} + 1}(1 - \tau_u)^{n-1} \right.$$
$$\left. + \frac{SD - \overline{D_{dd}} - 2}{SD - \overline{D_{dd}} + 1}(1 - \tau_{u2})^q \right)$$

(4.37)

In both expressions, the first term in the brackets corresponds to a success in the second backoff period of the superframe, the second term corresponds to a possible access from both deferred and non-deferred packets, while the third term corresponds to non-deferred packets only. The average value of β is therefore

$$\beta = \frac{n\tau_u \beta_u + \tau_{dtot}\beta_d}{n\tau_u + \tau_{dtot}}. \tag{4.38}$$

The probability γ that a packet will not experience collision. This probability can be calculated as the probability that there are no accesses to the medium by the background traffic during the period of one complete packet transmission time. (Note that both the current packet and all other packets that can cause a collision have succeeded in their respective CCA checks.) As before, the views from the uplink and downlink direction are different.

A collision can happen in $SD - \overline{D_{du}} + 1$ consecutive backoff periods, in case of an uplink transmission, or in $SD - \overline{D_{dd}} + 1$ consecutive backoff periods, in case of a downlink transmission, starting from the third backoff period in the superframe. The values of γ from the different viewpoints are

$$\gamma_u = \sum_{q=0}^{n-1} P_q \left(\frac{1}{SD - \overline{D_{du}} + 1}(1 - \tau_u)^{(D_{lu}+2+t'_{ack}(1)+G'_a(1))(n-1)}(1 - \tau_{dtot}) \right.$$
$$\left. + \frac{SD - \overline{D_{du}}}{SD - \overline{D_{du}} + 1}(1 - \tau_{u2})^{(D_{lu}+2+t'_{ack}(1)+G'_a(1))q}(1 - \tau_{dtot2}) \right)$$

$$\gamma_d = \sum_{q=0}^{n-1} P_q \left(\frac{1}{SD - \overline{D_{dd}} + 1}(1 - \tau_u)^{(D_{ld}+2+t'_{ack}(1)+G'_a(1))(n-1)} \right.$$
$$\left. + \frac{SD - \overline{D_{dd}}}{SD - \overline{D_{dd}} + 1}(1 - \tau_{u2})^{(D_{ld}+2+t'_{ack}(1)+G'_a(1))q} \right) \tag{4.39}$$

and its average value is

$$\gamma = \frac{n\tau_u \gamma_u + \tau_{dtot}\gamma_d}{n\tau_u + \tau_{dtot}}. \tag{4.40}$$

4.3 Probability Distribution for the Packet Service Time

The packet service time for a data packet is defined as the time interval between the moment the MAC layer is delivered the packet to transmit – i.e., the beginning of service – and the moment when the receipt of packet has been acknowledged by the destination node. This service time includes the overhead of the CSMA-CA algorithm, as well as re-transmissions, if necessary. For downlink packets, the service time is the sum of the service time for the data request packet and the service time for the actual data packet. In this case, if one of the packets involved in the downlink procedure is lost, the entire procedure is repeated until successful.

To obtain the probability distribution of the packet service time, we need the probability distribution of the service time for the packet queue at the network node. We begin the calculation by modeling the effect of freezing the backoff counter during the inactive period of the superframe. The probability that a backoff period is the last one within the active portion of the current superframe is $P_{last} = 1/SD$. The PGF for the effective duration of the backoff period is $B_{off}(z) = (1 - P_{last})z + P_{last}z^{BI-SD+1}B_{ea}(z)$ (this value includes the duration of the beacon frame). The PGF for the duration of i-th backoff attempt, where $i = 0 .. m$, is

$$B_i(z) = \sum_{k=0}^{W_i-1} \frac{1}{W_i} B_{off}^k(z) = \frac{B_{off}^{W_i}(z) - 1}{W_i(B_{off}(z) - 1)} \tag{4.41}$$

The transmission procedure will not start unless it is finished within the current superframe, as explained in Section 2.3. The probability that the remaining number of backoff periods is insufficient for the transmission is P_d (it will be calculated in Section 5.1). The PGFs for the number of backoff periods that are effectively 'wasted' in this manner are

$$B_{pd}(z) = \frac{1}{\overline{D_d}} \sum_{k=0}^{\overline{D_d}-1} z^k$$

$$B_{pr}(z) = \frac{1}{\overline{D_r}} \sum_{k=0}^{\overline{D_r}-1} z^k \tag{4.42}$$

for data and request packets, respectively.

Finally, the PGFs of the data packet transmission time are

$$T_{d1}(z) = B_{pd}(z)z^{(BI-SD)}B_{ea}(z)G_p(z)t_{ack}(z)G_a(z)$$
$$T_{d2}(z) = G_p(z)t_{ack}(z)G_a(z) \tag{4.43}$$

for deferred and non-deferred transmissions, respectively.

The PGFs $T_{r1}(z)$ and $T_{r2}(z)$ for deferred and non-deferred data request packet transmissions, respectively, can be derived in an analogous manner, except that $G_r(z)$ should be used instead of $G_p(z)$, and $B_{pr}(z)$ should be used instead of $B_{pd}(z)$.

4.3.1 Service time for uplink data packets

In this subsection we will derive the PGF $T_{ud}(z)$ for the duration of the uplink service time. For simplicity, let us introduce the probability that i-th backoff attempt is unsuccessful as $R_{ud,i} = 1 - P_{d,i} - (1 - P_{d,i})\alpha_u\beta_u$. (We assume that $R_{ud,-1} = 1$.)

We will also introduce several partial PGFs (i.e., PGF-like functions that do not evaluate to 1 when $z = 1$):

- The function that describes the time needed for the backoff countdown before a transmission attempt and the transmission attempt itself, as

$$P(z) = \sum_{i=0}^{m} \prod_{j=0}^{i} \left(B_j(z)R_{ud,j-1}\right) z^{2(i+1)} \left(P_{d,i}T_{d1}(z) + (1 - P_{d,i})\alpha_u\beta_u T_{d2}(z)\right) \tag{4.44}$$

- The function that describes the time for a successful transmission, together with the backoff countdown that precedes it, as $P_1(z) = \gamma_u P(z)$.

- The function that describes the time for an unsuccessful transmission attempt and, thus, requires the backoff procedure to be repeated, as $P_2(z) = (1 - \gamma_u)P(z)T_{ud}(z)$.

- Finally, the function that represents $m + 1$ unsuccessful backoff countdown iterations without a transmission attempt and requires that the backoff procedure is repeated, as

$$R_{ep}(z) = \prod_{j=0}^{m} B_j(z)z^{2(m+1)}R_{ud,j} \tag{4.45}$$

With this notation, the PGF of the uplink data packet service time $T_{ud}(z)$, accounting for the effects of backoffs, unsuccessful CCAs, transmission, deferral to the next superframe, and re-transmission, satisfies the equation

$$\begin{aligned}
T_{ud}(z) &= P_1(z) + P_2(z) \\
&\quad + R_{ep}(z)\left(P_1(z) + P_2(z)\right) \\
&\qquad + R_{ep}(z)\left(P_1(z) + P_2(z)\right) \\
&\qquad\quad + \ldots \\
&= (P_1(z) + P_2(z))\left(1 + R_{ep}(z) + R_{ep}^2(z) + R_{ep}^3(z)\ldots\right), \\
&= \frac{\gamma_u P(z) + (1 - \gamma_u)P(z)T_{ud}(z)}{1 - R_{ep}(z)}.
\end{aligned} \tag{4.46}$$

from which the value of $T_{ud}(z)$ may be obtained as

$$T_{ud}(z) = \frac{\gamma_u P(z)}{1 - R_{ep}(z) - (1 - \gamma_u)P(z)}. \tag{4.47}$$

The function $T_{ud}(z)$ would indeed be a valid PGF if $T_{ud}(1) = 1$. To show that the last condition holds, note that $P(1)$ represents the probability that one of the $m + 1$ backoff attempts is successful, i.e., that it will pass two CCAs. Also, note that $R_{ep}(1)$ represents the probability that none of the $m + 1$ backoff attempts is successful, which means that the node has to repeat the entire backoff procedure. Obviously, $R_{ep}(1) + P(1) = 1$, from which follows that $T_{ud}(1) = 1$, Q.E.D.

The first two moments of the uplink data packet service time can be obtained as $\overline{T_{ud}} = T_{ud}'(1)$ and $\overline{T_{ud}^{(2)}} = T_{ud}''(1) + T_{ud}'(1)$.

As noted earlier, our derivations are based on the assumption that fully reliable transmission is employed, i.e., the MAC sublayer will retry packet transmission until the acknowledgment is received. However, it is also possible to employ partially reliable transmission with at most *macMaxFrameRetries* = 3 repeated transmission attempts, as required by the standard (Section 2.4), instead. In this case, the calculation presented above will overestimate the packet service time under high loads. Still, an 802.15.4 cluster is more likely to operate under low to moderate loads and the error due to this difference is negligible in practice.

4.3.2   Service time for uplink request packets

The probability that the i-th backoff attempt to transmit a data request packet to the coordinator in order to initiate the extraction of downlink packets is unsuccessful is equal to $R_{ur,i} = 1 - P_{r,i} - (1 - P_{r,i})\alpha_u\beta_u$. (Note that the duration of the data request packets differs from that of the data packets.) The PGF-like function that describes the time for the backoff procedure and subsequent transmission attempt is

$$P^{ur}(z) = \sum_{i=0}^{m} \prod_{j=0}^{i} B_j(z) R_{ur,j-1} z^{2(i+1)} \left(P_{r,i} T_{r1}(z) + (1 - P_{r,i}) \alpha_u \beta_u T_{r2}(z) \right) \tag{4.48}$$

However, the data request packets may be blocked when the coordinator is busy executing the backoff procedure for the downlink packet towards another node (a more detailed analysis of this problem is presented in Section 5.4). In the presence of n nodes in the cluster, the probability of this type of blocking is

$$P^B = 1 - \frac{1}{1 + n \gamma_u \tau_{ur}(T_x + \overline{D_1} + R_x)}. \tag{4.49}$$

where $\overline{D_1}$ is the time needed for one transmission attempt, including up to $m+1$ backoff countdown iterations, in the downlink (this time is calculated in the next subsection), while T_x and R_x denote the time needed to switch the antenna from receiver to transmitter mode and back (which the standard limits the MAC constant *aTurnaroundTime*). Taking into account all these effects, the PGF of the service time for uplink data requests becomes

$$T_{ur}(z) = \frac{\gamma_u (1 - P^B) P^{ur}(z)}{1 - R_{epr}(z) - (1 - \gamma_u (1 - P^B)) P^{ur}(z)}, \tag{4.50}$$

where

$$R_{epr}(z) = \prod_{j=0}^{m} B_j(z) z^{2(m+1)} R_{ur,j} \tag{4.51}$$

As before, it can be shown that $T_{ur}(1) = 1$, which means that $T_{ur}(z)$ is indeed a PGF. The first two moments of the uplink request packet service time can be obtained as $\overline{T_{ur}} = T'_{ur}(1)$ and $\overline{T_{ur}^{(2)}} = T''_{ur}(1) + T'_{ur}(1)$.

4.3.3 Service time for downlink packets

The overall service time for downlink packets is the longest, since the extraction procedure comprises two distinct packet transmissions that follow the CSMA-CA algorithm, and one or both of them include acknowledgments. The data packet may suffer a collision, as can the data request packet, and the latter can be blocked at the coordinator; if any of these events occur, the entire procedure has to be repeated, starting with the re-transmission of the data request packet.

In order to calculate the service time for downlink packets, we need to introduce some auxiliary parameters and functions:

- The probability that one iteration of the downlink backoff procedure ends unsuccessfully, as $R_{dd,j} = 1 - P_{d,j} - (1 - P_{d,j}) \alpha_d \beta_d$. We assume that $R_{dd,-1} = 1$.

- The PGF-like function to describe the time needed to backoff countdown and the subsequent transmission attempt, as

$$P^{dd}(z) = \sum_{i=0}^{m} \prod_{j=0}^{i} B_j(z) R_{dd,j-1} z^{2(i+1)} (P_{d,i} T_{d1}(z) + (1 - P_{d,i}) \alpha_d \beta_d T_{d2}(z)) \tag{4.52}$$

- The PGF-like function to describe the time for a successful transmission together with the preceding uplink request and backoff countdown, as

$$P_{1d}(z) = T_{ur}(z)\gamma_d P^{dd}(z) \tag{4.53}$$

- The PGF-like function to describe the time for an unsuccessful transmission attempt which requires the entire backoff procedure to be repeated, as

$$P_{2d}(z) = T_{ur}(z)(1 - \gamma_d)P^{dd}(z)T_{dd}(z) \tag{4.54}$$

- Finally, the function which represents $m + 1$ unsuccessful backoffs without a transmission attempt, as

$$R_{epd}(z) = \prod_{j=0}^{m} B_j(z)z^{2(m+1)}R_{dd,j} \tag{4.55}$$

Note that a repetition of the entire backoff procedure, beginning with the data request packet, will be needed.

With these items in place, the PGF for time needed for one backoff and transmission attempt can be calculated as

$$D_1(z) = \frac{P^{dd}(z)}{P^{dd}(1)}, \tag{4.56}$$

since $D_1(1) = 1$ is a necessary condition for the function $D_1(z)$ to be a valid PGF.

The PGF for the downlink packet service time may be derived as

$$T_{dd}(z) = \frac{T_{ur}(z)\gamma_d P^{dd}(z)}{1 - R_{epd}(z) - T_{ur}(z)(1 - \gamma_d)P^{dd}(z)}, \tag{4.57}$$

and it can be shown that $T_{dd}(1) = 1$.

Finally, the first two moments of the downlink packet service time can be obtained as $\overline{T_{dd}} = T'_{dd}(1)$ and $\overline{T_{dd}^{(2)}} = T''_{dd}(1) + T'_{dd}(1)$.

4.3.4 The complete model and stability considerations

Equations (4.35), (4.37), and (4.39), which describe α, β and γ; Equations (4.26), (4.27), (4.28), which define access probabilities; Equation (4.49) for blocking probability at PAN coordinator; and the following equations which describe the offered load for uplink data, requests and downlink data: $\rho_{ud} = \lambda_u \overline{T_{ud}}$, $\rho_{ur} = \lambda_d \overline{T_{ur}}$, and $\rho_{dd} = \lambda_d \overline{T_{dd}}$, together form a complete system. This system can be numerically solved for a known number of nodes n, and uplink and downlink packet arrival rates per node, λ_u and λ_d.

A queueing system is said to be stable when the mean number of packets serviced is not smaller than the mean number of packets entered; if this is not the case, packets tend to pile up in the queue, and packet delays grow without bound. In an 802.15.4 cluster with uplink and downlink traffic, the overall stability requirement translates into the following conditions:

- The offered uplink load for each of the ordinary nodes, $\rho_{ud} + \rho_{ur}$, cannot exceed 1.

- The total offered downlink load at the coordinator, which is $\rho_{dtot} = n\rho_{dd}$, cannot exceed 1.

Adherence to these conditions is reflected in the access delays for uplink and downlink traffic. According to Takagi (1991), mean delay in the downlink queue for M/G/1 systems is

$$\overline{W_d} = \frac{\lambda_d \overline{T_{dd}^{(2)}}}{2(1 - \rho_{dtot})} \tag{4.58}$$

where $\overline{T_{dd}^{(2)}}$ denotes the second moment of the downlink data service time. However, the uplink traffic at an ordinary node may be considered to be serviced with two queues, the data request queue and the data packet queue, the former having higher priority than the latter. According to the analysis of priority queues (Takagi 1991), mean access delay is obtained as

$$\overline{W_u} = \frac{\lambda_u \overline{T_{ud}^{(2)}} + \lambda_d \overline{T_{ur}^{(2)}}}{2(1 - \rho_{ur})(1 - \rho_{ud} - \rho_{ur})}. \tag{4.59}$$

The Laplace transforms of the probability distributions for the uplink and downlink access times could be used as well. However, since we are concerned only with stability limits, mean values of the respective variables will suffice.

4.4 Performance of the Cluster with Bidirectional Traffic

We will now investigate the performance of an 802.15.4 cluster through some results obtained through analytical modeling described above. As before, the cluster operates in star topology, beacon enabled operating mode, using slotted CSMA-CA medium access mechanism. Although we have considered only the 2450 MHz PHY option with the maximum data rate of 250 kbps, the results hold for other PHY options as well, provided appropriate scaling is used. The superframe duration was controlled by $BO = SO = 0$, which means that the superframe contains 480 bytes, and there is no inactive period. The PHY and MAC headers had six and seven bytes, respectively, and all packets had a two-byte FCS (Section 2.9). Since the size of the data request packet is 16 bytes (IEEE 2006), we have rounded it to two backoff periods for simplicity. By the same token, the duration of the acknowledgment packet was set to one backoff period even though the correct size is 11 bytes. All other parameters were set to the default values specified by the standard.

We consider the scenario where each node sends packets to every other node with equal probability. Therefore, if the uplink packet arrival rate per node is λ_u, then each node receives data at the rate of $\lambda_d = \tau_{ud}\gamma_u$. We have fixed the data packet size to $G'_p(1) = 3$ backoff periods, and the packet arrival rate was varied between 1 arrival per minute to 240 arrivals per minute (4 arrivals per second).

The calculated offered loads for the downlink queue at the coordinator and the uplink queues at an individual node are shown in Figure 4.5. We clearly see that the offered load in the downlink direction is more critical with respect to stability, since it reaches the boundary value of one at smaller network sizes (i.e., much sooner) than its uplink counterpart. When this stability condition is exceeded, packet service times and access

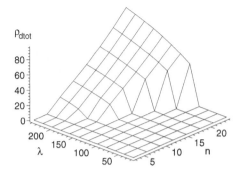

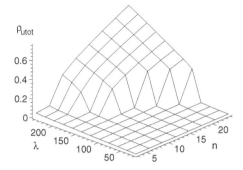

(a) Total offered load in the downlink direction.

(b) Offered load in the uplink direction (one node only).

Figure 4.5 Offered load. Adapted from J. Mišić, S. Shafi, and V. B. Mišić, 'Performance of a beacon enabled IEEE 802.15.4 cluster with downlink and uplink traffic,' *IEEE Trans. Parallel Dist. Syst.*, **17**(4): 361–376, © 2006 IEEE.

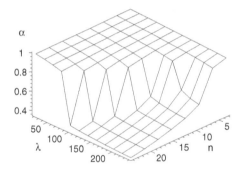

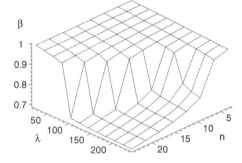

(a) Probability of success at the first CCA, α.

(b) Probability of success at the second CCA, β.

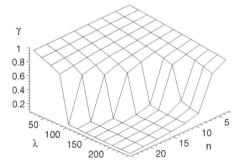

(c) Probability of successful transmission, γ.

Figure 4.6 Success probabilities. Adapted from J. Mišić, S. Shafi, and V. B. Mišić, 'Performance of a beacon enabled IEEE 802.15.4 cluster with downlink and uplink traffic,' *IEEE Trans. Parallel Dist. Syst.*, **17**(4): 361–376, © 2006 IEEE.

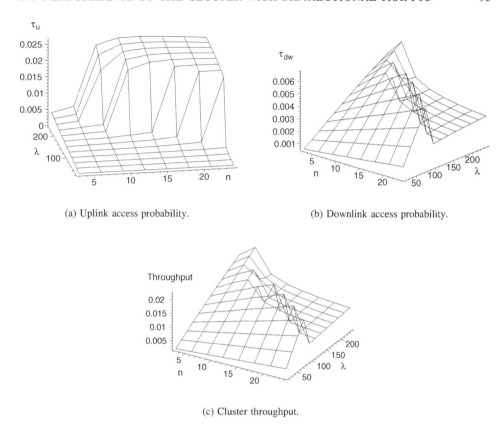

(a) Uplink access probability.   (b) Downlink access probability.

(c) Cluster throughput.

Figure 4.7 Performance of a cluster with uplink and downlink traffic. Adapted from
J. Mišić, S. Shafi, and V. B. Mišić, 'Performance of a beacon enabled IEEE 802.15.4
cluster with downlink and uplink traffic,' *IEEE Trans. Parallel Dist. Syst.*, **17**(4): 361–376,
© 2006 IEEE.

delays experience large growth. This leads us to the conclusion that an 802.15.4 cluster
should be used only for very light downlink communication. Fortunately, this is indeed
the case in many sensing and home networking applications, where the downlink traffic
consists of infrequent commands and, henceforth, presents a load much lower than that of
the uplink traffic. Uplink stability, on the other hand, limits the network size to at most
25 nodes or so, with packet arrival rates of at most three to four packets per second. This
stability region would be wider for lower intensity of downlink traffic.

   Figure 4.6 shows the probabilities α, β, and γ – i.e., the probability that the medium is
idle on the first CCA, the probability that the medium is idle on the second CCA, and the
probability of success. We present only the average of uplink and downlink values of all
three probabilities, as we have found that the uplink and downlink probabilities differ by
less than 10% under moderate and high loads. This may be explained by packet collisions
and blocking at the coordinator, which decrease the number of downlink packets to be
processed by the coordinator and reduce the impact of downlink transmissions. We note

that all three probabilities α, β, and γ reach lower (saturation) bounds at moderate loads for network size between 10 and 20 nodes. The lower bound for the success probability is close to zero, which means that virtually no packet is able to reach its destination when the network operates in this region.

Figure 4.7 shows the uplink and downlink access probabilities, as well as the throughput. The flattening of uplink access probability indicates that the onset of saturation region, in which case all accesses to the medium are contributed by the request packets that do not succeed. A rather dramatic decrease of downlink access probability for the coordinator is caused by the inability of the coordinator to receive any correct data requests due to collisions and blocking. The throughput also deteriorates rapidly when the cluster enters saturation.

5

MAC Layer Performance Limitations

All contention-based MAC protocols suffer from collisions which waste bandwidth and energy. Packets that experience collisions have to be retransmitted or are lost forever; if the application requires reliable transmission, explicit acknowledgments have to be used. Both of these reduce the maximum achievable throughput and impair network performance. In an 802.15.4 cluster operating in beacon enabled mode that uses slotted CSMA-CA access, collisions may occur in several scenarios. The most common among those is the 'standard' data packet collision: namely, when two or more nodes want to send their data, they may finish their random backoff countdown and attempt transmission at the same time. However, a number of other, context-specific scenarios that present performance risks can be identified in the slotted CSMA-CA MAC protocol, as defined in the 802.15.4 standard (IEEE 2006). While the exact amount of degradation caused by those risks depends on network and traffic parameters, in some cases it may rise to values that are sufficiently high to pose severe limitations on the throughput and packet delays. In this chapter, we analyze those performance risks and the scenarios in which they may occur; we quantify their impact; and we suggest minor modifications of the coordinator function that allows the network to handle higher traffic loads with improved performance.

5.1 Congestion of Packets Deferred to the Next Superframe

The 802.15.4 standard stipulates that all transactions must be contained within the active portion of a single superframe. As a result, some nodes may finish the random backoff countdown only to find that the remaining time in the active portion of the superframe does not suffice for the two CCAs, packet transmission, and the subsequent (optional) acknowledgment. In such cases, the transmission procedure is temporarily suspended until

Wireless Personal Area Networks Jelena Mišić and Vojislav B. Mišić
© 2008 John Wiley & Sons, Ltd

after the beacon frame of the next superframe. However, the random backoff countdown is not repeated, and all such nodes will undertake their CCAs immediately following the beacon frame. As each transmission must be preceded by the two CCAs, the channel is found idle in the first two backoff periods after the beacon frame. The nodes with deferred packets will then conclude that the channel is free, and start their transmission in the third backoff period – resulting in a collision. This scenario is schematically depicted in Figure 5.1.

It is worth noting that this problem is, by nature, deterministic: all packets deferred to the next superframe will collide, regardless of the previously chosen values for the random backoff countdown or the packet length. Furthermore, when acknowledged transfers are enabled, the lack of proper acknowledgment may result in another random backoff iteration (depending on the index of the current one and the current retry count) and, possibly, another collision. This contention problem becomes more pronounced for small superframe sizes (when, say, $SO = 1$ or 2) and large packet sizes, which are more likely to be deferred.

Note that the probability that a packet transmission will be deferred increases with the index of the current backoff countdown iteration. As the value of NB increases, so does BE and the corresponding range of the possible waiting time. Longer random backoff countdowns will finish closer to the end of the superframe, and thus increase the probability that the remaining time would not suffice for the two CCAs, packet transmission, and acknowledgment. Therefore, each collision increases the probability of deferral, thus leading to even more collisions later on. When the number of retries is limited (i.e., in case of

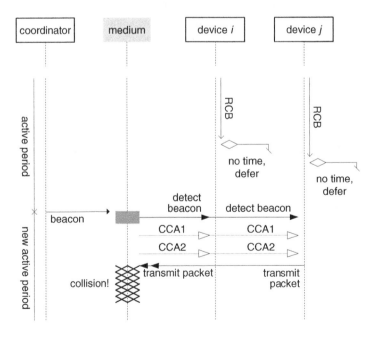

Figure 5.1 Congestion of packets deferred to the beginning of the superframe.

partially acknowledged transmission), the packet may even be dropped because the current backoff countdown retry count reaches its maximum value of five.

The probability of collisions of this type can be derived as follows. From Equation (4.41), the PGF for the duration of i-th backoff countdown phase can be calculated as

$$B_i(z) = \sum_{k=0}^{W_i-1} \frac{1}{W_i} B_{off}^k(z) \tag{5.1}$$

and the total duration of j consecutive unsuccessful backoff phases has the PGF of

$$B_{t,j}(z) = \prod_{i=0}^{j-1} B_i(z) z^{2(j-1)} \tag{5.2}$$

Being a product of polynomials, the last PGF can be expressed as the polynomial in z, where the coefficients are mass probabilities of $B_{t,j}(z)$. The largest exponent in this polynomial is

$$B_{T,j} = \sum_{i=0}^{j-1} (W_i - 1) \tag{5.3}$$

which corresponds to the largest duration of j consecutive backoff countdown attempts. The mass probability that j consecutive backoff countdown attempts will last for a total of m backoff periods is equal to

$$b_{t,j,m} = \left. \frac{d^m B_{t,j}(z)}{dz^m} \right|_{z=1} \tag{5.4}$$

Therefore, the $B_{t,j}(z)$ can be presented in the polynomial form as

$$B_{t,j}(z) = \sum_{m=0}^{B_{T,j}} b_{t,j,m} z^m \tag{5.5}$$

From Equation (4.18), the PGF for the time period between the data packet arrival and the end of the current superframe has the PGF of

$$S_1(z) = \sum_{k=0}^{SD-1} \frac{1}{SD} z^k \tag{5.6}$$

If the total transmission time of data packets is $\overline{D_d}$, the probability that the transmission will be deferred after j consecutive backoff countdown attempts is

$$P_{d,j} = \sum_{i=0}^{SD-1} \frac{1}{SD} \sum_{m=\max(0, i-\overline{D_d}+1)}^{\min(i, B_{T,j})} b_{t,j,m}. \tag{5.7}$$

5.2   Congestion after the Inactive Period

A similar problem occurs in 802.15.4 networks in which the superframe contains both an active and an inactive portion; however, its causes are different. In many applications, in particular in the area of sensor networks, the inactive portion of the superframe is used to conserve energy by switching off the radio subsystem. However, the requirements of the sensing application often necessitate that the sensing subsystem (which consumes much less energy) operates without interruption.

   In this setup, data packets may be collected during the inactive period of the superframe. Once the node wakes up to listen to the active portion of the superframe, it may find one or more data packets ready to be transmitted. All such nodes will start their CSMA-CA algorithms with the same initial value for *BE*, which is only three or two. Since the resulting range of countdown values in the first iteration is small, $0 .. 3$ or $0 .. 7$ only, the probability of two or more nodes choosing the same value is high. This scenario is schematically shown in Figure 5.2.

   Obviously, the probability of this congestion increases when the time interval during which the radio subsystem is inactive increases. However, the number of data packets that may be deferred in this manner depends on the capacity of the node buffer: too many packets will cause buffer overflow, and some of them may be lost. A more detailed discussion of activity management and related issues will be given in Chapter 6.

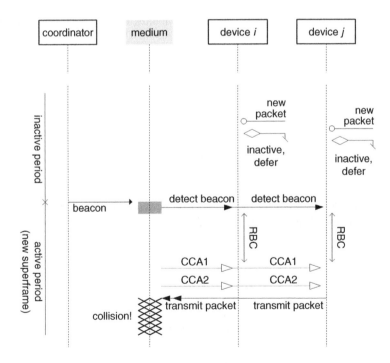

Figure 5.2 Congestion of packets collected during the inactive period of the superframe.

5.3 Congestion of Uplink Data Requests

Both of the scenarios described above apply to data packets regardless of the direction of transmission. In other words, both uplink and downlink data packets may suffer congestion in either of these scenarios. Unfortunately, congestion in 802.15.4 clusters is not limited to data packets only; it can also affect the transmission of data request packets from individual nodes to the coordinator.

One problematic scenario may occur when the coordinator has pending packets for two or more of the nodes. As explained in Section 2.5, the presence of downlink packets is announced through the beacon frame; a single beacon frame can accommodate at most seven destinations for such packets. The nodes that recognize their addresses in the beacon frame will immediately execute the CSMA-CA algorithm in order to transmit their data request packets. However, the range of countdown values in the first iteration is small, $0 . . 3$ or $0 . . 7$ only, and the probability of two or more nodes choosing the same value is high. After the countdown and the two required CCAs, such nodes may start their transmission simultaneously and collide. (A similar situation may occur in the second and subsequent iterations, but the corresponding probabilities are smaller.) This scenario is schematically shown in Figure 5.3.

It is worth noting that a data request packet may also collide with the packets collected during the inactive period, as well as with those deferred from the previous superframe, which leads to further performance deterioration.

To calculate the probability of collisions of this type, we have to distinguish between two possible cases. First, some request packets need to be repeated because of a collision, blocking by the coordinator (explained below), or because the subsequent downlink packet was not received due to a collision. In such cases, the node in question immediately repeats the backoff countdown and attempts to re-send the request packet. Such 'repeated' requests

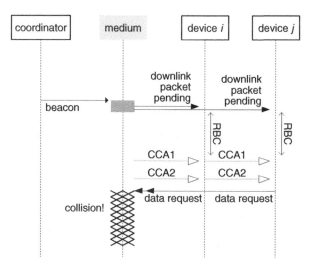

Figure 5.3 Congestion of uplink data requests.

may be assumed to be distributed uniformly over the duration of the superframe. Therefore, a repeated request packet with the transmission time of $\overline{D}_r$ behaves much like a data packet, and the probability that its transmission will be deferred is

$$P_{r,j}^r = \sum_{i=0}^{SD-1} \frac{1}{SD} \sum_{m=\max(0,i-\overline{D}_r+1)}^{\min(i,B_{T,j})} b_{t,j,m}. \tag{5.8}$$

The second case concerns new request packets that result directly from the corresponding announcements in the beacon; these behave in a different fashion. Namely, the target node will start the backoff countdown immediately after the beacon, which means that the arrivals of such request packets in fact coincide with the beginning of the superframe. Obviously the probability that their transmission will have to be deferred until the next superframe is low; but on the other hand, the probability that the transmission of such packets will collide with other packets which are delayed is non-negligible.

A node that learns about a pending downlink packet will immediately begin the random backoff countdown for the uplink request, unless it is not currently busy with an uplink data transmission. From Equation (4.13), the probability that a given request packet is a new one, is

$$P_n = \frac{\rho_{ur}(x^z + C_4^{ud} x_{0,0,0}^{ud} + C_4^{dd} x_{0,0,0}^{dd})}{A} \tag{5.9}$$

where, for simplicity, we have introduced an auxiliary variable

$$\begin{aligned} A &= \rho_{ur}(x^z + C_4^{ud} x_{0,0,0}^{ud} + C_4^{dd} x_{0,0,0}^{dd}) \\ &\quad + x_{0,0,0}^{ur}\left((1-\gamma_u)C_4^{ur} + C_9^{ur} + \gamma_u C_4^{ur} P^B\right) \\ &\quad + x_{0,0,0}^{dd}\left(C_4^{dd}(1-\gamma_d) + C_9^{dd}\right) \end{aligned} \tag{5.10}$$

The probability to start the backoff countdown immediately is $(1 - \rho_{ud})/W_0$, where W_0 is the size of the first backoff window. The probability that the data request packet in question is transmitted immediately after the beacon is $P_{r,j} = (1 - P_n)P_{r,j}^r + P_n(1 - \rho_{ud})/W_0$.

It is worth noting that this problem affects only the downlink traffic, and thus may lead to asymmetry of network performance in cases where both uplink and downlink traffic are present.

5.4 Blocking of Uplink Data and Data Requests

The delivery mechanism for downlink packets induces another, more subtle problem due to the fact that a typical 802.15.4 node is likely to have a single radio with half-duplex interface only. Such radios can, at any given moment, be in only one of the following states: transmit, receive, or turned off. In order to conserve energy, the radio should be turned off whenever possible, even during the random backoff countdown.

Now let us assume that the coordinator has received downlink packets for two or more nodes, and announced their presence through the beacon frame. When the destination nodes learn about the presence of a downlink packet, they attempt to send data request packets using the procedure from Figure 2.5. Upon successfully receiving first such request, the coordinator acknowledges it and begins the CSMA-CA procedure to send the data frame.

This procedure begins with a random backoff countdown, during which the coordinator does not listen actively to the medium and ignores any ongoing data transmission. Even during the subsequent CCAs, the coordinator will simply monitor the activity on the radio channel, rather than listen to the channel. (Active listening will be resumed only after the data packet is successfully transmitted, or the packet is dropped in case of partially acknowledged transfers.) As a result, any data request (or requests) that may have been sent during the countdown or the two CCAs will simply be ignored, even in the absence of collisions. This scenario is schematically depicted in Figure 5.4.

In this manner, all data requests except the first are effectively blocked, even in the absence of any collisions. All blocked requests will have to be retransmitted, which consumes bandwidth and increases the probability of further collisions. Furthermore, the blocking of requests will cause the pending downlink data transfers to be delayed, which in turn incurs the risk of buffer overflow at the coordinator and subsequent data loss.

The problem described here would not occur if the coordinator is able to send the downlink data frame 'piggybacked' onto the request acknowledgment frame, i.e., without CSMA-CA, as explained in Section 2.5. In that case, another node that attempts to send its data request packet will detect the ongoing transmission and back off; thus no data requests will be lost. However, this is possible only if: (a) the coordinator's radio subsystem meets the required performance levels, and (b) certain timing requirements are met; if this is not the case, the downlink packet will have to be sent using CSMA-CA.

Moreover, a single transaction that includes the data request, data request acknowledgment, downlink data packet, and the corresponding (optional) acknowledgment, is likely

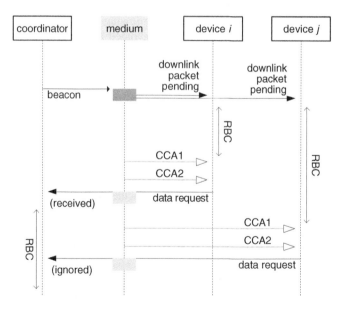

Figure 5.4 Some data request packets are ignored because the coordinator has already begun the countdown for another downlink packet.

to take a long time. Nodes that attempt to transmit their data requests or data packets during that time will be forced to back off, perhaps more than once, and thus experience performance degradation.

Ordinary data packets sent in the uplink direction during a random backoff countdown before a downlink transmission will be ignored as well. However, the uplink traffic will suffer less performance degradation than the downlink one, since the uplink transmission procedure involves fewer steps than its downlink counterpart.

5.5 Possible Remedies

Let us now outline some possible remedies for the bottlenecks described above. In doing so, we have strived to minimize the changes to the procedures and algorithms prescribed by the standard, as these would seriously limit the applicability of our work. Instead, we have focused on amending the issues that are not mentioned in the standard; in this manner, chip and device manufacturers can offer improved performance (as will be seen) while remaining compatible with the current specification.

Congestion and blocking of data request packets.   The congestion of data request packet contention can be reduced by limiting the number of announcements that a beacon frame can contain. A limit of, say, four or three announcements instead of the original seven would reduce the probability of collisions of this type; a limit of one announcement per beacon frame would completely eliminate them. This restriction may be easily enforced by the upper layers of the protocol stack, transparently to the MAC layer; any other remedy would require a change in the standard. However, it would lead to longer access delays for downlink packets and increased probability of overflowing the buffer at the coordinator.

The blocking of data request packets can be avoided (to some extent) by simply allowing the coordinator to listen to such packets throughout the countdown that leads to the transmission of acknowledgment packets. As the coordinator needs to perform the two CCAs before transmitting the acknowledgment packet, its radio may as well be switched on throughout the random backoff countdown; but the lower layers of the protocol stack should allow active listening during that time. Data request packets received during the countdown should be queued and acknowledged after the current request is acknowledged (or acknowledged and serviced, as appropriate). It should be noted that the small span of the first countdown iteration means that only one or two request packets may be received during the coordinator countdown, henceforth even small buffers would suffice for this purpose.

Note that the standard allows a node to decide 'whether to enable its receiver during idle periods', i.e., when no transmission is scheduled (IEEE 2006, Section 7.5.6.2). Therefore, the changes outlined above should not be difficult to implement: by changing the software that implements the 802.15.4 protocol stack, at best, accompanied by appropriate (small) changes in hardware, at worst.

Extending the receiver time-out.   However, the lack of immediate acknowledgment to second and later request packets might lead the nodes from which those packets were sent to conclude that those packets have been lost. In such cases, the originating nodes would initiate re-transmission, which would effectively cancel any improvement due to queueing and render the very concept of queueing data requests useless. This issue can be addressed by simply extending the receiver time-out for data request packets, which is determined by the MAC constant *macResponseWaitTime*. This constant can take any value from 5 to 64 symbols, with the default value of 32 symbols, which is obviously inadequate if the coordinator adopts the list-while-counting down approach, and an extended range should be used.

While this extension does require a change in the standard, the avoidance of some (or all) of the re-transmissions would substantially reduce the collision probability and improve the utilization of the medium, thus leading to improved throughput (and/or reduced power consumption) of the entire network.

Avoiding the congestion at the beginning of the superframe. Congestion of deferred transmissions immediately after the beginning of the superframe can also be avoided if the deferred packets do not attempt transmission immediately upon the beginning of the next superframe. Instead, some way must be found for the deferred transmissions to be 'spread' so as to reduce the probability of collisions and, consequently, increase the throughput. While this waiting may be accomplished by repeating the random backoff countdown after the beacon, it is unclear whether the current backoff exponent is to be used or not, and whether the retry index should be incremented (i.e., whether this is a valid countdown) or not. (Moreover, this would require appropriate changes to the standard.)

A better solution would be to let all such nodes wait for the time equal to the time needed to finish their corresponding transmissions: the two CCAs, the packet length of $\overline{Dd}$ backoff periods, and the subsequent (optional) acknowledgment. This wait should begin at the end of the random backoff countdown; it should be frozen during the inactive portion of the superframe (if any), and restarted immediately after the beacon frame. Since individual nodes finish their random backoff countdown at different times before the end of the active portion of the superframe, they will begin the transmission procedure (i.e., their CCAs) at different times. This scenario is shown in Figure 5.5.

The degree of 'randomness' provided in this manner suffices to result in nonzero probability that one node would succeed in transmitting its packets. (Note that this probability is exactly zero in the original scenario corresponding to the unmodified standard.) As before, this change could simply be accomplished by modifying the implementation of the 802.15.4 protocol stack, rather than by changing the standard itself.

If we adopt the latter approach, the Markov chain that corresponds to the corrected MAC is given in Figures 5.6 and 5.7. This chain may be analyzed along similar lines as the original one, but with the following differences:

$$\begin{aligned}
x_{0,1,0} &= x_{0,2,0}\alpha = x_{0,2,0}C_1 \\
x_{1,2,0} &= x_{0,2,0}(1 - \alpha\beta) = x_{0,2,0}C_2 \\
x_{0,0,0} &= x_{0,2,0}\alpha\beta = x_{0,2,0}C_3
\end{aligned} \tag{5.11}$$

The differences related to the delay line are

$$\begin{aligned}
P\{i, 2, 0, l | i, 2, 0\} &= 0, & i &= 0 .. m; l = 0 .. \overline{D_d} - 2 \\
P\{i, 2, 0, \overline{D_d} | i, 2, 0\} &= P_d, & i &= 0 .. m \\
P\{i, 2, 0, l - 1 | i, 2, 0, l\} &= 1, & i &= 0 .. m; l = 0 .. \overline{D_d} - 1
\end{aligned} \tag{5.12}$$

$$\sum_{l=0}^{\overline{D_d}-1} x_{i,2,0,l} = \frac{x_{0,0,0}C_2^i P_d \overline{D_d}}{C_3}$$

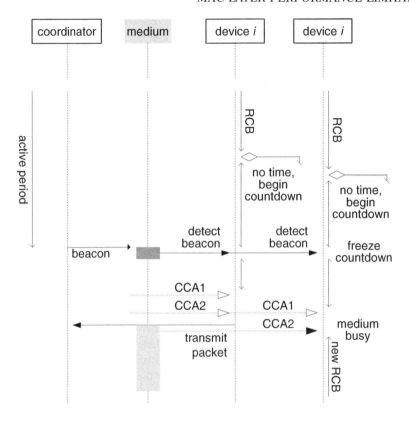

Figure 5.5 Congestion of packets can be avoided by introducing an extra countdown for the packets deferred to the beginning of the next superframe.

Using the balance Equation (3.6), which holds for this chain as well, we obtain $x_{0,0,0}$ as

$$
x_{0,0,0} = \cfrac{1}{\displaystyle\sum_{i=0}^{m} \frac{C_2^i(W_i+1)}{2C_3} + \frac{1-C_2^{m+1}}{1-C_2}\left(\overline{D_d} - 2 + \frac{C_1}{C_3} + \frac{\pi_0\beta}{1-P_0} + \frac{P_d\overline{D_d}}{C_3}\right) + \frac{C_2^{m+1}}{C_3}}
$$

(5.13)

The corresponding access probability can be obtained from Equations (5.13) and (3.8). Since, in this case, $\tau_1 = 0$, $\tau_2 = \tau$, $n_1 = 0$, and $n_2 = \overline{D_d}(n-1)\tau_2$, the probabilities that the first CCA, second CCA, and overall packet transmission are successful can be then obtained by substituting these values in Equations (3.11), (3.13), and (3.14) respectively.

Congestion after the inactive period. Note that the congestion after a prolonged inactive period can be avoided by turning off both the sensing and the radio subsystem. If this is unacceptable because of the requirements of the application, the duration of the inactive period should be limited. It also helps if a buffer of suitable size is available to store the packets received during the period of inactivity.

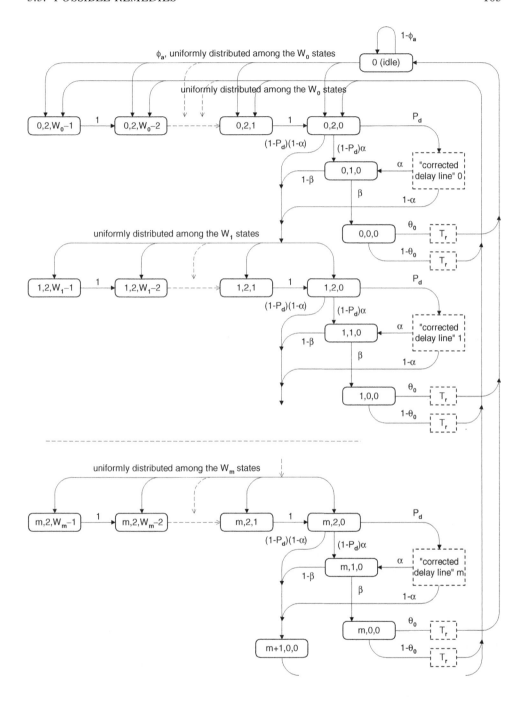

Figure 5.6 Markov chain model that avoids congestion of the packets deferred to the beginning of the next superframe.

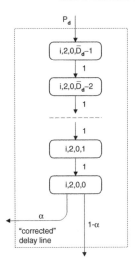

Figure 5.7 'Corrected' delay line for the extra countdown in Figure 5.6.

5.5.1 Performance of the improved MAC

Let us now investigate the performance of the improved MAC, analyzed with the aid of a custom-built simulator based on the object-oriented Petri net simulation engine Artifex by RSoft Design, Inc. (2003). We have considered a single 802.15.4 cluster operating in the ISM band at 2.4 GHz with raw data rate 250 kbps, using the parameter values of $SO = 0$ and $BO = 0$. The minimum value of backoff exponent *macMinBE* is set to 3 while the maximum value of the backoff exponent *aMaxBE* is set to 5. Partially acknowledged transfers are used, with the maximum number of backoff attempts set to 5, i.e., *macMaxCSMABackoffs* = 4.

We consider the scenario where each node sends packets to every other node with equal probability; the uplink packet arrival rate per node is λ_u, while each node receives data at the rate of $\lambda_d = \tau_{ud}\gamma_u$. The data packet size was fixed at 3 backoff periods, while the packet arrival rate was varied between 1 and 240 arrivals per minute. The duration of the MAC command frame for a data request and the acknowledgment frame were set to two and one unit backoff period respectively, for simplicity (the standard prescribes lengths of 16 and 11 bytes, respectively).

The results for the improved MAC were obtained by introducing all the modifications outlined above: deferred packets were made to wait for the period equal to the packet length before performing the two CCAs; the coordinator was allowed to listen to the medium and queue packet requests received during the random backoff countdown for the transmission of an already requested and acknowledged downlink packet; finally, the receiver time-out was extended to 660 backoff periods in order to allow data request acknowledgment packets to be received throughout the active portion of the superframe.

Figure 5.8 shows the probabilities of success for the first CCA, second CCA, and overall packet transmission, in the network with the original and improved MAC. Note that the probabilities that the channel is sensed to be idle at the first and second CCA (α and β, respectively) differ, for reasons explained in Section 3.1.

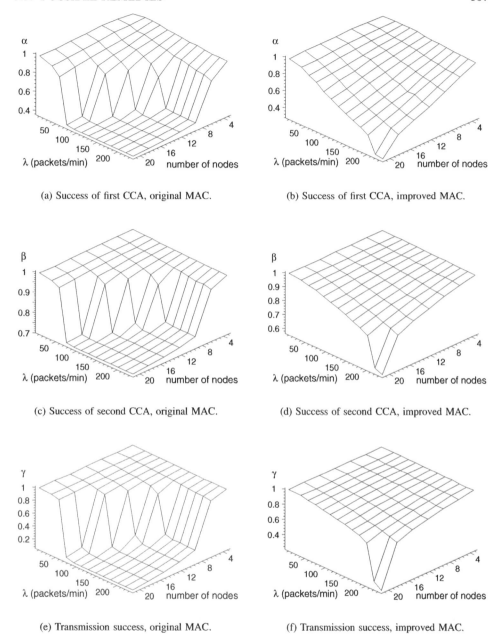

(a) Success of first CCA, original MAC.

(b) Success of first CCA, improved MAC.

(c) Success of second CCA, original MAC.

(d) Success of second CCA, improved MAC.

(e) Transmission success, original MAC.

(f) Transmission success, improved MAC.

Figure 5.8 Comparing the original MAC (left column) with the improved MAC (right column), in terms of success probabilities of the first CCA, second CCA, and packet transmission, respectively. Adapted from J. Mišić, S. Shafi, and V. B. Mišić, 'Performance limitations of the MAC layer in 802.15.4 low rate WPAN,' *Computer Communications*, **29**(13-14): 2534–2541, © 2006 Elsevier B. V.

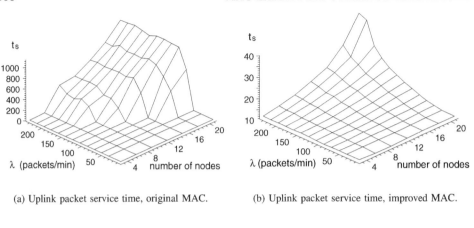

(a) Uplink packet service time, original MAC.

(b) Uplink packet service time, improved MAC.

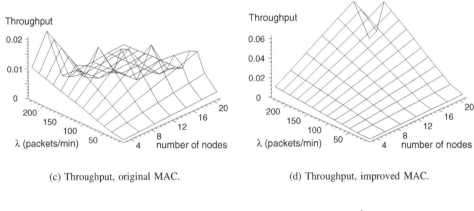

(c) Throughput, original MAC.

(d) Throughput, improved MAC.

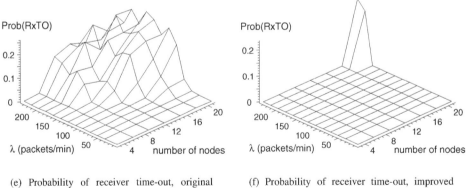

(e) Probability of receiver time-out, original
MAC.

(f) Probability of receiver time-out, improved
MAC.

Figure 5.9 Comparing the performance of the original MAC (left column) with that of the improved MAC (right column). Adapted from J. Mišić, S. Shafi, and V. B. Mišić, 'Performance limitations of the MAC layer in 802.15.4 low rate WPAN,' *Computer Communications,* **29**(13-14): 2534–2541, © 2006 Elsevier B. V.

As can be seen, the range in which the values of all those probabilities are close to one (which means that the network is, in effect, usable) is much wider under the improved MAC than it was for the original MAC. Furthermore, the onset of saturation – the regime in which the network is operating with reduced efficiency and throughput, as exemplified by the abrupt drop in success probabilities – occurs at much higher packet arrival rates and at larger cluster sizes under the improved MAC than under the original one, and it is less abrupt as well.

The improvements are further confirmed by the diagrams of mean packet service in the uplink direction, aggregate throughput (expressed as a percentage of the rated output of 250 kbps), and the probability of receiver time-out, shown in Figure 5.9. As can be seen, the uplink packet service time is drastically reduced with the improved MAC, while the probability of received timeout remains low in a much wider range of parameter values. Consequently, the maximum throughput has increased approximately threefold.

6

Activity Management through Bernoulli Scheduling

6.1 The Need for Activity Management

Wireless sensor networks are an increasingly important application area for WPANs such as IEEE 802.15.4 (Callaway, Jr. 2004; Gutiérrez et al. 2004). The primary objective of sensing applications running on those networks is to achieve and maintain the desired information throughput. This throughput, often referred to as the event sensing reliability, is defined as the mean number of packets received by the network sink in unit time (Sankarasubramaniam et al. 2003). In addition, many application scenarios require wireless sensor networks to operate with as little maintenance as possible due to high cost and, sometimes, infeasibility of maintenance activities. In such cases, maximization of network lifetime is often a secondary objective with no less importance than the primary one. This dual objective can only be achieved through activity management of both the individual nodes and the overall network.

Typically, the sensor lifetime can be extended by adjusting the frequency and ratio of active and inactive periods of sensor nodes (Sankarasubramaniam et al. 2003). This approach is supported in 802.15.4 networks that operate in beacon enabled mode, since the superframe (i.e., the time interval between two successive beacons) is divided into active and inactive periods, and individual nodes are free to switch to a power-saving mode during the latter; for obvious reasons, this is often referred to as 'sleeping'. Power saving typically involves turning off the radio subsystem which is the single greatest power consumer; in most hardware implementations, this will reduce power consumption by two to three orders of magnitude and allow prolonged operation on battery power. For example, a single AAA battery can power a radio transceiver that draws 10 mA (a fairly typical value for the current generation of sensor motes) for two years, provided that the duty cycle does not exceed 0.5% (Gutiérrez et al. 2004).

However, synchronized sleep in which individual nodes alternate between active and sleep periods in unison, means that the entire network is inactive for prolonged periods of time. In many applications this is not a problem: such is the case, for example, in environmental monitoring applications in agriculture and forestry, or in logistics applications. Other applications, such as surveillance, health care, and structural health monitoring, require continuous monitoring of relevant variables and events; letting the entire network sleep for any period of time is simply out of the question. In this case, activity management must be applied at the individual node level and each node will have its own activity pattern. As a result, at any given moment some of the nodes will be active while others will be sleeping. Obviously, the total number of nodes in the cluster must be larger than the minimum number needed to achieve the throughput constraint only. The simultaneous lifetime and throughput constraints are, then, achieved by independently adjusting two parameters:

- the mean number of active nodes will determine the information throughput (i.e., the traffic volume) that can be achieved in such a network;

- the ratio of active vs. inactive nodes, together with the energy budget of each node, will determine the lifetime of the network.

In line with the probabilistic character of the medium access mechanism, we propose a two-pronged approach to activity management using Bernoulli scheduling (Takagi 1991) and random sleep times; both mechanisms are applied at the individual node level. Under the Bernoulli scheduling policy, at the end of each packet transmission the MAC layer checks the node buffer to see if there are any more packets queued. If there are none, the node immediately goes to sleep by turning off as many of the hardware subsystems as possible (in particular, the radio subsystem which is the single greatest energy consumer); if there are one or more packets waiting to be transmitted, the node will decide either to service the next packet or go to sleep, with the probabilities of P_{ber} and $1 - P_{ber}$, respectively. In both cases, the sleep time is a random variable that follows a predefined probability distribution. By adjusting the probability P_{ber} and the appropriate parameters of the chosen sleep time distribution, independent control of both the node utilization (which, in turn, affects the number of active nodes) and the duration of sleep periods is made possible.

6.2 Analysis of Activity Management

Let us now evaluate the performance of this approach using a queueing theoretic approach, assuming for the moment that both P_{ber} and P_{sleep} are known. Our main assumptions are similar to those used in Chapters 3 to 5. We assume that the probability distribution of the packet service time at the MAC layer is described by the Probability Generating Function (PGF) of $T_t(z)$ derived earlier in Section 3.4. The corresponding Laplace-Stieltjes Transform (LST), $T_t^*(s)$, can be obtained by substituting the variable z with e^{-s} in the expression for $T_t(z)$. Also, let the PDF (probability distribution function) of the packet service time be denoted with $T_{dt}(x) = \text{Prob}[T_{dt} \leq x]$, while the corresponding probability density function (pdf) is denoted with $t_{dt}(x)$. Mean duration of the packet service time is, then, $\overline{T_t} = T_t'(1)$.

Let the cluster consist of the coordinator and n ordinary nodes; as before, the cluster is organized in a star topology and it operates in beacon enabled mode with slotted CSMA-CA medium access. Let each device have an input queue implemented with a buffer capable of storing up to L packets. Let the packet arrival process to the node i (where $1 \leq i \leq n$) be modeled as a Poisson process with the arrival rate of λ. The input queue is serviced by the server which consists of the MAC and PHY layers of the 802.15.4 protocol stack. The server has two states: active state, in which it services packets from the input queue, and sleep state. When the server is in the active state, it follows the Bernoulli scheduling policy, as follows:

- if the input queue is non-empty after a successful packet transmission, the server proceeds to service the next packet from the queue with the probability P_{ber}, or begins the sleep state with the probability $1 - P_{ber}$;

- if the input queue is empty upon successful packet transmission or upon returning from the sleep state, the server immediately begins a new sleep.

The operation of the node is schematically shown in Figure 6.1.

Since the actual buffer has finite capacity, some form of buffer management must be employed. When the server is active, a newly arrived packet that finds a buffer filled to its capacity is simply dropped. When the server is in the sleep state, a slightly different policy is utilized. In this policy, often referred to as 'push-out' policy, a newly arrived packet is always admitted to the buffer. If the buffer was full at that time, admission is accomplished by discarding the packet at the head of the buffer and moving all other packets in the buffer for one position so as to make space for the newly arrived packet. In this manner, the server that returns from the sleep state will always find the most recent packets in its buffer, which is beneficial in many sensing applications.

The sleep periods during which the server is unavailable to service any packets may be considered as server vacations, and appropriate analysis techniques may be used (Takagi 1991). As to the choice of sleep time distribution, a simple solution is to use geometric distribution which is controlled with a single adjustable parameter. Let us denote this

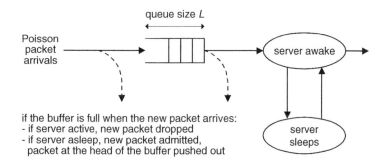

Figure 6.1 Queueing/vacation model for a single node. Adapted from J. Mišić, S. Shafi, and V. B. Mišić, 'Maintaining reliability through activity management in an 802.15.4 sensor cluster,' *IEEE Trans. Vehicular Technology* **55**(3): 779–788, © 2006 IEEE.

parameter with P_{sleep}; the duration of the vacation period may, then, be expressed as

$$V(z) = \sum_{k=1}^{\infty} (1 - P_{sleep}) P_{sleep}^{k-1} z^k \tag{6.1}$$

$$= \frac{(1 - P_{sleep})z}{1 - zP_{sleep}}$$

and its mean duration is $\overline{V} = V'(1) = 1/(1 - P_{sleep})$.

Note that the server immediately takes a new vacation (i.e., goes to sleep again) if the buffer is empty upon returning from vacation. In this manner, most of the idle waiting is eliminated; the remaining part is due to the manner in which the cluster operates. Namely, in a beacon enabled 802.15.4 cluster, a node that returns to the active state with a non-empty buffer will have to synchronize with the beacon prior to undertaking any packet transmission. The PGF for this synchronization time is

$$D_1(z) = \frac{1 - z^{SD}}{SD(1 - z)} \tag{6.2}$$

However, we have shown that the packets received during the inactive period will experience much higher collision probability due to the congestion caused by other similar packets (Section 5.2), as well as the packets received during the previous active period but deferred to the next superframe (Section 5.1) and, possibly, data request packets in case the cluster has downlink traffic (Section 5.3). Therefore, some means of reducing this congestion would be beneficial, as the reduced collision rate will lead to improved power efficiency, and, ultimately, to an increase in cluster lifetime – which is the original objective of activity management. One such approach was described in Section 5.5 above, and we will use a similar procedure here. Namely, when a node returns to the active state to find that the buffer contains at least one packet, it will not undertake the regular CSMA-CA algorithm for uplink transmission immediately upon synchronizing with the beacon. Instead, it will wait for additional D_2 backoff periods before starting the countdown, where D_2 is a uniformly distributed random number in the range between 0 and SD, the duration of the active portion of the superframe. If the beacon interval is divided into active and inactive parts of equal duration (i.e., if $BI = 2SD$), the PGF for this time is equal to the PGF for the time to synchronize with the beacon: $D_2(z) = D_1(z)$.

Let us now analyze the operation of system, starting from the Markov points that correspond to the moments of packet departure from the node and the moments when the server wakes up (i.e., ends its vacation). Let $V^*(s)$ denote the LST of the vacation time, with the corresponding PDF of $V(x)$ and pdf of $v(x)$. Let d_k, a_k and f_k denote the probability of exactly k packet arrivals to the node buffer during the set-up time, packet service time, and during the sleep time, respectively. These probabilities amount to

$$s_k = \int_0^{\infty} \frac{(\lambda x)^k}{k!} e^{-\lambda x} d(x) dx$$

$$a_k = \int_0^{\infty} \frac{(\lambda x)^k}{k!} e^{-\lambda x} t_{dt}(x) dx \tag{6.3}$$

$$f_k = \int_0^{\infty} \frac{(\lambda x)^k}{k!} e^{-\lambda x} v(x) dx$$

We also note (Takagi 1991) that the PGF for the number of packet arrivals to the node buffer during the packet service time and sleep time, respectively, are

$$S(z) = \sum_{k=0}^{\infty} s_k z^k = \int_0^{\infty} e^{-x\lambda(1-z)} d(x) = D^*(\lambda - z\lambda)$$

$$A(z) = \sum_{k=0}^{\infty} a_k z^k = \int_0^{\infty} e^{-x\lambda(1-z)} dt_t(x) = T_t^*(\lambda - z\lambda) \qquad (6.4)$$

$$F(z) = \sum_{k=0}^{\infty} f_k z^k = \int_0^{\infty} e^{-x\lambda(1-z)} v(x) = V^*(\lambda - z\lambda)$$

When the LSTs of the packet service time and the server vacation time are known, the probabilities s_k, a_k and f_k can be obtained as

$$s_k = \frac{1}{k!} \frac{d^k S(z)}{dz^k}\Big|_{z=0}$$

$$a_k = \frac{1}{k!} \frac{d^k A(z)}{dz^k}\Big|_{z=0} \qquad (6.5)$$

$$f_k = \frac{1}{k!} \frac{d^k F(z)}{dz^k}\Big|_{z=0}$$

As defined above, P_{ber} denotes the probability that the service will continue after the departure of the packet from the head of the buffer, provided the buffer is non-empty at that time. Let π_k and q_k denote the steady state probabilities that there are k packets in the device buffer immediately upon a packet departure and after returning from vacation, respectively. Then, the steady state equations for state transitions are

$$q_0 = (q_0 + \pi_0) f_0$$

$$q_k = (q_0 + \pi_0) f_k + (1 - P_{ber}) \sum_{j=1}^{k} \pi_j f_{k-j}, \qquad 1 \le k \le L - 1$$

$$q_L = (q_0 + \pi_0) \sum_{k=L}^{\infty} f_k + (1 - P_{ber}) \sum_{j=1}^{L-1} \pi_j \sum_{k=L-j}^{\infty} f_k$$

$$\pi_k = \sum_{j=1}^{k+1} q_j \sum_{l=0}^{k-j+1} (s_l + a_{k-j+1-l}) + P_{ber} \pi_j a_{k-j+1}, \qquad 0 \le k \le L - 2 \qquad (6.6)$$

$$\pi_{L-1} \sum_{j=1}^{L} q_j \sum_{k=L-j}^{\infty} \sum_{l=0}^{k} (s_l + a_{k-l}) + P_{ber} \sum_{j=1}^{L-1} \pi_j \sum_{k=L-j}^{\infty} a_k$$

$$\sum_{k=0}^{L} q_k + \sum_{k=0}^{L-1} \pi_k = 1$$

The probability distribution of the device queue length at the time of packet departure π_i, $i = 0 .. L - 1$, and at the time of the return from vacation q_i, $i = 0 .. L$, can be found by solving the system of equations given above. The system can be simplified by noting that

$$\sum_{k=L-j}^{\infty} f_k = 1 - \sum_{k=0}^{L-j-1} f_k.$$

From the system (6.6), the probability that an arbitrary Markov point is followed by a vacation can be obtained as

$$P_v = q_0 + \pi_0 + (1 - P_{ber}) \sum_{k=1}^{L-1} \pi_k \tag{6.7}$$

By the same token, the probability that an arbitrary Markov point is followed by a packet service is $1 - q_0 - \pi_0$. Then, the mean distance between two successive Markov points is

$$\eta = P_v \overline{V} + \sum_{j=1}^{L} q_j (\overline{D} + \overline{T_t}) + (1 - P_v - \sum_{j=1}^{L} q_j) \overline{T_t} \tag{6.8}$$

where

$$1 - P_v = \sum_{k=1}^{L} q_k + P_{ber} \sum_{k=1}^{L-1} \pi_k.$$

We will also need the probability that the radio subsystem of a node is active, which can be obtained as

$$\eta = \frac{\eta - P_v \overline{V}}{\eta} \tag{6.9}$$

The offered load to the system is $\rho = \lambda \overline{T_t}$, while the probability that the packet will not be admitted to the node buffer due to insufficient space is

$$P^B = 1 - \frac{\eta}{\rho}, \tag{6.10}$$

Given that the total number of nodes is n, the mean number of active nodes is

$$n_{on} = \sum_{k=0}^{n} k \binom{n}{k} \eta^k (1 - \eta)^{n-k} \tag{6.11}$$

For simplicity, we assume that the MAC layer operates with fully reliable transfer, in which case the event sensing reliability per node (i.e., the number of packets that a node delivers to the coordinator per time unit) is equal to the admitted packet rate per second:

$$r = \frac{\eta}{\overline{T_t} t_{boff}} = \frac{\lambda}{t_{boff}} (1 - P^B) \tag{6.12}$$

where t_{boff} denotes the duration of the unit backoff period (IEEE 2006).

The probability distribution of packet service time in the MAC layer depends on the packet arrival rate per node, packet length, the number of nodes, backoff window sizes and the way how the decision to transmit is made. We will derive these dependencies in the following sections.

6.3 Analysis of the Impact of MAC and PHY Layers

The assumptions about packet sizes are the same as those made in earlier chapters, i.e., the PGF of the data packet length be $G_p(z) = z^k$, and the mean data packet size is $G'_p(1) = k$

backoff periods. The PGF of the time interval between packet transmission and subsequent acknowledgment is $t_{ack}(z) = z^2$. $G_a(z) = z$ stands for the PGF of the acknowledgment duration. We assume that acknowledgments are used to ensure fully reliable transfer, as defined in Section 3.1.1. The PGF for the total transmission time of the data packet will be denoted with $D_d(z) = z^2 G_p(z) t_{ack}(z) G_a(z)$.

The PGF of the duration of the beacon frame is denoted as $B_{ea}(z) = z^2$. We assume that the beacon uses short device addresses of four bytes each and does not contain GTS announcements or beacon payload. The superframe is assumed to have active period only, since the nodes go to sleep for prolonged periods of time, as determined by the value of P_{sleep}. We will also assume that the superframe duration has the minimum allowed value of $SD = aBaseSuperframeDuration$, or 48 backoff periods; the extension of the model to accommodate different lengths of the superframe is straightforward.

The discrete-time Markov chain for the 802.15.4 CSMA-CA algorithm executed by a node, with the inclusion of sleep and synchronization states, is presented in Figures 6.2 and 6.3. Each node is assumed to work in non-saturated regime. The boxes labeled as *D1* and *D2* correspond to the beacon synchronization time upon returning from the sleep state and the additional countdown that reduces congestion at the beginning of the super-frame, respectively; the duration of those time intervals was calculated in Section 6.2. The 'delay line' boxes model the case in which a packet transmission is deferred to the next superframe because the remaining time within the current superframe does not suffice for two CCAs, packet transmission, and subsequent acknowledgment; the probability of this event is $P_d = \overline{D_d}/SD$. Finally, the T_r boxes model the time taken by packet transmission.

We assume that this Markov chain, together with the higher level structure into which it is incorporated, has a stationary distribution (Bianchi 2000). The state of the device at unit backoff period boundaries synchronized with the beacon is defined by the process $\{i, c, k, d\}$, where:

- $i \in (0..m)$ is the index of current backoff attempt, where m must be less than the MAC constant *macMaxCSMABackoffs*, the default value of which is 5.

- $c \in (0, 1, 2)$ is the index of the current Clear Channel Assessment (CCA) phase.

- $k \in (0..W_i - 1)$ is the value of backoff counter, with W_i being the size of backoff window in i-th backoff attempt. The minimum window size is $W_0 = 2^{macMinBE}$, while other values are equal to $W_i = W_0 2^{min(i, 5-macMinBE)}$ (by default, *macMinBE* = 3).

- $d \in (0..\overline{D_d} - 1)$ denotes the index of the state within the delay line mentioned above; in order to reduce notational complexity, it will be shown only within the delay line and omitted in other cases.

From the analysis of $M/G/1/K$ queues with vacations and non-zero set-up time, we find that, upon a successful packet transmission, the system will enter the vacation state with the conditional probability

$$P_c = (\pi_0 + (1 - P_{ber}) \sum_{k=1}^{L-1} \pi_k) / \sum_{i=0}^{L-1} \pi_i,$$

Conditioning follows from the fact that the term in the numerator is the probability that a Markov point corresponds to a packet departure and the vacation starts after the packet

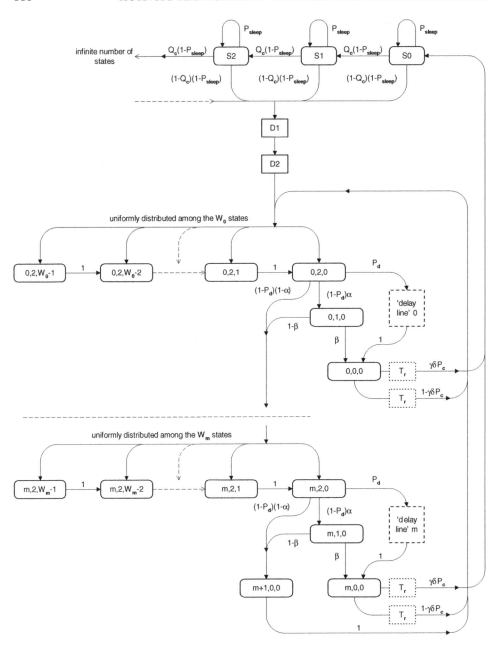

Figure 6.2 Markov chain model in the presence of sleep periods. Adapted from J. Mišić, S. Shafi, and V. B. Mišić, 'Maintaining reliability through activity management in an 802.15.4 sensor cluster,' *IEEE Trans. Vehicular Technology* 55(3): 779–788, © 2006 IEEE.

departure, whereas the sum in the denominator denotes the probability that the Markov point corresponds to a packet departure.

By the same token, the conditional probability that the Markov point corresponds to a return from the vacation and the queue is empty at that moment is

$$Q_c = q_0 / \sum_{i=0}^{L} q_i.$$

If we denote the probability that the node will switch to the sleeping state as $\mathrm{Prob}(In)$, the probabilities of the vacation states become

$$
\begin{aligned}
s_0 &= \frac{\mathrm{Prob}(In)}{1 - P_{sleep}} \\
s_i &= s_{i-1} Q_c, \qquad i = 1..\infty
\end{aligned}
\tag{6.13}
$$

and the sum of probabilities of all inactive states becomes

$$S_{tot} = \sum_{i=0}^{\infty} s_i = \frac{\mathrm{Prob}(In)}{(1 - P_{sleep})(1 - Q_c)},$$

which allows us to re-write the formula for utilization from Equation (6.9) in the form $\eta = 1 - S_{tot}$.

The sum of probabilities of synchronization states in which the node is awake but does not attempt transmission, is $w_{tot} = S_{tot}(1 - Q_c)(1 - P_{sleep}) = \mathrm{Prob}(In)SD$.

Using the transition probabilities indicated in Figures 6.2 and 6.3, we can derive the relationships between the state probabilities and solve the Markov chain. We use our standard notation for success probabilities: α and β are the probabilities that the medium is idle on first and second CCA, respectively; γ is the probability that the transmission does not experience a collision; and δ is the probability that the transmission is not corrupted by noise. Also, we will omit l whenever it is zero, and introduce auxiliary variables C_1, C_2, C_3 and C_4 through

$$
\begin{aligned}
x_{0,1,0} &= x_{0,2,0}(1 - P_d)\alpha = x_{0,2,0}C_1 \\
x_{1,2,0} &= x_{0,2,0}(1 - P_d)(1 - \alpha\beta) = x_{0,2,0}C_2 \\
x_{0,0,0} &= x_{0,2,0}((1 - P_d)\alpha\beta + P_d) = x_{0,2,0}C_3 \\
C_4 &= \frac{1 - C_2^{m+1}}{1 - C_2}
\end{aligned}
\tag{6.14}
$$

Then, by substituting $\mathrm{Prob}(In) = x_{0,0,0}\gamma\delta P_c C_4$, we obtain

$$
\begin{aligned}
S_{tot} &= x_{0,0,0}C_4 \frac{P_c\gamma\delta}{(1 - P_{sleep})(1 - Q_c)} \\
w_{tot} &= x_{0,0,0}\, SD\, C_4 P_c\gamma\delta
\end{aligned}
\tag{6.15}
$$

The sum of all probabilities in the Markov chain must be equal to one:

$$
\begin{aligned}
S_{tot} + w_{tot} + \sum_{i=0}^{m}\sum_{k=0}^{W_i-1} x_{i,2,k} + (\overline{D_d} - 2)\sum_{i=0}^{m} x_{i,0,0} + \\
\sum_{i=0}^{m} x_{i,1,0} + x_{m+1,0,0} + \sum_{i=0}^{m}\sum_{l=0}^{\overline{D_d}-1} x_{i,2,0,l} = 1
\end{aligned}
\tag{6.16}
$$

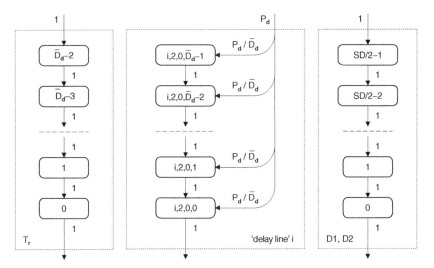

Figure 6.3 Delay lines for the Markov chain of Figure 6.2. Adapted from J. Mišić, S. Shafi, and V. B. Mišić, 'Maintaining reliability through activity management in an 802.15.4 sensor cluster,' *IEEE Trans. Vehicular Technology* **55**(3): 779–788, © 2006 IEEE.

which gives the value for $x_{0,0,0} = 1/N_{0,0,0}$ where

$$N_{0,0,0} = C_4 \left(\overline{D_d} + \frac{C_1}{C_3} + \frac{P_c \gamma \delta}{(1 - P_{sleep})(1 - Q_c)} + SD\, P_c \gamma \delta + \frac{P_d(\overline{D_d} - 1)}{2C_3} \right)$$
$$+ \sum_{i=0}^{m} \frac{C_2^i(W_i + 1)}{2C_3} + \frac{C_2^{m+1}}{C_3}$$

(6.17)

The total probability to access the medium is

$$\tau = \sum_{i=0}^{m} x_{i,0,0} = x_{0,0,0}C_4$$

(6.18)

However, we should distinguish between the probability τ_1, which describes medium access in the case where a transmission is deferred to the next superframe due to insufficient time, and the probability τ_2, which describes regular medium access in the current superframe:

$$\tau_1 = \frac{P_d}{C_3} \tau$$
$$\tau_2 = \left(1 - \frac{P_d}{C_3} \right) \tau$$

(6.19)

The values of success probabilities at the first and second CCAs, α and β, and the probability of successful transmission, γ, are identical to those derived earlier in Section 3.2.1, while the probability distribution for the packet service time is identical to the one derived in Section 3.4.

The impact of the PHY layer deserves a brief comment. As explained in Section 2.1, the 802.15.4 standard provides several PHY options in three RF bands. For simplicity, and because future sensing applications will likely require as much bandwidth as possible, let us consider only the 2450 MHz PHY option. In this case, a 16-ary quasi-orthogonal modulation technique is employed where, in each symbol period, 'four information bits are used to select one of 16 nearly orthogonal pseudo-random noise chip sequences to be transmitted' (IEEE 2006). The resulting chip sequence is then modulated onto the carrier using offset quadrature phase shift keying (O-QPSK). Since the chip rate is 2 Mcps and the raw data rate is 250 kbps, the processing gain is 8, which gives the maximum supported ratio of bit energy to the noise power spectral density as $E_b/N_0 = 8$. According to Garg et al. (1998), the resulting bit error rate is

$$BER = Q\left(\sqrt{\frac{E_b}{N_0}}\right)$$

where $Q(u) \approx e^{-u^2/2}(\sqrt{2\pi}u)$, $u \gg 1$.

Therefore, in the absence of interference we may expect a BER value slightly below 10^{-4}. As significant interference may be expected in the ISM band, which hosts Bluetooth and 802.11 wireless networks as well as other devices, it is, then, more realistic to expect that the BER will have a value of 10^{-3}, which will result in the PER value of $PER = 1 - (1 - BER)^X$, where X is the packet length expressed in bits. This is confirmed by the claim that the Packet Error Rate (PER) value of less than 1% is expected on packets with the length of 20 octets (IEEE 2006, section 6.1.7, Table 4). (The packet length includes MAC and PHY level headers.) Since the transmission includes both the data packet and subsequent acknowledgment packet, the probability that it is not corrupted by noise is

$$\delta = (1 - BER)^{X_d + X_a} \tag{6.20}$$

where X_d and X_a are lengths of the data and acknowledgment packets, respectively. Both of these lengths are expressed in bits, and both include the MAC and PHY headers.

6.4 Controlling the Event Sensing Reliability

The activity management problem we are trying to solve can succintly be defined as tuning activity patterns of individual nodes in such a way that the aggregate event sensing reliability (i.e., the total throughput received by the cluster coordinator) and the individual node utilization are kept at their predefined values, or as close to those values as possible. As the activity pattern of a node is determined by the probability P_{ber} and the parameter P_{sleep}, the activity problem can be defined in more formal terms: namely, we need to find (the algorithm to adjust) the values of P_{ber} and P_{sleep} that maintain the aggregate event sensing reliability and the node utilization at the given values of $\mathcal{R}$ and $\mathcal{U}$, respectively:

$$\text{find } P_{ber}, P_{sleep} \text{ such that } r \geq \mathcal{R} \text{ and } \eta \leq \mathcal{U} \tag{6.21}$$

It is safe to assume that the cluster coordinator is aware of the number of nodes and packet arrival rates. Namely, every node that participates in the cluster has to be explicitly admitted by the coordinator (Section 2.8). Moreover, every packet received by the coordinator carries the source address, which makes it simple to estimate the packet arrival rates. We can also assume that the coordinator knows the required reliability $\mathcal{R}$ as well as the packet size, both of which are set by the sensing application.

But before we define our design task and outline the solution, let us investigate the behavior of the cluster with Bernoulli activity scheduling policy and geometrically distributed sleep periods. To that end, we have considered a star topology cluster operating in beacon enabled mode with CSMA-CA medium access. The cluster employs the 2450 MHz PHY option with the raw data rate of 250 kbps. The superframe size was controlled by $BO = SO = 0$, which gives the superframe duration of 480 bytes. All other MAC parameters were set to their default values. Bit error rate was set to a rather high value of $BER = 10^{-3}$.

The packet size has been fixed at $G'_p(1) = 3$ backoff periods, while individual node buffers had a fixed size of $L = 2$ packets. The coordinator is assumed to run the actual sensing application, so that its buffer size is effectively infinite. Buffer management was performed according to the policy outlined in Section 6.2 (Figure 6.1).

In this setting, we have solved the system formed by Equations (6.18), (6.15), (3.11), (3.13), (3.14), and (6.20), with the goal of finding the values of P_{sleep} needed to maintain the aggregate event sensing reliability at $\mathcal{R} = 7$ packets per second. The packet arrival rate was set to one packet per second for each node, while the cluster size and probability P_{ber} were kept as independent variables.

The calculated values for individual node utilization are shown in Figure 6.4. For convenience, the plane that corresponds to the utilization value of $\mathcal{U} = 0.005$ (i.e., 0.5%) is superimposed on the diagram on the left; the reason for choosing this particular value of utilization was mentioned earlier in Section 6.1.

Apart from the obvious conclusion that node utilization depends on both cluster size (i.e., the number of nodes) and probability P_{ber}, the following observations can be made from the diagrams. First, if the value of utilization η is required to remain below a predefined limit, the number of nodes has to exceed a certain threshold. This threshold is depicted in Figure 6.4 as the intersection of the main surface and the plane that corresponds to

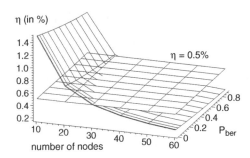

Figure 6.4 Node utilization under controlled reliability. Adapted from J. Mišić, S. Shafi, and V. B. Mišić, 'Maintaining reliability through activity management in an 802.15.4 sensor cluster,' *IEEE Trans. Vehicular Technology* **55**(3): 779–788, © 2006 IEEE.

the utilization of 0.005. The threshold is variable since its exact position depends on the probability P_{ber}: lower values of P_{ber} correspond to higher threshold values, as more nodes are needed to maintain the utilization below the limit, and vice versa.

Second, the boundary value $P_{ber} = 0$ corresponds to the so-called 1-limited service, a well-known scheduling policy for single-server queues (Takagi 1991). In this case, a node will service only a single packet from the queue (or none, if the queue is empty) before going to sleep. The threshold value is the highest in this case, slightly above 3 in our example.

Third, for cluster sizes close to the threshold, the utilization U stays below the predefined limit only for values of the probability P_{ber} close to 1. Note that, when $P_{ber} = 1$, the node stays awake (and services the packets in its buffer) as long as its buffer contains packets to send; this policy is known as exhaustive service (Takagi 1991).

Exhaustive service policy is able to reduce utilization because the overhead incurred by the synchronization with the beacon and the additional waiting time that reduces congestion at the beginning of the superframe are, in fact, shared among several packets. Alternatively, exhaustive service may be used to reduce the number of nodes while maintaining the node utilization below the predefined limit. Still, this number must be above the threshold which, in our example, is around 20 when $P_{ber} = 1$.

Although it may seem that our activity management problem can simply be solved by setting the probability to $P_{ber} = 1$, this may not be the best solution in all scenarios. Namely, the packet payload is, more often than not, large enough to accommodate the results of a single sensing measurement. (In our example, the packet size is 3 backoff periods or 60 bytes; quite a few numeric values can fit therein without a problem.) If exhaustive service is used, several packets may be sent in quick succession, i.e., without any sleep period between them. However, those measurements are taken at the same physical location, and the correlation between the measured values, in both spatial and temporal sense, may be quite high. Depending on the requirements of the sensing applications, the usability of highly correlated measurements could be diminished; in extreme cases, they might be considered redundant and, therefore, useless.

Consequently, a more general solution is needed; this policy, together with the details outlining its implementation in practice, is explained in the next section.

6.5 Activity Management Policy

The activity management problem, as defined above, aims at maintaining the aggregate event sensing reliability (i.e., the total throughput received by the cluster coordinator) and the individual node utilization at some predefined values. The control parameters that need to be adjusted to achieve this are the probability P_{ber} and the geometric distribution parameter P_{sleep}. The packet arrival rate per node is assumed to be known, as it will likely be determined by the requirements of the sensing application.

Since the boundary values for the probability P_{ber} correspond to different threshold values for the number of nodes in the cluster, the idea is to utilize the entire range for P_{ber} for control purposes. We start with a given number of nodes in the cluster, which must exceed the threshold for $P_{ber} = 0$. The value of P_{sleep} that achieves the desired event sensing probability is calculated from the system of equations defined above, and all nodes are made aware of it. Therefore, when the cluster begins operating, the node activity patterns are determined by the calculated value of P_{sleep} and the probability of $P_{ber} = 0$.

In time, the nodes will consume energy from their batteries and the batteries will begin to die. Since the activity patterns are random, the nodes will die one by one, at different times, and the value of P_{sleep} has to be adjusted accordingly in order to maintain the desired event sensing reliability.

When the number of nodes decreases below the threshold that corresponds to $P_{ber} = 0$, further decrease of sleep time is no longer useful since packet arrivals are assumed to follow a Poisson distribution. Hence the probability P_{ber} must be increased. (In our example, Figure 6.4, this situation occurs when the number of the nodes that are still alive drops below 30.)

Finally, when the number of nodes drops below the threshold that corresponds to $P_{ber} = 0$, the cluster is no longer able to maintain the desired event sensing reliability and it ceases to fulfill its intended function. Since the remaining nodes have almost exhausted their batteries as well, this is effectively the end of the cluster's operational life.

The system of equations resulting from our analytical model can be used to obtain the minimum cluster size needed to achieve predefined values for $\mathcal{R}$ and $\mathcal{U}$. It can also be used to estimate the cluster lifetime, provided the operational characteristics of the power source are known, as there is an obvious tradeoff between the cluster size (and, of course, the capacity of node power source) and its lifetime, and virtually any desired lifetime may be achieved by scaling the cluster size accordingly.

Centralized control.   A straightforward way to implement the policy described above is to use centralized control hosted at the cluster coordinator. In this case, the coordinator would calculate the initial values of P_{sleep} and P_{ber} (the preferred value for the latter being zero) which are needed to achieve the desired event sensing reliability and node utilization. The calculated values can be periodically broadcast in the beacon frame, or perhaps in each beacon frame, so that all nodes in the cluster can adjust their activity pattern accordingly. The values of P_{sleep} and P_{ber} have to be recalculated upon every change in cluster size, caused by the death of a node and detected through received time-outs at the coordinator. At first, the recalculation would attempt to increase the probability P_{ber} but not P_{sleep}; the latter would need to be changed only when P_{ber} reaches its maximum value of one.

The performance of this approach is demonstrated in Figures 6.5 and 6.6. As in the previous experiment, the aggregate event sensing reliability was set to $\mathcal{R} = 7$ packets per second, while all time intervals are shown as multiples of the unit backoff period.

Figure 6.5 shows the aggregate event sensing reliability, the individual (per-node) event sensing reliability, and the mean number of active nodes, as functions of the number of nodes in the cluster and the packet arrival rate. Note that the vertical range of Figure 6.5(a) has been expanded to demonstrate that the algorithm maintains the event sensing reliability within about 1% of the desired value of $\mathcal{R} = 7$ packets per second – if the usual range of 0 to, say, 8 packets per second had been used, the surface would be almost perfectly flat. Due to the control mechanism, the reliability per node depends on the number of nodes, but not on the packet arrival rate, as per Figure 6.5(b). Finally, Figure 6.5(c) shows how the average number of active nodes changes with the cluster size and (to a smaller extent) the value of P_{ber}.

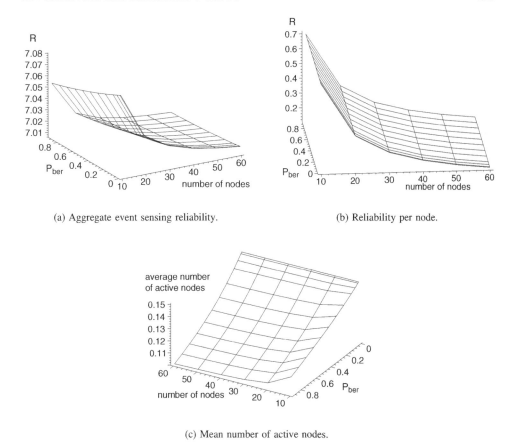

(a) Aggregate event sensing reliability.

(b) Reliability per node.

(c) Mean number of active nodes.

Figure 6.5 Network performance and node activity under controlled reliability. Adapted from J. Mišić, S. Shafi, and V. B. Mišić, 'Maintaining reliability through activity management in an 802.15.4 sensor cluster,' *IEEE Trans. Vehicular Technology* **55**(3): 779–788, © 2006 IEEE.

Figure 6.6 show the performance of activity management from the viewpoint of an individual node: Figure 6.6(a) shows the mean sleep duration (which is determined by P_{sleep}, while Figure 6.6(b) shown the mean duration of the busy period. As can be seen, the mean sleep duration is very long, of the order of tens of thousands of backoff periods; at the same time, the busy periods are extremely short, typically just one superframe in which a packet is sent and the preceding synchronization period.

To summarize, these results show that the control mechanism described above is capable of maintaining the event sensing reliability and the individual node utilization. However, it requires the entire system of equations to be solved every time the cluster size changes, which may not be feasible due to the limited computational capabilities of individual nodes, including the cluster coordinator. Consequently, a different and computationally more efficient implementation is needed.

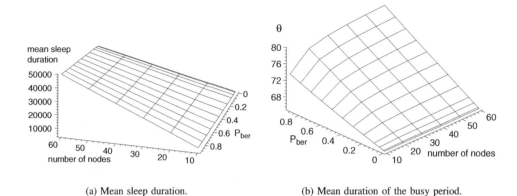

(a) Mean sleep duration.

(b) Mean duration of the busy period.

Figure 6.6 Node activity under controlled reliability. Adapted from J. Mišić, S. Shafi, and V. B. Mišić, 'Maintaining reliability through activity management in an 802.15.4 sensor cluster,' *IEEE Trans. Vehicular Technology* **55**(3): 779–788, © 2006 IEEE.

Distributed control. Fortunately, such an implementation can readily be devised by distributing the calculations between the cluster coordinator and the individual nodes in the cluster, in a way which does not strain the computational and memory resources of either of them. We assume that the number of nodes in the cluster is sufficiently high so as to maintain individual node utilization below the prescribed limit of $\mathcal{U}$, at least in the initial phase when all the nodes are alive. Furthermore, the coordinator is aware of the required event sensing reliability $\mathcal{R}$, and it knows how many nodes are still alive. This may be accomplished through a suitable timeout mechanism; if the node does not transmit anything before the timeout expires, the coordinator may assume that it is dead. The duration of the timeout may be determined by the coordinator itself, or the node itself may provide that information before going to sleep.

In this setup, the responsibility of the coordinator would be to monitor the aggregate event sensing reliability r and calculate the individual node utilization η_d required to maintain the actual event sensing reliability at its desired value, based on the number of active nodes. The value of η_d is then made known through the beacon frame. The sleep time and the probability P_{ber} for an individual node are determined by the node itself, making use of the fact that the mean sleep time, thanks to its geometric probability distribution, is $t_{boff}/(1 - P_{sleep}) = 1/r$, where t_{boff} is the duration of the unit backoff period, and r is the event sensing reliability.

Initially, the node assumes that $P_{ber} = 0$ and sets $P_{sleep} = 1 - rt_{boff}$; it is also aware of the desired value for the utilization η_0 by listening to the beacon frames when active. The node continuously monitors its own utilization by finding the ratio of its busy time and the current sleep time. The values of P_{ber} and P_{sleep} are then recalculated as follows:

$$P_{ber} = \begin{cases} 0, & \text{for } \eta \leq \eta_d \\ 1 - \dfrac{\eta_d}{\eta}, & \text{otherwise} \end{cases}$$

$$P_{sleep} = 1 - rt_{boff}(1 + P_{ber}) \tag{6.22}$$

The actual sleep time is obtained as a random number drawn from a geometric distribution with the parameter P_{sleep}.

Note that, as long as the actual utilization is below the desired value, P_{ber} can be left unchanged at zero; otherwise it is modified to reflect the discrepancy between the actual and desired values of the node utilization.

Once the probability P_{ber} reaches 1, which indicates that the number of nodes that are alive has reached the minimum number needed for the required event sensing reliability, the node continues to use $P_{ber} \approx 1$ until it dies.

To demonstrate the effectiveness of this control mechanism we have built a simulator of the 802.15.4 cluster, focusing on the MAC and PHY layers of the network protocol stack, using the object-oriented Petri Net-based simulation engine Artifex by RSoft Design, Inc. (2003). The simulator modeled a cluster with 50 nodes, each of which had the power budget of 100,000 unit backoff periods. For each backoff period in which a node has had its radio subsystem turned on, this number was decremented. Strictly speaking, this is a simplification – power consumption usually differs between transmit and receive modes (Moteiv Corporation 2006) – but the error thus incurred is small and irrelevant in the current context. (We assume that the coordinator, which functions as the network sink that collects all data from the cluster, has unlimited energy at its disposal.) We have used 1-limited scheduling, in which $P_{ber} = 0$ all the time, as the benchmark against which the performance of the adaptive scheduling is to be assessed.

In this setup, we have measured aggregate event sensing reliability and node utilization as functions of time; the resulting diagrams are shown in Figure 6.7. Since the time range is large and all node deaths would occur in a relatively short interval near the end of the diagram, we have used the number of the nodes that are still alive as the independent variable.

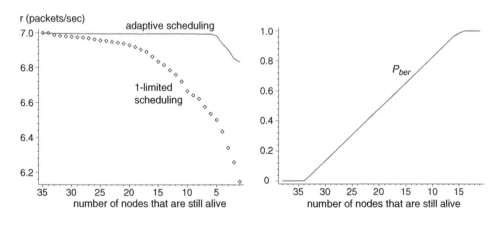

(a) Aggregate event sensing reliability r (diamonds correspond to 1-limited scheduling).

(b) Mean value of the probability P_{ber}.

Figure 6.7 Performance of adaptive scheduling vs. the number of nodes that are still alive. Adapted from J. Mišić, S. Shafi, and V. B. Mišić, 'Cross-layer activity management in a 802.15.4 sensor network,' *IEEE Communications Magazine* **44**(1): 131–136, © 2006 IEEE

As can be seen from Figure 6.7(a), adaptive scheduling succeeds in maintaining the event sensing realibility close to the predefined value of $\mathcal{R} = 7$ packets per second, in a larger range of current cluster sizes than 1-limited scheduling. Figure 6.7(b) shows how the probability P_{ber} changes as the number of nodes in the cluster decreases toward zero.

In both these diagrams, little difference may be observed between adaptive and pure 1-limited scheduling until the cluster is left with about 34 nodes. From that point on, the performance of adaptive scheduling departs from that of 1-limited scheduling. With further node deaths, the mean value of P_{ber} increases almost linearly and reaches the boundary value of one when only five or four nodes are left alive; however, the cluster has already ceased to provide the required event sensing reliability at that time.

Obviously, the use of adaptive Bernoulli scheduling extends the useful lifetime of the network; the question is by how much. The use of redundant nodes to extend the lifetime of the cluster may be described by the so-called lifetime gain. In the ideal case (i.e., disregarding the waste due to collisions, and assuming that the cluster does not enter the saturation regime), the use of n nodes with the packet arrival rate of λ each to achieve the aggregate reliability of $\mathcal{R}$ should result in cluster lifetime of $n \cdot (n\lambda/\mathcal{R})$ times the individual node lifetime. In our example, the ideal lifetime gain is about 357.

The actual lifetime can be observed by combining the information about acceptable event sensing reliability from Figure 6.7(a) with the information about time when those nodes die, which is shown in Figure 6.8. If we assume that the 'useful' life of the cluster, when 1-limited scheduling is used, ends when about 35 nodes remain alive; and that the cluster in which adaptive scheduling is used remains useful until about seven nodes are left, the corresponding lifetime gains are around 349 and 355.2, respectively. Therefore,

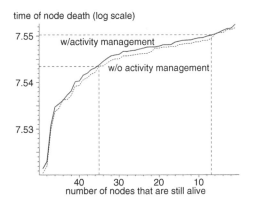

Figure 6.8 Individual node lifetimes, expressed as the time of death, vs. number of nodes that are still alive; vertical axis uses logarithmic scale for clarity. Adapted from J. Mišić, S. Shafi, and V. B. Mišić, 'Maintaining reliability through activity management in an 802.15.4 sensor cluster,' *IEEE Trans. Vehicular Technology* **55**(3): 779–788, © 2006 IEEE.

the useful lifetime of the network has been extended by about six node lifetimes (which is around 1.77%) when adaptive scheduling is used.

These results vividly show that adaptive activity management can be used to extend the useful lifetime of the 802.15.4 cluster. It is capable of achieving graceful degradation of cluster performance, and it can be implemented in a distributed, computationally efficient manner.

7

Admission Control Issues

7.1 The Need for Admission Control

Admission control is tightly coupled to the concept of Quality of Service (QoS). In some application scenarios, wireless PANs are expected to provide a certain throughput; in others, delay constraints have to be respected. Achieving and maintaining the QoS objectives is rather difficult, in particular when the medium access protocol uses some form of the CSMA-CA mechanism where collisions may substantially and unpredictably affect transmission efficiency. A general solution to the above problem may be found in the mechanism of admission control described by Yang and Kravets (2005), in which a central authority – most likely, a network coordinator or cluster head – admits or rejects requests by individual nodes to join the network or cluster.

The most common form of admission control is a contention-based mechanism which strives to maintain high utilization in the network under consideration, while simultaneously minimizing contention. This goal may be achieved through simple means such as limiting the number of nodes or devices in the cluster, but more complicated schemes that use a combination of traffic characteristics and Quality-of-Service (QoS) indicators, such as delay and throughput, have been proposed as well. The fact that fairness is another desirable goal further complicates the design of suitable admission control techniques.

It is worth noting that admission control techniques have long been studied as a promising approach to solving the generic problem of improving bandwidth utilization while maintaining the desired QoS indicators within prescribed limits. The problem was first identified in cellular networks (i.e., mobile telephony), where traffic is predominantly composed of voice calls with well-known and standardized traffic characteristics, and issues related to admission control can be treated in isolation from issues related to PHY and other layers. As a result, a number of proposals for admission control in this environment have been described.

More recently, ad hoc and sensor network applications have received much less attention on account of their higher complexity caused by users' mobility and energy considerations. Moreover, admission control in wireless ad hoc and sensor networks is usually treated together with other aspects of the Medium Access Control (MAC) layer functionality, often under the generic label of 'resource allocation' or 'resource management'.

Time-Division Multiple Access (TDMA) techniques, bearing a strong resemblance to cellular networks, probably offer the simplest way in which admission control can be enforced. The main challenge, in this case, is to calculate the impact of the traffic generated by a node which requests admission, while actual bandwidth allocation is conducted by the central controller that allocates a certain number of time units to individual nodes. For example, this approach can be used in Bluetooth piconets, where all communications are synchronized to, and controlled by, the piconet master (Mišić et al. 2004a). In other types of networks synchronization may be much more difficult to achieve, which means that TDMA-based approach may not be a viable option for implementing admission control.

Admission control is much more complex in CSMA-based ad hoc networks where all nodes, once admitted, have to contend for medium access. Effective admission control necessitates the availability of a suitable traffic monitoring and shaping (policing) mechanism, which is used in two ways. First, the node which requests admission must advertise its traffic parameters, hence the need for monitoring. Second, the node which gets admitted to the network must be able to maintain its traffic rate within the allocated portion of the total available bandwidth, hence the need for traffic shaping. This is a common theme in most, if not all, related proposals: the Admission Control and Dynamic Bandwidth Management scheme (Shah et al. 2005), the Contention-Aware Admission Control (CACP) (Yang and Kravets 2005), both of which are described in the context of IEEE 802.11 networks, as well as in the schemes for peer-to-peer networks (Saxena et al. 2003, 2004).

Sensor networks pose a slightly different set of challenges. In most cases, sensor nodes are not mobile; but on the other hand, sensor nodes are often required to operate on battery power for prolonged periods of time. What this means is that the network topology is again nonstationary – not because of mobility, but because of the fact that sensor nodes will eventually exhaust their power source and cease functioning. Energy efficiency becomes one of the main factors that affect the design of the network and dictate the choice of communication technology and operating regime, if not *the* main such factor. It is worth noting that energy efficiency is important both for individual nodes and for the network as a whole. In this case, bandwidth utilization may simply be regulated through a sleeping mechanism that is applied to the large number of nodes present in the cluster after being admitted by the cluster coordinator. In this case, each node is free to undertake the transmission of its packets (provided there are some) whenever it wakes up from sleep; this sleep may be coordinated in a centralized manner (i.e., through explicit commands sent by the cluster coordinator) or in a distributed fashion, where each node goes to sleep of its own volition, based on some aggregate information broadcast by the cluster coordinator. An example of the latter solution was originally proposed in Mišić et al. (2005d).

It is also possible to combine the two approaches described above into a single scheme. In this cases, nodes go to sleep, but are not automatically allowed to transmit when they wake up. Instead, they first check with the cluster coordinator whether they are permitted to transmit; if not, they go to sleep again. The cluster coordinator, on the other hand, will only admit (and allow transmission by) the number of nodes that allows the network to perform the sensing task without excessive collisions. A suitable admission scheme can be developed based on the activity management model of Mišić et al. (2005d). However, this model is rather complex and requires numerical solution of a large number of non-linear equations. While this solution is acceptable for off-line performance analysis, it is virtually intractable when admission decisions have to be made in real time.

In order to find a viable solution for the sensor network environment, we have developed a simplified and approximate cluster model that can be repeatedly solved using limited computational resources available to the cluster coordinator. The model uses number of nodes, average packet arrival rate, and packet length as independent variables; from these, it calculates packet service time, which is then used as the major admission condition. This simplified cluster model is developed using the results of simulation experiments conducted on clusters with a variable number of nodes, assuming that the packet arrival rates are not symmetric but follow a uniform distribution instead, with small to moderate deviations from the average value.

7.2 Performance under Asymmetric Packet Arrival Rates

In order to develop the admission control algorithm, we have simulated an IEEE 802.15.4 cluster using the object-oriented Petri net simulation engine Artifex by RSoft design, Inc. RSoft Design, Inc. (2003). The goal of this analysis is to investigate the impact of different parameter values on cluster performance, in particular the effect of the asymmetry of the load originating from individual sensors, and to identify parameters that might be suitable candidates for the implementation of an admission control scheme.

For these simulations, we assumed that the network operates in the ISM band at 2.4 GHz, with the maximum data rate of 250kbps. The cluster had the coordinator and $n = 15$ and 25 ordinary sensor nodes, respectively. Packet arrival rates at each of the ordinary nodes were uniformly distributed in the range of $(1 - \delta)\lambda$ to $(1 + \delta)\lambda$, where the variation span δ took values from 0.1 to 0.8, and λ was the arrival rate averaged over all nodes. (For example, the load variation of $\delta = 0.5$ corresponds to uniformly distributed arrival rates in the range between 0.5λ and 1.5λ.) We assume the packet size is fixed at 90 bytes, which includes all PHY and MAC layer headers. Ordinary nodes had buffers that can hold $L = 3$ packets. All other parameters were set to default values prescribed in the 802.15.4 standard IEEE (2006).

In this setup, we have measured the throughput and the service time. The corresponding values are shown in top and middle rows of Figure 7.1 as functions of the average packet arrival rate for the entire cluster λ and the load variation span δ. Note that the mean packet service time does include the effect of packet retransmissions: namely, if the sending node does not receive an acknowledgment from the cluster coordinator, it will assume that the packet is lost due to a collision. The sending node will then repeat the transmission until the proper acknowledgment is received.

From Figure 7.1, two important observations can be made. First, cluster performance is predominantly dependent on the total traffic load in the cluster, but virtually independent of the load asymmetry, i.e., the load variation among individual cluster nodes. This observation holds for both throughput and mean packet service time, under a rather wide range of individual node traffic.

The second observation is that the cluster effectively goes into saturation beyond a certain traffic load. Saturation is caused by the CSMA-CA algorithm which allows collision

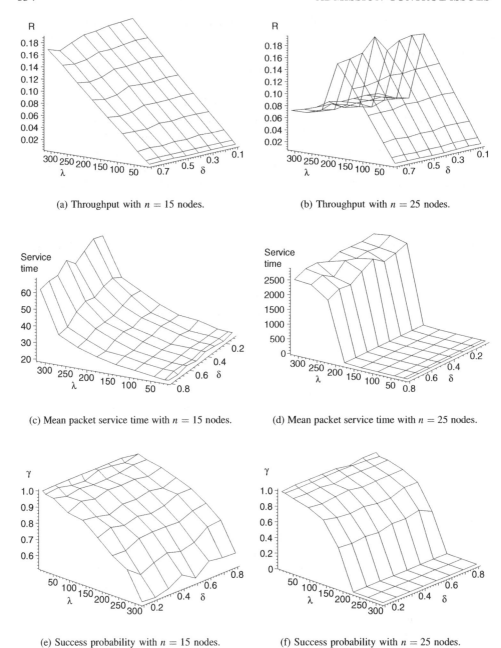

(a) Throughput with $n = 15$ nodes.

(b) Throughput with $n = 25$ nodes.

(c) Mean packet service time with $n = 15$ nodes.

(d) Mean packet service time with $n = 25$ nodes.

(e) Success probability with $n = 15$ nodes.

(f) Success probability with $n = 25$ nodes.

Figure 7.1 Cluster performance under asymmetric traffic: cluster throughput (top row), mean packet service time (middle row), and probability of successful transmission (bottom row). Adapted from J. Mišić, S. Shafi, and V. B. Mišić, 'Real-time admission control in 802.15.4 sensor clusters,' *Int. J. Sensor Networks*, **1**(1): 34–40, © 2006 Inderscience Publishers.

of packets sent by different nodes. As long as the number of nodes is small ($n = 5, 15$), the overall range of traffic loads in the diagrams is small and the cluster does not saturate.

However, in the cluster with $n = 25$ nodes, the probability of collisions is much higher. When the traffic load exceeds a certain level, the collisions become prevalent. When this happens, the majority of the bandwidth is taken up by retransmissions and the overall cluster efficiency decreases. At the same time, the sensor nodes are forced to reject or drop newly arrived packets, since the input buffers are occupied by the packets awaiting retransmission.

The cumulative effect of those phenomena is that the throughput experiences a sharp decrease, Figure 7.1(b), while the packet service time shows a large increase due to packet retransmission, Figure 7.1(d).

Note that collisions, effectively, waste bandwidth; the presence of a large number of collisions means that the nodes are using up their energy resources without actually managing to send useful data. Therefore, the number of collisions must be kept as low as possible in order to conserve energy. A suitable activity management algorithm was presented and analyzed in Chapter 6.

In order to verify that the aforementioned effects are indeed caused by overwhelming number of collisions, we have also measured the probability of successful transmission γ, which is shown in the bottom row of Figure 7.1. As can be seen, the success probability is over 0.9 in the cluster with five nodes (note that seemingly large variations are due to the reduced vertical range for γ), and drops to about 0.55 in the cluster with 15 nodes. However, when saturation occurs in the cluster with 25 nodes, the success probability virtually drops to zero, as shown in Figure 7.1(f), which confirms our analysis.

7.3 Calculating the Admission Condition

The analysis presented above can be summarized as follows. First, the main factor affecting performance is the *total cluster load*, rather than the degree of asymmetry among the nodes in the cluster. Consequently, the analysis results, and admission schemes derived thereof, for the case of symmetric cluster traffic can safely be utilized for clusters with asymmetric traffic.

Interestingly enough, similar conclusion were made in the context of Bluetooth piconets Mišić et al. (2004a); however, the underlying mechanism that leads to this independence is quite different since Bluetooth uses TDMA.

Second, the main objective of the admission control in an 802.15.4 cluster is to prevent the onset of the saturation condition.

In view of this, the actual admission algorithm is quite simple. The cluster coordinator monitors the operation of the cluster, in particular the traffic load. When the new node requests admission, the cluster coordinator calculates whether the added traffic will cause the cluster to move into saturation. If the answer is negative, the new node is admitted to the cluster.

The next step is to find a suitable indicator variable, and to devise the actual algorithm for the necessary calculation. In an earlier work, the performance of the network under symmetric load was considered (Mišić et al. 2004b), and it was found that the forthcoming saturation regime is reliably predicted by any of the following conditions:

- packet service times which are longer than a certain value (approx. 50 backoff periods);

- transmission success probability lower than about 70%;

- finally, access probability at the node larger than a certain value (around 0.005).

While any of these descriptors can be used for the purpose of admission control, the packet service time appears to be the most accurate indicator of the overall cluster load. Therefore, we will focus on calculating or approximating the packet service time as the main indicator of the cluster operating regime. A detailed but complex analytical model for an 802.15.4 cluster with uplink traffic is presented in Figure 7.3. But before we proceed, let us first introduce the necessary notation.

Let the PGF of the data packet length be $G_p(z) = \sum_{k=2}^{12} p_k z^k$, where p_k denotes the probability of the packet size being equal to k backoff periods or $10 \cdot k$ bytes. Then, the mean data packet size is $G'_p(1)$ backoff periods.

The assumptions about packet sizes are the same to those made in Chapter 3, i.e. the PGF of the data packet length be $G_p(z) = z^k$, and the mean data packet size is $G'_p(1) = k$ backoff periods. The PGF of the time interval between packet transmission and subsequent acknowledgment is $t_{ack}(z) = z^2$. $G_a(z) = z$ stands for the PGF of the acknowledgment duration. The PGF for the total transmission time of the data packet will be denoted with $D_d(z) = z^2 G_p(z) t_{ack}(z) G_a(z)$, while its mean value is $\overline{D_d} = 2 + G'_p(1) + t'_{ack}(1) + G'_a(1)$.

Let us denote the probabilities that the medium is idle on first and second CCA with α and β, respectively, and the probability that the transmission is successful with γ. Note that the first CCA may fail because of a packet transmission in progress (originating from another device), and this particular backoff period may be at any position with respect to that packet. The second CCA, however, fails only if some other device has just started its transmission – i.e., this must be the *first* backoff period of that packet. Since the corresponding probabilities differ, we need two different variables.

The simplified probability that the medium is busy at the first CCA is

$$1 - \alpha = \frac{1 - (1 - \tau)^{n-1}(G'_p(1) + G'_a(1))}{\overline{D_d}} \tag{7.1}$$

where τ denotes the access probability – i.e., the probability that the packet will be transmitted. (γ denotes the probability that the transmission will be successful.)

If the cluster operates in non-saturated regime, we are able to ignore the potential congestion effects at the beginning of superframe due to delayed transmissions from previous superframe as described in Chapter 5. In this case, the access probability τ is uniform throughout the superframe and the above expression holds.

The probability that the medium is idle on the second CCA for a given node is, in fact, equal to the probability that neither one of the remaining $n - 1$ nodes has started a transmission in that backoff period. Then the probability of success of second CCA becomes:

$$\beta = (1 - \tau)^{n-1}. \tag{7.2}$$

By the same token, the probability of success of the whole transmission attempt becomes:

$$\gamma = \beta^{\overline{D_d}} \tag{7.3}$$

The PGF of the time needed to conduct one transmission attempt including backoffs is

$$\mathcal{A}(z) = \frac{\sum_{i=0}^{m} \prod_{j=0}^{i} \left(B_j(z) R_{ud} \right) z^{2(i+1)} \left(\alpha \beta T_d(z) \right)}{\sum_{i=0}^{m} \prod_{j=0}^{i} R_{ud} \alpha \beta} \tag{7.4}$$

where

- $T_d(z) = G_p(z) t_{ack}(z) G_a(z)$ represents the transmission and ACK time without the backoff procedure;

- $R_{ud} = 1 - \alpha\beta$ is the probability that one CCA will not be successful;

- $B_j(z) = \sum_{k=0}^{W_j-1} \frac{1}{W_j} z^k = \frac{z^{W_i} - 1}{W_j(z-1)}$ represents the PGF for the duration of j-th transmission attempt; and

- $0 \ldots W_i - 1$ represents the backoff counter value in i-th transmission attempt. If the battery saving mode is not turned on then $W_0 = 7$, $W_1 = 15$, $W_2 = W_3 = W_4 = 31$.

If we assume that transmission will eventually succeed in five attempts, which is reasonable under light to moderate load, the probability distribution of the packet service time follows the geometric distribution and its PGF is

$$\begin{aligned}
T(z) &= \sum_{k=0}^{\infty} (\mathcal{A}(z)(1-\gamma))^k \mathcal{A}(z)\gamma \\
&= \frac{\gamma \mathcal{A}(z)}{1 - \mathcal{A}(z) + \gamma \mathcal{A}(z)}
\end{aligned} \tag{7.5}$$

In this case, mean packet service time can simply be written as

$$\overline{T} = T'(1) = \frac{\mathcal{A}'(1)}{\gamma} \tag{7.6}$$

The period between two consecutive packet service times also depends on the state of the input buffer of the device in question. If the buffer is not empty after packet departure (when positive acknowledgment is received), then the next service time will start immediately; if the buffer is found to be empty after a successful packet transmission, the next service period starts when the next packet arrives. If the probability that the buffer is empty after a packet transmission is denoted with π_0, the mean time period between two packet departures is $\overline{P} = \overline{T} + \frac{\pi_0}{\lambda_i}$, and the period between two transmission attempts by the same device is

$$p = \overline{A} + \frac{\pi_0 \gamma}{\lambda_i} \tag{7.7}$$

In Chapter 3, probability distribution of the length of device queue was derived using the M/G/1/K queueing model coupled with a Markov chain model of the 802.15.4 MAC algorithm. Due to the large computational complexity of the model, the probability π_0 could be obtained only through elaborate numerical calculations. While this approach is appropriate for offline performance evaluation, it is definitely unsuitable for real-time admission control. The limited computational resources of sensor nodes necessitates a much simpler solution.

In order to arrive at a suitable algorithm, we have made two simplifying approximations. First, the expression for the mean time for a transmission attempt was simplified by retaining just the first two members of the Taylor series expansion around point $\tau = 0$. The accuracy of the approximate solution was found to be quite sufficient. Second, we have assumed that the product $\pi_0 \gamma$ is approximately constant and equal to 0.2.

Then, the approximate period between transmission attempts is

$$\widetilde{p} = \widetilde{A} + \frac{0.2}{\lambda} \tag{7.8}$$

In this case the access probability for node becomes:

$$\widetilde{\tau} = \frac{1}{\widetilde{p}}. \tag{7.9}$$

and we can solve Equation 7.9 algebraically for $\widetilde{\tau}$ using n, λ and $\overline{D_d}$ as parameters. Thus, we can obtain approximate values for $\widetilde{A}$, γ, and $\widetilde{St}$ as functions of n, λ and $\overline{D_d}$. The latter two are shown in Figure 7.2, for $\overline{D_d} = 9$.

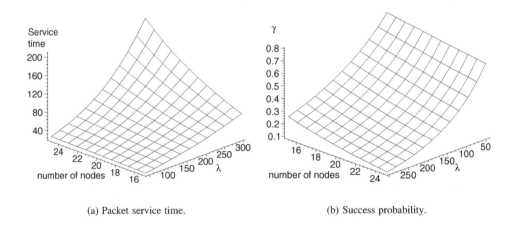

(a) Packet service time.   (b) Success probability.

Figure 7.2 Approximation of performance indicators. Adapted from J. Mišić, S. Shafi, and V. B. Mišić, 'Real-time admission control in 802.15.4 sensor clusters,' *Int. J. Sensor Networks*, **1**(1): 34–40, © 2006 Inderscience Publishers.

7.4 Performance of Admission Control

In order to verify the performance of the admission control, we have augmented the simulator with the calculation of approximate expressions $\tilde{\tau}$, $\tilde{\gamma}$, and $\widetilde{St}$. The cluster was set with variable number of nodes, starting with $n = 15$ nodes and adding one node at a time. The packet arrival rate of each node is featured both in symmetric and in asymmetric approach. In symmetric approach, we assumed the packet arrival rate $\lambda = 2$ packets per second for each and every node. In asymmetric case, each of the nodes featured a variable packet arrival rate λ for $i = 1 .. n$, where the load variation span δ took the value of 0.5, and the average packet arrival rate λ was set to two packets per second. In both cases, the packet size was set to nine backoff periods.

After every 120 seconds, a new node applied for admission. The cluster coordinator then performed the approximate calculations outlined above to obtain the approximate packet service time. The node is admitted based on a desirable admission condition. The admission condition is set as the smallest duration of an 802.15.4 superframe – 48 unit backoff periods. The mean packet service time below 48 backoff periods means that most of the packets will be processed within a single superframe, which is certainly desirable from the standpoint of energy conservation. The algorithm executed by the coordinator when a new node requests admission is shown below.

The calculated and measured values for packet service time under symmetric and asymmetric traffic are shown in Figure 7.3(a) and Figure 7.3(b), respectively. Short horizontal lines show approximate values of service times as calculated by the algorithm, while circle points show measured values (averaged over a five-second time window) after admission. The superimposed line at $T = 48$ is set as a desirable admission condition.

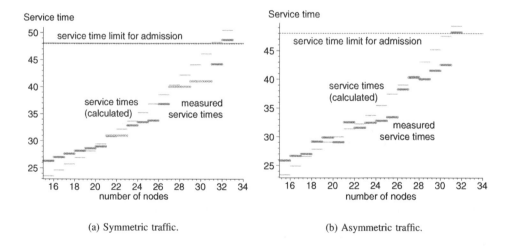

(a) Symmetric traffic. (b) Asymmetric traffic.

Figure 7.3 Measured and calculated service times of individual nodes. Adapted from J. Mišić, S. Shafi, and V. B. Mišić, 'Real-time admission control in 802.15.4 sensor clusters,' *Int. J. Sensor Networks*, **1**(1): 34–40, © 2006 Inderscience Publishers.

Algorithm 1: Admission policy.

Data: number of existing nodes n ;
packet arrival rates λ_i ;
new node packet arrival rate λ'
Result: admission decision

1   calculate the new number of nodes $n' = n + 1$;
2   **if** *traffic load is asymmetric* **then**
3 | calculate the new average arrival rate $\lambda_{avg} = (\lambda_1 + \lambda_2 + ...\lambda_n + \lambda')/n'$;
4   **else**
5 | calculate $\lambda_{avg} = \lambda$;
6   obtain approximate packet service time $\widetilde{St}$ using n' and λ_{avg} ;
7   **if** $\widetilde{St} \leq 48$ **then**
8 | allow the new device to join ;
9 | update $n = n'$, $\lambda_{n+1} = \lambda'$;
10   **else**
11 | reject admission ;
12 | retain existing value of n ;

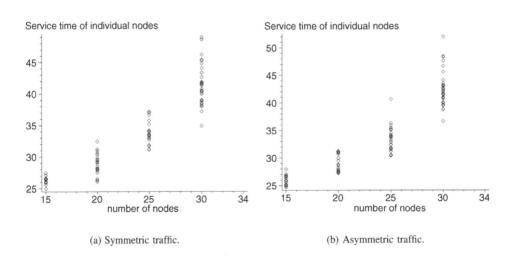

(a) Symmetric traffic.   (b) Asymmetric traffic.

Figure 7.4 Packet service time distribution of individual nodes. Adapted from J. Mišić, S. Shafi, and V. B. Mišić, 'Real-time admission control in 802.15.4 sensor clusters,' *Int. J. Sensor Networks*, **1**(1): 34–40, © 2006 Inderscience Publishers.

We observe a very close match between calculated and simulated values for both symmetric and asymmetric traffic. We also observe that the admission condition is same in both cases, allowing 31 nodes to be admitted with an average packet arrival rate $\lambda = 2$ packets per second. Interestingly, measured values of packet service time do not increase

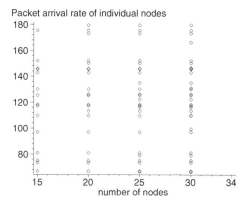

Figure 7.5 Distribution of packet arrival rates of individual nodes under asymmetric traffic.

all the time with increasing number of nodes in asymmetric case – sometimes simulation results with higher number of nodes fall slightly behind the simulation results with lower number of nodes (Figure 7.4). The reason for such behaviour will be clear if we observe Figure 7.5, which shows the distribution of packet arrival rates among the nodes in the case of asymmetric traffic. Although we assume that the distribution of load variation is uniform in the range of 50% around the mean, the number of samples (which is between 15 to 31) is not sufficiently high enough to generate a steady average arrival rate λ. As a result, when a new node is admitted, the new average arrival rate may vary within a few percent of the average arrival rate λ. Thus, after admitting a new node, the new measured service time is affected by the current number of nodes as well as the new average arrival rate. For example, the average service time generated by a cluster with 20 nodes, each of which generates traffic at a rate of $\lambda = 2.03$ packets per second, may be slightly greater than the service time generated by the cluster with 21 nodes with $\lambda = 1.98$ packets per second arrival rate.

Part II Summary and Further Reading

In Part II, we have investigated the performance of single-cluster 802.15.4 networks operating in beacon enabled, slotted CSMA-CA medium access mode with start topology. We have considered clusters operating with uplink traffic only (Chapter 3), as well as clusters with both uplink and downlink traffic (Chapter 4). We have described possible performance risks that stem from the definitions of the official 802.15.4 standard (Chapter 5). We have also discussed activity management and lifetime maximization (Chapter 6), and admission control (Chapter 7). While most of the results presented here were obtained through queueing theoretic analysis, simulation results are also available in a number of earlier papers (Mišić et al. 2005c), (Mišić et al. 2006e), (Mišić et al. 2006f), (Mišić et al. 2006d), and (Mišić et al. 2006g).

A number of authors have also focused on the performance of single-cluster 802.15.4 networks. Among the earliest reports were a brief but rather exhaustive simulation study by Zheng and Lee (2006), who developed an ns-2 simulation module for 802.15.4, and similar study by Lee (2005), which exhibits a notable experimental bias. Other, more focused studies described the performance of the cluster (Pollin et al. 2006) adjustment of various MAC layer parameters (Koubaa et al. 2007a; Neugebauer et al. 2005; Ramachandran et al. 2007), and general performance evaluation.

Some proposals include extensions or modifications to the protocol specified in the standard with the goal of optimizing the performance; the list includes papers by Kim (2006), Kim and Choi (2006), Kim et al. (2006a), Kwon and Chae (2006), Xing et al. (2006), and Cheng and Bourgeois (2007), among others.

Energy efficiency and activity management are typically investigated in the context of wireless sensor networks, where unattended operation and lifetime maximization are of primary importance. Relevant work includes papers by Lu et al. (2004), Timmons and Scanlon (2004), Bougard et al. (2005), Howitt et al. (2005), Mirza et al. (2005), Park et al. (2005), Cho et al. (2006), Koubaa et al. (2006a), and Athanasopoulos et al. (2007).

Issues related to the operation of 802.15.4 networks at the physical layer, including coexistence with other wireless networks in the same RF range, were investigated by Howitt and Gutiérrez (2003), Golmie et al. (2005), Shin et al. (2005), and (in part) by Petrova et al. (2006), while Myoung et al. (2007) investigate the coexistence problem from the viewpoint of a collocated 802.11b wireless LAN. Antonopoulos et al. (2006) provide a comparison of 802.15.4 and 802.11b standards, which is somewhat atypical because of the difference in data rates.

It is worth noting that a number of authors have considered the operation of a single-cluster 802.15.4 network in non-beacon enabled, unslotted CSMA-CA medium access mode with peer-to-peer topology, most notably Kim et al. (2006b), Latré et al. (2005), and Latré et al. (2006). Overall, this mode of operation has not received much attention, most likely on account of its remarkable similarity with the ubiquitous 802.11 protocol.

Part III

Multi-cluster Networks

8

Cluster Interconnection with Master-Slave Bridges

Individual 802.15.4 clusters with separate coordinators can be interconnected to form larger networks. Communication between individual clusters is accomplished through shared nodes or bridges that switch between different clusters in some predefined order. When such a device is present in a particular cluster, it collects the data to be delivered to destinations in other clusters, and/or delivers the data from other clusters to the destinations in the current cluster.

The manner in which the bridge operates when present in a cluster, the duration of bridge residence in each cluster, and the manner in which the switch-over times are arranged, are determined by the bridge scheduling algorithm. All three of these sub-problems heavily depend on the communication technology used and, in particular, on the characteristics of the MAC layer, but the latter two are much harder to solve than the first one.

Fortunately, the 802.15.4 standard provides a simple yet effective mechanism for bridge switching, provided individual clusters operate in beacon enabled, slotted CSMA-CA mode. Namely, when the superframe has both an active and an inactive period, no communication takes place during the inactive period. The bridge (or bridges) may, then, switch to another cluster during the inactive period, and return to the cluster for the next active period. The problem of bridge scheduling is thus reduced to the problem of aligning the superframes (and their active periods) of two or more clusters which is much simpler to solve. The 802.15.4 standard offers a ready-made solution to this problem through the provisions for creating and operating a so-called multi-cluster tree (Section 2.8, p. 34). In this arrangement, several clusters operate using a single RF channel with identical values for superframe duration and beacon interval, but with their active periods distributed over the beacon interval in a non-overlapping fashion. In this manner, a multi-cluster tree with two or more levels can be formed and maintained with ease. An example of a four-level tree is shown in Figure 8.1.

It is also possible to build a multi-cluster tree with fewer levels, as is the case with the topology shown in Figure 8.2. In this topology, which will be the focus of the analysis in this

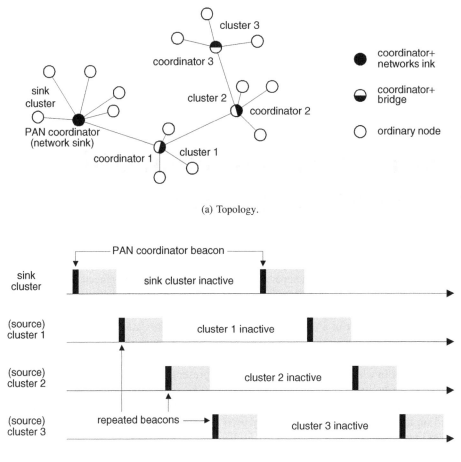

(a) Topology.

(b) Timing of active and inactive periods.

Figure 8.1 A multi-cluster, multi-level tree with four clusters.

chapter, the network consists of one parent cluster and κ child clusters. The parent cluster is controlled by the PAN coordinator, while the child clusters have their own coordinators. Both the parent cluster and the child clusters have a number of ordinary nodes as well. We assume that the network runs a sensing application in which ordinary nodes send their data to the network sink. Obviously, the PAN coordinator is in the best position to act as that sink and it is, thus, the ultimate destination for all traffic. Since child cluster coordinators have to repeat the beacon frames received from the PAN coordinator after specified delays, they must be within the transmission range of the PAN coordinator. Consequently, they are in the best position to act as bridges, i.e., to collect the data sent by the nodes in their respective clusters and subsequently deliver it to the PAN coordinator. By analogy with the Bluetooth technology, the bridge that acts as the coordinator in another cluster will be referred to as a master-slave bridge (Mišić and Mišić 2005, Chapter 10).

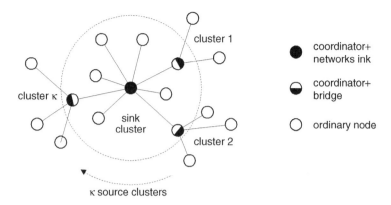

Figure 8.2 A two-level multi-cluster tree with the sink cluster and κ source clusters.

For simplicity, the PAN coordinator will be referred to as the sink, and its cluster will be referred to as the sink cluster. Child clusters will be referred to as the source clusters, and the terms 'coordinator' and 'bridge' will be used according to the role the node in question plays in the current context.

Since all clusters operate in beacon enabled mode and utilize slotted CSMA-CA medium access, there are two possible ways in which the bridge can deliver its data to the sink:

- The bridge may use CSMA-CA, in which case it will compete for medium access with ordinary nodes.

- The bridge may also use contention-free access through GTS, which have to be negotiated beforehand with the coordinator of the sink cluster.

In the discussions that follow, we will refer to those modes as the CSMA-CA and GTS mode, respectively.

8.1 Analysis of Bridge Operation

Let us now investigate the operation of the bridge in more detail. In the analysis that follows, we assume that all traffic flows in the uplink direction. Ordinary nodes in source clusters deliver their traffic to their respective coordinators, which, upon switching to the sink cluster, then deliver it to the sink (i.e., the coordinator of the sink cluster). This traffic will occasionally be referred to as remote traffic in the sink cluster. Ordinary nodes in the sink cluster deliver their data directly to the sink; this traffic will be referred to as local traffic in the sink cluster.

Since the bridges are acting as coordinators of their respective source clusters, they can spend only a single active period in the sink cluster; once that active period ends, the bridges return to their clusters and resume the coordinator role. The reason for this restriction is simple: a cluster operating in beacon enabled mode relies on the periodic beacon frames sent by its coordinator (Section 2.3). If a beacon frame is missing, no communication can take place; in case *aMaxLostBeacons* = 4 beacons are lost, the nodes will assume that the coordinator is lost forever and the cluster essentially ceases to exist. Therefore, the coordinator should not be absent from its cluster for prolonged periods; restricting its absence to the inactive period of a single superframe in the source cluster ensures that the nodes in that cluster suffer no service interruption.

All bridges are assumed to operate in the same manner, either CSMA-CA or GTS. However, regardless of the manner of bridge operation, ordinary nodes in all clusters use the CSMA-CA access mode.

If the bridge operating in the CSMA-CA mode has a packet to transmit to the sink, but is unable to do so because the remaining time in the active portion of the sink cluster superframe is insufficient for the two CCAs, data packet transmission, and (optional) acknowledgment packet, we assume that the bridge will freeze its backoff counter and leave the sink cluster. The backoff countdown will be resumed upon returning to the sink cluster for the next superframe. Obviously, this procedure requires a certain level of functionality beyond the one prescribed by the standard; but so does the bridging operation itself.

Let us now consider a single source cluster and calculate the amount of traffic that reaches the sink cluster. Let the source cluster contain n ordinary nodes, each with the packet arrival rate of λ. As the cluster performance depends mainly on the aggregate packet arrival rate, rather than on the spread of it among individual nodes (Mišić et al. 2006b), the assumption that the node arrival rate is uniform does not affect the validity of our analysis.

We assume that the packet arrival process is Poisson and that the packet may arrive at any time. As the input buffer at the source has a limited size of L packets, some of the packets that arrive may be rejected because the buffer is already filled to capacity. Packets may be dropped at the coordinator/bridge as well, since its input buffer has a finite capacity of L_{bri} packets; however, this loss may be compensated for if acknowledged transfer is used. Let us denote the blocking probabilities at an ordinary node in the source cluster and at the bridge as P_{src}^B and P_{bri}^B, respectively. (For clarity, variables pertaining to the ordinary nodes in the source cluster, ordinary nodes in the sink cluster, and the bridge, will be distinguished through subscripts *src*, *snk*, and *bri*, respectively.)

Since the number of nodes n is relatively large and packet blocking by the bridge, collisions with other packets, and noise corruption are non-correlated events, we may assume that packet arrival process to the bridge is also Poisson, but with the arrival rate of λ_{bri}.

Non-acknowledged transfer. In this case, the cluster coordinator does not send acknowledgments regardless of whether the packet has been successfully received or not. Without acknowledgments, packets sent to the bridge but lost due to noise, collisions, or blocking at the bridge buffer will not be re-transmitted. Total traffic admitted in the source cluster is $n\lambda(1 - P_{src}^B)$. Total traffic offered to the bridge is $\lambda_{bri} = n\lambda(1 - P_{src}^B)\gamma_{src}\delta_{src}$, where γ_{src} denotes the probability that the packet did not suffer a collision, while δ_{src} denotes the probability that packet is not corrupted by the noise. (The latter probability may be calculated if the bit error rate *BER* at the PHY level and the packet length in bytes $\overline{G_p}$ are known, as shown in Section 3.2.) Finally, the total traffic that the bridge will offer to the sink is $\lambda_{bri} = n\lambda(1 - P_{src}^B)\gamma_{src}\delta_{src}P_{bri}^B$, since some packets that have been successfully received may be blocked in case the bridge buffer is full. A simplified representation of the source cluster operation in case of non-acknowledged transfer is shown in Figure 8.3(a).

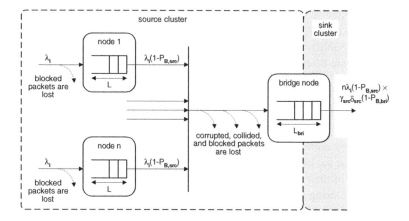

(a) Non-acknowledged transfer.

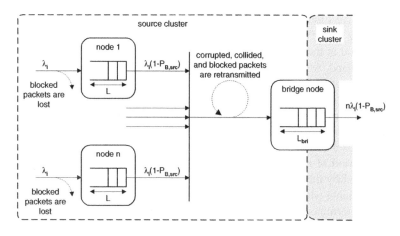

(b) Acknowledged transfer.

Figure 8.3 Queueing model of the bridging process between source and sink cluster. Adapted from J. Mišić and C. J. Fung, 'The impact of master-slave bridge access mode on the performance of multi-cluster 802.15.4 network,' *Computer Networks* **51**: 2411–2449, © 2007 Elsevier B.V.

Acknowledged transfer. In this case, the bridge acknowledges all packets that are successfully received and admitted to the bridge buffer. Lack of acknowledgment is interpreted as the sign that the packet was not received due to noise corruption, collision, or blocking at the input buffer; the sending node will then repeat the packet up to a times before finally giving up. (The default value of a, as defined in the standard, is *macMaxFrameRetries* $= 5$.) Repeated packets contribute to an increase in overall traffic volume and also increase the collision rate. A simplified representation of the source cluster operation in case of acknowledged transfer is shown in Figure 8.3(b). The amount of traffic admitted in the source cluster is $n\lambda(1 - P_{src}^B)$. From the equality $n\lambda(1 - P_{src}^B)P_a = \lambda_{bri}(1 - P_{bri}^B)$, where P_a denotes the probability that packet will be successfully transmitted within a attempts, the packet arrival rate offered by the bridge may be obtained as

$$\lambda_{bri} = \frac{n\lambda(1 - P_{src}^B P_a)}{1 - P_{bri}^B} \tag{8.1}$$

Bridge and ordinary nodes as servers.   In order to obtain blocking probabilities at an ordinary node and at the bridge, P_{src}^B and P_{bri}^B, we need the probability distributions of the number of packets in their buffers at certain points of time. For an ordinary node, packets arrive at any time. During the active part of the superframe, queued packets are sent to the bridge; packets that arrive during the inactive part of the superframe are queued (provided the buffer capacity of L packets is not exceeded) but not transmitted. From a queueing theoretic viewpoint, such a node can be considered as server which serves a random number of packets, as determined by the duration of the packet service time at the MAC and PHY layers, in exhaustive mode during the active part of the superframe, but takes a vacation during the inactive part. Therefore, we can analyze the buffer of ordinary node at the packet departure times and at the end of its inactive superframe time, as shown in Figure 8.4.

On the other hand, when the bridge resides with the source cluster, it receives packets (up to the capacity of its input buffer of L_{bri}) but does not transmit them. Queued packets are sent when the bridge switches to the sink cluster; during that time, no new packets are received. From a queueing theoretic standpoint, when the bridge is present in the source cluster, it acts as a server on vacation; when it is present in sink cluster, it acts as a gated server that services a certain number of packets – those that were present in the buffer at the end of the period of residence in the source cluster (bridge vacation). The number of packets transmitted by the bridge depends on its operating mode: in the CSMA-CA mode, it is a random variable which depends on local traffic in the sink cluster; in the GTS mode, it is a constant determined by the size of the allocated GTS.

We note that ordinary nodes in the sink cluster act in a manner similar to that of the nodes in the source cluster. We assume that there is no packet blocking at the sink (i.e., the sink buffer has infinite capacity), but the presence of the bridge affects the performance of local traffic in the ways that will be discussed below.

Buffer losses at an ordinary node.   Let us begin the analysis by noting that the LST of the packet service times $T_{t,src}^*(s)$ and $T_{t,snk}^*(s)$ can be obtained by substituting the variable z with e^{-s} in the corresponding PGFs

$$T_{t,src}(z) = \sum_{k=0}^{\infty} p_{t,src}(k)z^k$$
$$T_{t,snk}(z) = \sum_{k=0}^{\infty} p_{t,snk}(k)z^k$$

The derivation of these PGFs is a bit involved, which is why it is deferred to Section 8.6 at the end of this chapter. The PGFs for the duration of multiples of packet service

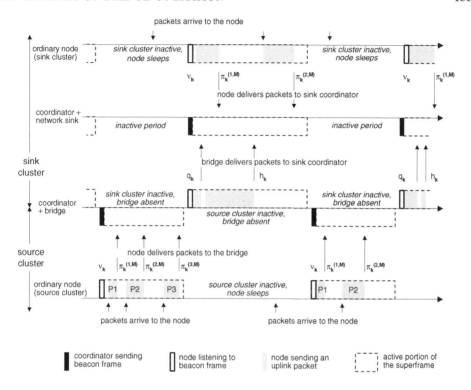

Figure 8.4 Timing and activities of ordinary nodes and the bridge in the source cluster. Adapted from J. Mišić and C. J. Fung, 'The impact of master-slave bridge access mode on the performance of multi-cluster 802.15.4 network,' *Computer Networks* **51**: 2411–2449, © 2007 Elsevier B.V.

times are

$$T_{t,src}^n(z) = \sum_{k=0}^{\infty} p_{nt,src}(k)z^k \quad n = 2, 3, \ldots$$

where the required mass probabilities can be found by matching the coefficients $p_{nt,src(k)}$ with the corresponding terms of $T_{t,src}^n(z)$, expressed in polynomial form. Also, let the corresponding pdfs be denoted by $t_{dt,src}(x)$ and $t_{dt,snk}(x)$, respectively.

We also need the probability distribution of the number of packets which can be served during the active part of the superframe. In the source cluster, this probability distribution is discrete with a few mass probabilities only, since the packet service time has some minimal value dictated by the MAC layer. Therefore, the maximum number $\widehat{M}$ of packets that can be served in one service period (i.e., during the active portion of a single superframe) is $\widehat{M} = \lfloor SD/T_{min,src} \rfloor$, where SD is duration of the active superframe part and $T_{min,src}$ is the minimum service time for a packet, which includes the minimum backoff duration, two backoff periods for the CCAs, actual packet transmission time, and the time to receive the acknowledgment, if any. The mass probabilities for the number of packet transmissions in active superframe part can be derived as

$$pm(0) = \sum_{k=SD+1}^{\infty} p_{t,src}(k)$$

$$pm(1) = (1 - pm(0)) \sum_{k=SD+1}^{\infty} p_{2t,src}(k)$$

$$pm(2) = (1 - pm(0) - pm(1)) \sum_{k=SD+1}^{\infty} p_{3t,src}(k) \tag{8.2}$$

$$\dots$$

$$pm(\widehat{M}) = 1 - \sum_{i=0}^{\widehat{M}-1} pm(i)$$

It is possible that no packet is served in a given service period, $pm(0) > 0$, even through the buffer is not empty; this may happen when several successive backoff and/or transmission attempts fail due to high traffic load in the cluster.

The probability of k packet arrivals during a single packet service time is

$$a_k = \int_0^{\infty} \frac{(\lambda x)^k}{k!} e^{-\lambda x} t_{dt,src}(x) dx \tag{8.3}$$

and the corresponding PGF is

$$A_{src}(z) = \sum_{k=0}^{\infty} a_k z^k = \int_0^{\infty} e^{-x\lambda(1-z)} t_{dt,src}(x) dx = T^*_{t,src}(\lambda - z\lambda) \tag{8.4}$$

Conversely, when the LST of the packet service time is known, the probability a_k can be obtained as

$$a_k = \frac{1}{k!} \frac{d^k A_{src}(z)}{dz^k} \bigg|_{z=0} \tag{8.5}$$

By the same token, the duration of the inactive part of the superframe has a LST of

$$V^*_{const}(s) = e^{-s(BI-SD)} \tag{8.6}$$

The PGF for the number of packet arrivals to an ordinary node in the source cluster during that time is

$$F_{const}(z) = V^*_{const}(\lambda - z\lambda) \tag{8.7}$$

and the probability of k packet arrivals to the buffer during the vacation is

$$f_k = \frac{1}{k!} \frac{d^k F_{const}(z)}{dz^k} \bigg|_{z=0} \tag{8.8}$$

Let $\pi_k^{(m,M)}$ denote the steady state probability that there are k packets in the buffer immediately after m-th packet service in the service period in which a total of M packets are served. (Obviously, M is a random variable which depends on the traffic intensity in the cluster.) Also, let v_k denote the steady state probability that there are k packets left in the buffer at the end of the inactive part of the superframe. Sample moments when $\pi_k^{(m,M)}$ and v_k are evaluated for one node in each cluster are shown in Figure 8.4. Then, the packet queue at an ordinary node in the source cluster, at the beginning of the superframe and after packet departures from the queue, can be described with the following steady state equations:

$$
v_k = pm(0) \sum_{j=1}^{k} v_j f_{k-j} + \left(v_0 + \sum_{M=2}^{\widehat{M}} \sum_{m=1}^{M-1} \pi_0^{(m,M)} \right) f_k
$$

$$
+ \sum_{M=1}^{\widehat{M}} \sum_{j=0}^{k} \pi_j^{(M,M)} f_{k-j}, \qquad\qquad 0 \le k \le L-1
$$

$$
v_L = pm(0) \sum_{j=1}^{L} v_j \sum_{k=L-j}^{\infty} f_k + \left(v_0 + \sum_{M=2}^{\widehat{M}} \sum_{m=1}^{M-1} \pi_0^{(m,M)} \right) \sum_{k=L}^{\infty} f_k
$$

$$
+ \sum_{M=1}^{\widehat{M}} \sum_{j=0}^{L-1} \pi_j^{(M,M)} \sum_{k=L-j}^{\infty} f_k
$$

$$
\pi_k^{(1,M)} = pm(M) \sum_{j=1}^{k+1} v_j a_{k-j+1}, \qquad 0 \le k \le L-2; \ 1 \le M \le \widehat{M}
$$

$$
\pi_{L-1}^{(1,M)} = pm(M) \sum_{j=1}^{L} v_j \sum_{k=L-j}^{\infty} a_k, \qquad\qquad 1 \le M \le \widehat{M}
$$

$$
\pi_k^{(m,M)} = \sum_{j=1}^{k+1} \pi_j^{(m-1,M)} a_{k-j+1},
$$

$$
\qquad\qquad 0 \le k \le L-2; \ 2 \le m \le M; \ 1 \le M \le \widehat{M}
$$

$$
\pi_{L-1}^{(m,M)} = \sum_{j=1}^{L-1} \pi_j^{(m-1,M)} \sum_{k=L-j}^{\infty} a_k
$$

(8.9)

The probabilities that describe the queue status must add up to one:

$$
\sum_{k=0}^{L} v_k + \sum_{M=1}^{\widehat{M}} \sum_{m=1}^{M} \sum_{k=0}^{L-1} \pi_k^{(m,M)} = 1 \qquad\qquad (8.10)
$$

The probability distribution of the device queue length at the time of packet departure can be found by solving the system (8.9) together with normalization Equation (8.10).

In order to find the buffer blocking probability at an arbitrary time, we need to find the average time period between Markov points that correspond to end of vacation time and packet departure times. Since the probability of a vacation period after an arbitrary Markov point is

$$
v_{src} = \sum_{k=0}^{L} v_k,
$$

the average distance between two consecutive Markov points is

$$
\eta_{src} = v_{src}(BI - SD) + (1 - v_{src})T'_{t,src}(1), \qquad\qquad (8.11)
$$

where $T'_{t,src}(1)$ denotes the average packet service time for an ordinary node in the source cluster. The carried load, in this case, can be determined as

$$\rho' = \frac{(1 - v_{src})T'_{t,src}(1)}{\eta_{src}} \tag{8.12}$$

Given that the offered load for an ordinary node is $\rho = \lambda T'_{t,src}(1)$, the blocking probability for the buffer at the node is

$$P^B_{src} = 1 - \frac{\rho'}{\rho} = 1 - \frac{(1 - v_{src})}{\lambda \eta_{src}} \tag{8.13}$$

The probability that the node buffer is empty at the end of the service period is

$$\pi_{0,src} = \sum_{M=1}^{\widehat{M}} \sum_{m=1}^{M} \pi_0^{(m,M)} \tag{8.14}$$

while the probability that the node buffer is empty at an arbitrary time is

$$P_{0,src} = \frac{\pi_{0,src}}{\lambda \eta_{src}} \tag{8.15}$$

Although the impact of the bridge behavior does not explicitly appear in Equations (8.9), (8.10) and (8.13), it does affect it through the probability distribution of the packet service time (to be derived in Section 8.6).

The blocking probability at the buffer of an ordinary node in the sink cluster can be determined in an analogous manner, which is why the corresponding Markov points in Figure 8.4 are labeled in the same manner as those for an ordinary node in the source cluster. However, the packet service time for an ordinary node in the sink cluster must be calculated in a different manner because the sink does not block packets and because of the presence of the bridge; this calculation will also be shown in Section 8.6 .

Blocking at the bridge.   The number of packets in the bridge buffer has to be evaluated at different Markov points. However, Markov points after a packet departure are irrelevant because there are no packet arrivals to the bridge during its service period in the sink cluster; we only need to consider Markov points at the end of the bridge vacation, i.e., when it switches to the sink cluster, and after every service period in the sink cluster, as shown in Figure 8.4. To that end, we need to find the probability distribution of the number of packets transmitted during the service period (i.e., the bridge residence in the sink cluster). Let us assume for the moment that the probability distribution for the service time of packets sent by the bridge in the sink cluster can be described with the PGF of

$$T_{t,bri}(z) = \sum_{k=0}^{\infty} p_{t,bri}(k)z^k \tag{8.16}$$

The correct expression will be derived in Section 8.6 . Then, the PGFs for the duration of two, three, and more packet service times are

$$T^n_{t,bri}(z) = \sum_{k=0}^{\infty} p_{nt,bri}(k)z^k, \quad n = 2, 3, \ldots \tag{8.17}$$

where the required mass probabilities can be found by equating the coefficients of the PGFs $T_{t,bri}^i(z)$, where $i = 1 .. \widehat{B} - 1$, to the matching terms in the polynomial form of the corresponding LST. The label $\widehat{B}$ denotes the maximum number of packets that are served during a single service period (active part of the sink's superframe). This number is finite and equal to the ratio of the active superframe size in sink cluster and minimum packet service time:

$$\widehat{B} = \left\lfloor \frac{BI - SD}{T_{t,bri}'(1)} \right\rfloor \tag{8.18}$$

However, the packet service time for bridge transmissions is longer than the corresponding time experienced by an ordinary node in the source cluster, since the bridge has to compete against ordinary nodes for access to the medium. The mass probabilities for the number of bridge packet transmissions B during the bridge residence in the sink cluster can be derived as

$$pb(0) = \sum_{k=SD+1}^{\infty} p_{t,bri}(k)$$

$$pb(1) = (1 - pb(0)) \sum_{k=SD+1}^{\infty} p_{2t,bri}(k)$$

$$pb(2) = (1 - pb(0) - pb(1)) \sum_{k=SD+1}^{\infty} p_{3t,bri}(k) \tag{8.19}$$

$$\cdots$$

$$pb(\widehat{B}) = 1 - \sum_{i=0}^{\widehat{B}-1} pb(i)$$

Since packets can arrive to the bridge only when the bridge is present in the source cluster, which lasts for exactly SD backoff periods, the probability of k packet arrivals to the bridge can be expressed as

$$u_k = \frac{(\lambda_{bri}SD)^k}{k!} e^{-\lambda_{bri}SD} \tag{8.20}$$

Note that these considerations apply only if the bridge is operating in the CSMA-CA mode; when the bridge operates in the GTS mode, the number of packets delivered to the sink cluster is constant in every superframe.

Let us q_k and h_k denote the probability that the bridge buffer contains k packets at the beginning and at the end of the service period, respectively. Consider the bridge service period where B packets can be transmitted. Then, the state of the bridge buffer at relevant Markov points is determined by

$$h_0 = \sum_{B=1}^{L_{bri}} pb(B) \sum_{k=1}^{B} q_k$$

$$h_k = \sum_{B=1}^{L_{bri}-k} pb(B)q_{k+B}, \qquad 1 \leq k \leq L_{bri} - 1 \tag{8.21}$$

$$q_k = q_0 u_k + pb(0) \sum_{j=1}^{k} q_j u_{k-j} + \sum_{j=0}^{k} h_j u_{k-j}, \quad 0 \leq k < L_{bri} - B$$

$$q_k = q_0 u_k + pb(0) \sum_{j=1}^{k} q_j u_{k-j} + \sum_{j=0}^{L_{bri}-B} h_j u_{k-j}, \quad L_{bri} - B + 1 \leq k < L_{bri}$$

$$q_{L_{bri}} = q_0 \sum_{l=L_{bri}}^{\infty} u_l + pb(0) \sum_{j=0}^{L_{bri}} q_j \sum_{l=L_{bri}-j}^{\infty} u_l \qquad (8.21)$$

$$+ \sum_{j=0}^{L_{bri}-B} h_j \sum_{l=L_{bri}-j}^{\infty} u_l$$

As always, buffer occupancies at all Markov points should add up to one:

$$\sum_{k=0}^{L_{bri}} q_k + \sum_{k=0}^{L_{bri}-B} h_k = 1 \qquad (8.22)$$

By solving the system (8.21) together with the Equation (8.22), we obtain the probability distribution of the bridge buffer occupancy at relevant Markov points.

The blocking probability at the bridge buffer is

$$P_{bri}^B = \frac{q_{L_{bri}}}{\sum_{i=0}^{L_{bri}} q_i} \qquad (8.23)$$

which critically affects the performance of the source cluster under acknowledged transfer, since blocked packets have to be retransmitted until acknowledged.

8.2   Markov Chain Model for a Single Node

Let the PGF of the data packet length be $G_p(z) = z^k$, where the packet size is equal to k backoff periods. In case of non-acknowledged transfer, the PGF for the total transmission time of a data packet is $D_d(z) = z^2 G_p(z)$, where the term z^2 models the two CCAs. In case of acknowledged transfer, we need to include the time to receive the acknowledgment: the time interval between the packet transmission and subsequent acknowledgment and the acknowledgment packet itself, the PGFs of which are $t_{ack}(z) = z^2$ and $G_a(z) = z$, respectively, as explained in Section 3.3. Then, the PGF for the total transmission time of the data packet under acknowledged transfer will be $D_d(z) = z^2 G_p(z) t_{ack}(z) G_a(z)$, while its mean value is $\overline{D_d} = 2 + G'_p(1) + t'_{ack}(1) + G'_a(1)$. The PGF of the duration of the beacon frame is denoted as $B_{ea}(z) = z^2$, which assumes that the beacon contains no payload and no GTS announcements.

The discrete-time Markov chain that models the behavior of ordinary nodes in both clusters, as well as the behavior of the bridge node, is shown in Figure 8.5 and 8.6. Although this chain is virtually identical to the one presented in Section 3.3 and shown in Figure 3.4 and 3.2, we repeat it here for convenience; of course, the values of various

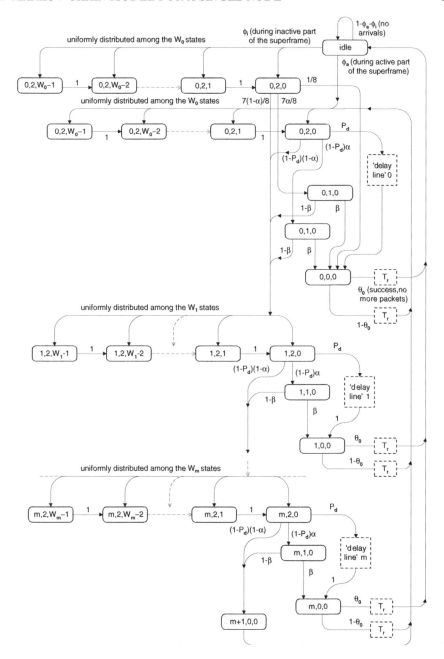

Figure 8.5 Markov chain model that applies to an ordinary node in both source and sink clusters, as well as to the bridge. Adapted from J. Mišić and C. J. Fung, 'The impact of master-slave bridge access mode on the performance of multi-cluster 802.15.4 network,' *Computer Networks* **51**: 2411–2449, © 2007 Elsevier B.V.

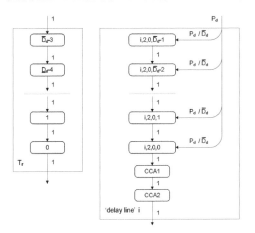

Figure 8.6 Delay lines for Figure 8.5. Adapted from J. Mišić and C. J. Fung, 'The impact of master-slave bridge access mode on the performance of multi-cluster 802.15.4 network,' *Computer Networks* **51**: 2411–2449, © 2007 Elsevier B.V.

transition probabilities will differ. We assume that this Markov chain has a stationary distribution. The state of the node at backoff unit boundaries is defined by the process $\{i, c, k, d\}$, where

- $i \in (0 .. m)$ is the index of current backoff attempt (m is a constant defined by MAC with default value 4).

- $c \in (0, 1, 2)$ is the index of the current Clear Channel Assessment (CCA) phase.

- $k \in (0 .. W_i - 1)$ is the value of backoff counter, with W_i being the size of backoff window in i-th backoff attempt. The minimum window size is $W_0 = 2^{macMinBE}$, while other values are equal to $W_i = W_0 2^{min(i, \, 5-macMinBE)}$ (by default, $macMinBE = 3$).

- $d \in (0 .. \overline{D_d} - 1)$ denotes the index of the state within the delay line mentioned above; in order to reduce notational complexity, it will be shown only within the delay line and omitted elsewhere.

The Markov chain from Figure 3.4, is general in the sense that the probabilities of leaving the transmission state θ_0 and leaving the idle state ϕ_a, ϕ_i are generally labeled; in order to model the bridge, source node or sink node under acknowledged or non-acknowledged transfer, they need to take actual values. θ_a and θ_i denote the probability that a new packet will arrive to the node which is currently in the idle state (i.e., it has no packets in its buffer) during the active and inactive part of the superframe, respectively. θ_0 is the probability that the node buffer is empty after successful packet transmission. For clarity, we will first solve the general model and then substitute the values specific to bridge, source, and sink nodes, respectively.

We note that the part of the Markov chain that models the first backoff phase actually has two components. The sub-chain connected to the idle state with the probability ϕ_i represents the case in which a new packet arrives to the empty buffer during the inactive portion of the superframe. In that case, the first backoff countdown will start immediately after the beacon and the value of the backoff counter will be in the range $0 .. W_0 - 1$. The corresponding states in the sub-chain will be denoted as $x^s_{i,c,k}$. On the other hand, if the packet arrives to a node during the active portion of the superframe, the backoff countdown will start at a random position within the superframe; those states will be denoted as $x_{i,c,k}$. As these two cases will have different effect on the behavior of the medium, they have to be modeled separately.

From the Markov chain, we define the probability to access the medium as

$$\tau = \sum_{i=0}^{m} x_{i,0,0}.$$

Then, the probability that the node will switch into the idle state is $\tau\theta_0$. After setting the balance equation for the idle state, the probability of being in the idle state is obtained as

$$\text{Prob(idle)} = P_z = \frac{\tau\theta_0}{\phi_a + \phi_i}.$$

If we further consider the output from the idle state, and set the balance equations for the first backoff phase after the idle state started during the inactive part of the superframe, we obtain

$$x^s_{0,2,W_0-k} = P_z\frac{k\phi_i}{W_0}, \ 1 \le k \le W_0 \tag{8.24}$$

The first backoff phase in the active part of the superframe may begin at the following moments: upon the arrival of a packet during the idle state, upon a packet transmission (but regardless of the transmission success), or after the last unsuccessful backoff phase. Let us denote the state after the last unsuccessful backoff phase with $x_{m+1,0,0}$ (this is done for clarity only, and has no physical meaning whatsoever). The state probabilities of the first backoff phase started in the active superframe part are represented as $x_{0,2,k}$, $0 \le k < W_0$. The input probability for that set of states is

$$U_a = P_z\phi_a + \tau(1-\theta_0) + x_{m+1,0,0} = \tau\left(1 - \theta_0\frac{\phi_i}{\phi_i + \phi_a}\right) + x_{m+1,0,0} \tag{8.25}$$

By setting the balance equations we obtain

$$x_{0,2,W_0-k} = k\frac{U_a}{W_0}, \ 1 \le k \le W_0 \tag{8.26}$$

A similar approach can be applied to derive $x_{i,2,k}$ for backoff attempts $i = 2, 3, \ldots m$.

Using the transition probabilities indicated in Figure 8.5 and 8.6, we can derive the relationships between the state probabilities and solve the Markov chain. For brevity, we will omit index d whenever it is zero, and introduce the auxiliary variables C_1, C^s_1, C_2,

C_2^s, C_3 and C_3^s through the following equations:

$$x_{0,1,0} = x_{0,2,0}(1 - P_d)\alpha + x_{0,2,0}^s \frac{7\alpha}{8} = x_{0,2,0}C_1 + x_{0,2,0}^s C_1^s$$

$$x_{1,2,0} = x_{0,2,0}(1 - P_d)(1 - \alpha\beta) + x_{0,2,0}^s \frac{7}{8}(1 - \alpha\beta) = x_{0,2,0}C_2 + x_{0,2,0}^s C_2^s$$

$$= \tau \left(\theta_0 \frac{\phi_i}{\phi_i + \phi_a} C_2^s + \left(1 - \theta_0 \frac{\phi_i}{\phi_i + \phi_a} \right) C_2 \right) + C_2 x_{m+1,0,0} \qquad (8.27)$$

$$x_{0,0,0} = x_{0,2,0}((1 - P_d)\alpha\beta + P_d) + x_{0,2,0}^s \left(\frac{7\alpha\beta}{8} + \frac{1}{8} \right)$$

$$= x_{0,2,0}C_3 + x_{0,2,0}^s C_3^s$$

where α and β denote the success probability for the first and second CCA, respectively. From the expressions that describe the Markov chain we obtain

$$\begin{aligned} x_{i,1,0} &= C_1 C_2^{(i-1)} x_{1,2,0}, & i &= 1 .. m \\ x_{i,2,0} &= C_2^{(i-1)} x_{1,2,0}, & i &= 1 .. m \\ x_{i,0,0} &= C_3 C_2^{(i-1)} x_{1,2,0}, & i &= 1 .. m + 1 \\ \sum_{d=0}^{\overline{D_d}-1} x_{i,2,0,d} &= x_{i,2,0} C_2^i (\overline{D_d} - 1)/2 \end{aligned} \qquad (8.28)$$

Of course, the sum of all probabilities in the Markov chain must be equal to one:

$$U + \sum_{k=0}^{W_0-1} x_{0,2,k}^s + x_{0,1,0}^s + \sum_{i=0}^{m} \sum_{k=0}^{W_i-1} x_{i,2,k} + \sum_{i=0}^{m} x_{i,0,0}(\overline{D_d} - 2) + $$

$$+ \sum_{i=0}^{m} x_{i,1,0} + x_{m+1,0,0} + \sum_{i=0}^{m} \sum_{d=0}^{\overline{D_d}-1} x_{i,2,0,d} = 1 \qquad (8.29)$$

which has to be solved for τ, the probability of successful transmission, averaged over the duration of the active portion of the superframe. However, no access is possible in the first two, as well as in the last $\overline{D_d} - 1$ backoff periods, and the access probability has to be scaled accordingly.

By the same token, packets that arrive to an idle node during the inactive portion of the superframe can only access the medium (i.e., be transmitted) between the third backoff period after the beacon and the $W_0 + 2$-th backoff period of the superframe. (The reader may recall that $W_0 = 2^{macMinBE}$ denotes the minimum window size for the random backoff countdown.) From the solution of the Markov chain, $x_{0,2,0}^s$ is the probability that medium access is possible in any backoff period; again, this value needs to be scaled to the time interval in which access can occur. Since the initial value for the countdown is chosen at random between 0 and $W_0 - 1$, the probability of access in the third backoff period after the beacon is

$$\tau_{3,1} = x_{0,2,0}^s \frac{1}{W_0} \frac{SM - 2}{W_0}, \qquad (8.30)$$

where $SM = SD - \overline{D_d} + 1$ is the duration of the part of the superframe wherein either CCAs or actual packet transmission can occur; by extension, $SM - 2$ denotes the time interval in which packet transmission can occur. The probability that the access will occur in some other (fourth to $W_0 + 2$-th) backoff period after the beacon is

$$\tau_{3,2} = x_{0,2,0}^s \frac{W_0 - 1}{W_0} \alpha\beta \frac{SM - 2}{W_0 - 1} \tag{8.31}$$

The reason for separating $\tau_{3,1}$ from $\tau_{3,2}$ is that the former overlaps with the transmissions deferred from the previous superframe due to insufficient time. The probability to access the medium in this case is $SM - 2$ times the value averaged over the entire superframe, as it can happen only in the third backoff period after the beacon. Therefore, the success probability is

$$\tau_1 = (SM - 2)\frac{P_d}{C_3} \left(\tau - x_{0,2,0}C_3^s\right)$$
$$\tau_2 = \left(1 - \frac{P_d}{C_3}\right)(\tau - x_{0,2,0}C_3^s) \tag{8.32}$$

for deferred and non-deferred packets, respectively.

8.2.1 Case 1: ordinary node in the source cluster

Let us now discuss the performance of specific nodes in the network, starting with ordinary nodes in the source cluster.

Non-acknowledged transfer. The idle state of the Markov chain is reached when the buffer is empty after transmission, regardless of whether the packet suffered the collision or blocking by the bridge. In this case, $\theta_{0,src} = \pi_{0,src}$, which was derived in Equation (8.14) above. Since the packet arrival rate λ is small, the probability of no arrivals during a single backoff period can be approximated with the first term of the corresponding Taylor series, $exp(-\lambda) \approx 1 - \lambda$. Then, the probability of non-zero packet arrivals (in which case the node will leave the idle state) is λ which further gives $\phi_{i,src} = P_{sync}\lambda$ and $\phi_{a,src} = (1 - P_{sync})\lambda$, where $P_{sync} = 1 - 2^{SO-BO}$ is the conditional probability that a packet has arrived to an idle node during the inactive portion of the superframe.

Acknowledged transfer. In this case, the idle state is reached only if the node buffer is empty after the transmission, the transmission was successful, and the packet was accepted by the bridge. As the bridge has a finite buffer, a packet sent by an ordinary node can be rejected even though the transmission itself was successful. Therefore,

$$\theta_{0,src} = \frac{\gamma_{src}\delta_{src}(1 - P_{bri}^B)}{P_{a,src}}\pi_{0,src},$$

where

$$P_{a,src} = \sum_{i=0}^{a-1}(1 - \gamma_{src}\delta_{src}(1 - P_{bri}^B))^i \gamma_{src}\delta_{src}(1 - P_{bri}^B)$$
$$= 1 - (1 - \gamma_{src}\delta_{src}(1 - P_{bri}^B))^a$$

denotes the probability that the transfer is completed within a attempts. Furthermore, the probabilities of leaving the idle state are $\phi_{i,src} = P_{sync}\lambda$ and $\phi_{a,src} = (1 - P_{sync})\lambda$ (since the packet arrival rate at an ordinary node in the source cluster is λ).

8.2.2   Case 2: bridge in the CSMA-CA mode

Non-acknowledged transfer.   The probability to enter the idle state of the Markov chain is

$$\theta_{0,bri} = \frac{h_0}{\displaystyle\sum_{i=0}^{L_{bri}-B} h_i}$$

Packets can arrive to the bridge only when it is present in the source cluster, and the packet arrival rate during this period is

$$\lambda_{bri} = n\lambda(1 - P^B_{src})\gamma_{src}\delta_{src}.$$

Therefore, the probabilities to leave the idle state are $\phi_{a,bri} = 0$ and $\phi_{i,bri} = \lambda_{bri}$.

Acknowledged transfer.   In this case, the probability to enter the idle state of the Markov chain is

$$\theta_{0,bri} = \frac{h_0}{\displaystyle\sum_{i=0}^{L_{bri}-B} h_i} \cdot \frac{\gamma_{bri}\delta_{bri}}{P_{a,bri}}$$

where δ_{bri} denotes the probability that a packet transmitted by the bridge is not corrupted by noise, and

$$P_{a,bri} = \sum_{i=0}^{a-1}(1 - \gamma_{bri}\delta_{bri})^i \gamma_{bri}\delta_{bri}$$

is the probability that transfer is completed within a attempts. Packet arrival to the bridge is possible only when the bridge is present in the source cluster, in which case the packet arrival rate to the bridge queue is

$$\lambda_{bri} = n\frac{\lambda(1 - P^B_{src})P_{a,bri}}{1 - P^B_{bri}}.$$

Finally, the probabilities to leave the idle state are $\phi_{i,bri} = \lambda_{bri}$ and $\phi_{a,bri} = 0$.

8.2.3   Case 3: ordinary node in the sink cluster

Under non-acknowledged transfer, the idle state is reached when the node buffer becomes empty regardless of the success of the transmission. In that case, the probabilities to enter the idle state are $\theta_{0,snk} = \pi_{0,snk}$, $\phi_{i,snk} = P_{sync}\lambda$, and $\phi_{a,snk} = (1 - P_{sync})\lambda$.

Under acknowledged transfer, an ordinary node in the sink cluster can reach the idle state of the chain only if its buffer remains empty after a successful transmission, which means that

$$\theta_{0,snk} = \frac{\gamma_{snk}\delta_{snk}}{P_{a,snk}}\pi_{0,snk}$$

and $\delta_{snk} = \delta_{bri}$. The idle state is left upon the arrival of a new packet, which happens with the probability of $\phi_{a,snk} = (1 - P_{sync})\lambda$ and $\phi_{i,snk} = P_{sync}\lambda$ in the active and inactive state, respectively.

Now, the bridge may operate in either CSMA-CA or GTS mode. While ordinary nodes in the sink cluster use CSMA-CA medium access regardless of the bridge access mode, and the Markov chain for both modes looks virtually the same, there are two subtle differences that need to be accounted for:

- When the bridge operates in the CSMA-CA mode, ordinary nodes in the sink cluster must compete with the bridge to access the medium. As the bridge delivers the traffic from the entire source cluster (of which it is the coordinator), the arrival rate from the bridge will be much higher than that of an ordinary node in the sink cluster.

- When the bridge operates in the GTS mode, its transmissions are confined to the CAP and do not interfere with those from ordinary nodes in the sink cluster. However, the active portion of the superframe in the sink cluster is shorter because the CAP is present, and contention among ordinary nodes will increase.

The exact details of these interactions will be seen in the following.

8.3 Performance of the Network

In the next two sections we discuss the performance of the network with a sink cluster and κ source clusters (and, consequently, κ coordinator/bridges). We assume that the network uses the 2450 MHz PHY option, in which case the raw data rate is 250kbps, *aUnitBackoffPeriod* has 10 bytes, and *aBaseSlotDuration* has 30 bytes. The superframe size in both clusters was controlled with $SO = 0$, $BO = 1$; as the value of *aNumSuperframeSlots* is 16, the *aBaseSuperframeDuration* is exactly 480 bytes. The minimum and maximum values of the backoff exponent, *macMinBE* and *aMaxBE*, were set to three and five, respectively, while the maximum number of backoff attempts was five. Other parameter values were set as shown in Table 8.1.

In each case we begin with a comparison of CSMA-CA vs. GTS bridge mode under acknowledged transfer, and then repeat the comparison using non-acknowledged transfer. In order to make a fair comparison, the size of the GTS is chosen to accommodate the throughput of the bridge operating in the CSMA-CA mode at a certain point, typically at $n = 30$ nodes in each cluster and the arrival rate of $\lambda = 3$ packets per second per ordinary node; provisions were also made for a second GTS for the acknowledgment packets, where appropriate.

Table 8.1 Parameters used to model the behavior of the network

number of nodes in each cluster, n	5–30
packet arrival rate, λ	30–240 packets per minute per node
packet size	3 backoff periods
buffer size at an ordinary node, L	2 packets
buffer size at the bridge, L_{bri}	6 packets
superframe size, SD	480 bytes
maximum number of re-transmissions, a	5
bit error rate, BER	10^{-4}

Each component of the model (i.e., source cluster, sink cluster and a bridge) was described by separate set of equations. For the network in which the bridge uses the CSMA-CA mode, the source cluster is modeled through Equations (8.13), (8.14), (8.15), (8.29), (8.39), (8.43), and (8.47); the sink cluster uses Equations (8.13),(8.14), (8.15), (8.29), (8.39), (8.43), and (8.49); finally, the bridge is modeled through Equations (8.1), (8.19), (8.21), (8.22), (8.23), (8.29), (8.39), (8.44), and (8.48).

For the network in which the bridge uses the GTS mode, the source cluster is modeled through Equations (8.13), (8.14), (8.15), (8.29), (8.39), (8.43), and (8.47); the sink cluster uses Equations (8.13), (8.14), (8.15), (8.29), (8.39), (8.46), and (8.50) with superframe duration SD_{gts} (and the interval SM_{gts}) modified to account for the presence of the CAP; finally, the bridge is modeled through Equations (8.1), (8.21), (8.22) and (8.23), since the number of packets delivered to the sink cluster is constant in every service period.

The resulting systems of equations were numerically solved under varying network size and packet arrival rates. Analytical processing was done using Maple 10 from Waterloo Maple, Inc. (2005). The results can be summarized as follows.

8.4   Network with a Single Source Cluster/Bridge

Acknowledged transfer, CSMA-CA bridge.   Let us first consider the CSMA-CA access mode with acknowledged transmissions. Figure 8.7 show the packet blocking probability and throughput at different points of observation in the network, as functions of individual cluster size and packet arrival rate per node. As can be seen, the packet loss probability at all three observation points sharply increases when the network size and packet arrival rate exceed a certain threshold, while the throughput experiences a noticeable dip. These effects signal that the network enters the saturation state in which the efficiency is drastically reduced and few packets manage to reach the sink.

For the chosen parameter values, saturation occurs in the source cluster first, due to the high bridge blocking rate. The sink cluster is still operating and the bridge still manages to deliver some data, albeit at a slower rate than before, partly because its input flow is reduced on account of saturation in the source cluster.

As we increase the values of n and λ, the local load in the sink cluster increases and ultimately the sink cluster enters the saturation regime as well. When this happens, the bridge effectively ceases to operate: it is not very successful in delivering the data to the sink, and it cannot accept new data from the source cluster because its buffer is full most of the time.

Overall, in the case of the CSMA-CA access mode we can distinguish between three operation regimes: in the first, both clusters operate normally; in the second, the source cluster is saturated but the sink cluster still operates normally; and in the third, both clusters are saturated. The three regimes can best be observed in Figure 8.7(c).

Increasing the bridge buffer may seem as a way to improve performance. Indeed, when the bridge buffer size is increased, a point is reached when the saturation occurs in the sink cluster first, due to its higher load and reduced blocking at the bridge. However, saturation in the sink cluster occurs at a lower traffic load than is the case when $L_{bri} = 6$; furthermore, the onset of saturation in the sink cluster immediately brings the source cluster into saturation as well, since the bridge buffer cannot be emptied. Therefore, little overall improvement is gained in this manner, if any at all. This conclusion is corroborated by

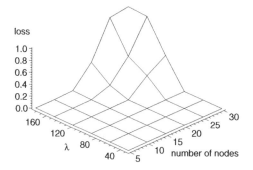

loss

(a) Blocking at an ordinary node in the source cluster.

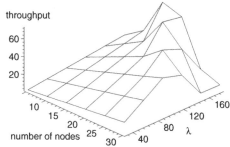

throughput

(b) Throughput from the source cluster to the bridge.

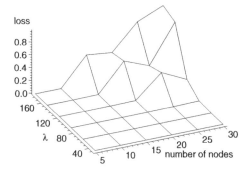

loss

(c) Blocking at the bridge in the source cluster.

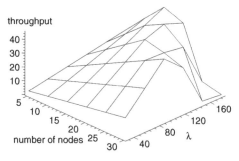

throughput

(d) Throughput from the bridge (i.e., source cluster) to the sink.

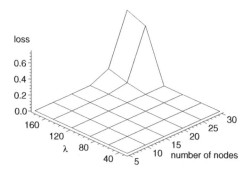

loss

(e) Blocking at an ordinary node in the sink cluster.

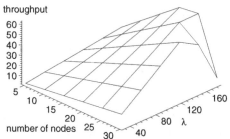

throughput

(f) Throughput from the sink cluster to the sink.

Figure 8.7 Blocking probability and throughput under acknowledged transfer in the network with one source cluster and CSMA-CA bridge. Adapted from J. Mišić and C. J. Fung, 'The impact of master-slave bridge access mode on the performance of multi-cluster 802.15.4 network,' *Computer Networks* **51**: 2411–2449, © 2007 Elsevier B.V.

comparing the traffic load offered to the bridge, to the one actually admitted and delivered, shown in Figures 8.7(b) and 8.7(d), respectively. Namely, the bridge packet service time depends on the local traffic in the sink cluster, and so does the amount of traffic from the source cluster which is actually admitted and delivered by the bridge. When the local traffic load in the sink cluster is high, the bridge blocking probability increases, in which case the increase of the bridge buffer size will not improve the overall bridge performance.

Overall, the use of the CSMA-CA access mode for the bridge brings undesirable coupling between the interconnected clusters. Furthermore, the onset of saturation in either cluster, which occurs at larger network sizes and high values of the packet arrival rate, degrades the performance of both clusters and the entire network. While the choice of parameter values is ultimately affected by the requirements of the sensing application, some limit on the product of the cluster size and packet arrival rate should be observed in order to avoid saturation, or at least to defer its onset as much as possible. The results presented here should serve as useful guidelines in choosing those values.

Acknowledged transfer, GTS bridge.   Packet blocking probability and aggregate throughput for the network with one source cluster and one bridge in the GTS mode are shown in Figure 8.8 as functions of individual cluster size and per-node packet arrival rate.

Since the CSMA bridge with acknowledged transmissions manages to deliver about one packet per second to the sink (slightly more under low traffic, but less than that under high traffic), we have set one uplink GTS of three backoff periods for data packets, and one downlink GTS of one backoff period for acknowledgment packets. (From the queueing theoretic viewpoint, this means that $pb(1) = 1$ and $pb(i) = 0$, for $i \neq 1$.) As the number of packets delivered to the sink cluster at low loads is higher for CSMA-CA bridge than for the GTS one, more uplink GTSs could have been used; however, we wanted to obtain comparable throughput at higher loads. Furthermore, the performance of local traffic in the sink cluster would suffer because the CAP would be much shorter.

In the chosen setup, the bridge that uses GTS access will block more packets in the source cluster, thus pushing the source cluster closer to saturation than in the case of the CSMA-CA bridge, as can be seen by comparing Figure 8.7(c) with Figure 8.8(c). Overall, the network with a GTS bridge achieves about 15% higher throughput that its CSMA-CA counterpart and reaches this value faster as well, as can be observed from Figure 8.8(b), as opposed to Figure 8.7(b).

The picture looks different under high traffic loads. Namely, under the GTS access mode the bridge succeeds in performing the bridging function (to a certain level) at higher loads than under the CSMA-CA access. As a result, the saturation condition for the remote traffic in the sink cluster appears much less abrupt, Figure 8.8(d), than in the case of CSMA-CA bridge access, Figure 8.7(d).

Under the GTS access mode, remote traffic delivered to the sink coordinator through the bridge is decoupled from the local traffic. Under moderate to high source traffic, the number of packets delivered by the bridge remains essentially constant up to the point where the number of collisions in the source cluster becomes so high that the input flow to the bridge decreases, as can be observed from Figures 8.8(d) and 8.8(b). This effect is confirmed by the data shown in Figure 8.8(c), where the bridge blocking probability never exceeds 60%, and even drops beyond certain traffic intensity, due to the decreased input

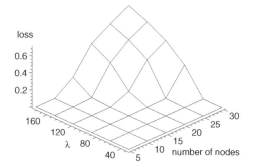
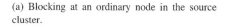
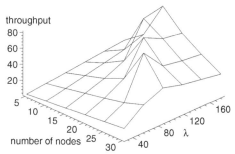

(a) Blocking at an ordinary node in the source cluster.

(b) Throughput from the source cluster to the bridge.

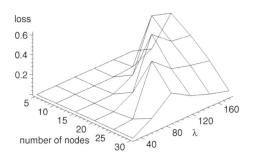
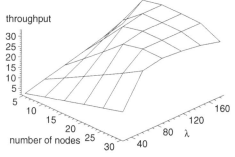

(c) Blocking at the bridge in the source cluster.

(d) Throughput from the bridge (i.e., source cluster) to the sink.

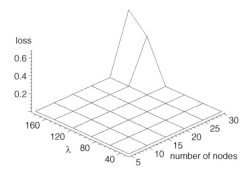
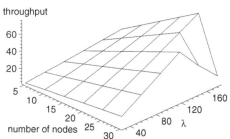

(e) Blocking at an ordinary node in the sink cluster.

(f) Throughput from the sink cluster to the sink.

Figure 8.8 Blocking probability and throughput under acknowledged transfer in the network with one source cluster and GTS bridge. Adapted from J. Mišić and C. J. Fung, 'The impact of master-slave bridge access mode on the performance of multi-cluster 802.15.4 network,' *Computer Networks* **51**: 2411–2449, © 2007 Elsevier B.V.

and constant output flow. Note that the shape of these dependencies is rather different from those obtained under the CSMA-CA access mode, shown in Figure 8.7(c).

All the while, the performance of the local traffic in the sink cluster does not differ much between the CSMA-CA and GTS access modes. In the former case, the ordinary nodes in the sink cluster compete for access to the medium with the bridge. In the latter, the bridge does not compete but the presence of GTS slots decreases the time available for access, effectively shortening the active portion of the superframe for the ordinary nodes in the sink cluster. Consequently, the performance indicators for the local traffic in the sink cluster do not change much.

As in the case of the CSMA-CA access mode, increasing the size of the bridge buffer will slightly reduce congestion at low traffic load in the source cluster. Under moderate and high loads, the increase in bridge buffer size does not improve the maximum bridge throughput and, therefore, cannot significantly defer the onset of saturation in the source cluster.

Non-acknowledged transfer, CSMA-CA bridge. In non-acknowledged mode, packets can be lost due to noise, collisions, or blocking. While this mode may not be an acceptable option for a PAN, it seems acceptable for many sensing applications where data sent by the nodes close to each other are correlated both spatially and temporally. As explained earlier, end-to-end loss rates are

$$
\begin{aligned}
E_{loss} &= 1 - (1 - P_{src}^B)\gamma_{src}(1 - P_{bri}^B)\gamma_{bri} \\
L_{loss} &= 1 - (1 - P_{snk}^B)\gamma_{snk}
\end{aligned}
\tag{8.33}
$$

for the traffic originating in the source and sink cluster, respectively.

Figure 8.9 shows packet blocking probability and throughput at different points of observation in the network, as functions of individual cluster size and packet arrival rate per node. The cluster size and packet arrival rates were varied in the same range as for acknowledged transfer. As can be seen, throughput values are much higher than their counterparts in the network where acknowledged transfer is used; as a result, blocking probability is lower, and the network does not show signs of the saturation regime. This does not mean than the network will never enter saturation – it just means that saturation does not occur anywhere in the range of independent variables shown.

Non-acknowledged transfer, GTS bridge. From the diagrams in Figures 8.9(b), 8.9(d), and 8.9(f), the CSMA-CA bridge manages to deliver about three packets to the sink in each superframe (more under low traffic, but less under high traffic. Therefore, three uplink GTSs of three backoff periods each were allocated to the GTS bridge ($pb(3) = 1$, $pb(i) = 0$, $i \neq 3$). Together with the required interframe spacing, this means that 12 backoff periods are reserved as the CFP at the end of the superframe. Figure 8.10 shows packet blocking probability and throughput at different points of observation in the network, as functions of individual cluster size and packet arrival rate per node. As can be seen, the throughput at various observation points in the network is similar to that obtained under the CSMA-CA bridge access, which should come as no surprise since the matching of throughput was, after all, our design goal.

However, the bridge does manage to deliver more packets to the sink, and so do the ordinary nodes in the sink cluster, Figures 8.10(d) and 8.10(f), respectively. In both cases,

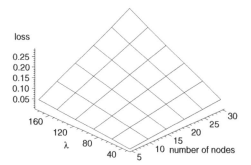

(a) Blocking at an ordinary node in the source cluster.

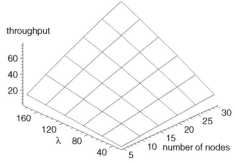

(b) Throughput from the source cluster to the bridge.

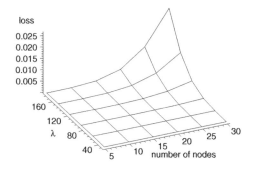

(c) Blocking at the bridge in the source cluster).

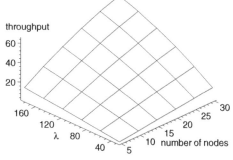

(d) Throughput from the bridge (i.e., source cluster) to the sink.

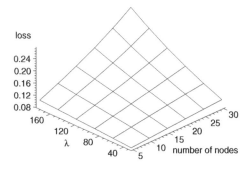

(e) Blocking at an ordinary node in the sink cluster.

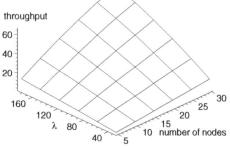

(f) Throughput from the sink cluster to the sink.

Figure 8.9 Blocking probability and throughput under non-acknowledged transfer in the network with one source cluster and CSMA-CA bridge. Adapted from J. Mišić and C. J. Fung, 'The impact of master-slave bridge access mode on the performance of multi-cluster 802.15.4 network,' *Computer Networks* **51**: 2411–2449, © 2007 Elsevier B.V.

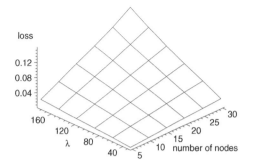

(a) Blocking at an ordinary node in the source cluster.

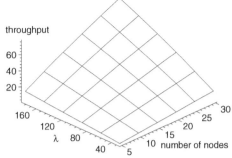

(b) Throughput from the source cluster to the bridge.

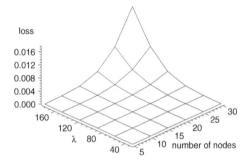

(c) Blocking at the bridge in the source cluster.

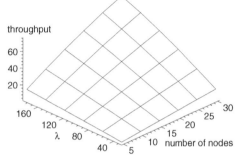

(d) Throughput from the bridge (i.e., source cluster) to the sink.

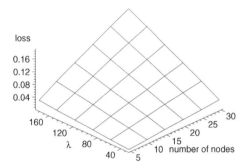

(e) Blocking at an ordinary node in the sink cluster.

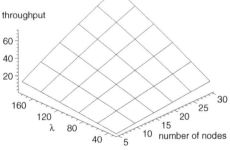

(f) Throughput from the sink cluster to the sink.

Figure 8.10 Blocking probability and throughput under non-acknowledged transfer in the network with one source cluster and GTS bridge. Adapted from J. Mišić and C. J. Fung, 'The impact of master-slave bridge access mode on the performance of multi-cluster 802.15.4 network,' *Computer Networks* **51**: 2411–2449, © 2007 Elsevier B.V.

the improvement stems from the elimination of contention between bridge and local traffic in the sink cluster, despite the shortening of the CAP because of the presence of GTSs. This is further corroborated by reduced blocking probabilities in the sink cluster, Figure 8.10(e).

The improvement in bridge throughput is reflected onto the blocking probabilities at the bridge, Figure 8.10(c), and, by extension, onto the blocking probabilities at ordinary nodes, Figure 8.10(a). Overall, the blocking probabilities in the network with the GTS bridge are smaller by about one-third than their counterparts in the network with a CSMA-CA bridge.

Direct comparison between the acknowledged and non-acknowledged case in the network with the GTS bridge is hard since the bridge was allocated a single GTS in the former, but three of them in the latter. Nonetheless, there are no signs of saturation in the case of non-acknowledged transfer; again, this holds only for the range of independent variables shown in Figure 8.10.

8.5 Network with Two Source Clusters/Bridges

Acknowledged transfer, CSMA-CA bridges. Let us first consider the network with two CSMA-CA bridges, under acknowledged transfer. Figure 8.11 shows the maximum achievable packet rates and packet blocking probability, respectively, as functions of the cluster size and per-node packet arrival rate. Upon comparing these results with those obtained with the network with a single source cluster, Figure 8.7, we may conclude the following.

Local traffic in the sink cluster, Figure 8.11(f), experiences slightly lower throughput than in the network with one source cluster only, Figure 8.7(f). This may be attributed to the increase in the average duration of backoff countdowns due to increased contention in the sink cluster.

However, the throughput from either of the source clusters that reaches the sink, which is shown in Figure 8.11(d), suffers a significant decrease. Namely, the presence of two bridges increases contention, which increases the probability that neither of the bridges may empty its buffer within a single residence period; as a result, the probability that both bridges (and, perhaps, some of the ordinary nodes in the sink cluster as well) will attempt a deferred transmission immediately after the beacon frame is much higher. This produces a cumulative feedback that limits the throughput to only about 75% of the value achieved in the network with a single source cluster, Figure 8.7(d).

As the consequence of increased contention among bridges, their packet service time increases, as does the occupancy of their respective buffers, and they block more of the traffic coming from their respective source clusters. This can be verified by comparing the diagrams of the blocking probability for ordinary nodes in the source cluster and the bridge, shown in Figures 8.11(a) and 8.11(c).

As before, three distinct operating regions may be identified in Figure 8.11(c). On account of increased blocking by the bridge, the source clusters enter saturation much sooner than in the network with only one source cluster. The wedge in the bridge blocking probability stems from the reduced input flow toward the bridge caused by saturation of ordinary nodes, while the sink cluster is still operating out of saturation in that range of n and λ. Note that the throughput from the bridge in the sink cluster, Figure 8.11(d), is steady in that range, albeit lower than its peak value. But once the sink cluster saturates, the bridge ceases to serve any incoming packets; its blocking probability goes up, while

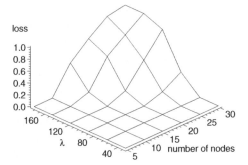
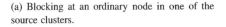

(a) Blocking at an ordinary node in one of the source clusters.

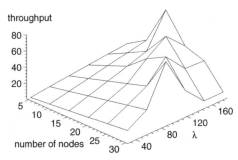

(b) Throughput from one of the source clusters to its bridge.

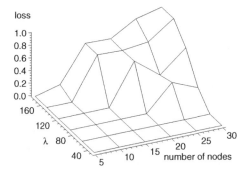

(c) Blocking at the bridge in one of the source clusters).

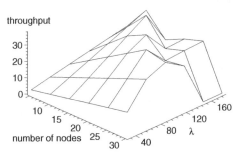

(d) Throughput from one of the bridges to the sink.

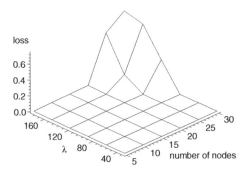

(e) Blocking at an ordinary node in the sink cluster.

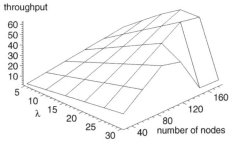

(f) Throughput from the sink cluster to the sink.

Figure 8.11 Blocking probability and throughput under acknowledged transfer in the network with two source clusters, each with a CSMA-CA bridge. Adapted from J. Mišić and C. J. Fung, 'The impact of master-slave bridge access mode on the performance of multi-cluster 802.15.4 network,' *Computer Networks* **51**: 2411–2449, © 2007 Elsevier B.V.

the throughput drops precipitously. The throughput originating in the sink cluster itself, Figure 8.11(f), also drops abruptly.

Acknowledged transfer, GTS bridges. When bridges use GTS access, source clusters are decoupled from each other and, thus, able to attain higher throughput than was possible in the previous case. As in the previous case, each bridge was allocated one uplink and one downlink GTS. Therefore, when we compare Figure 8.12 with Figure 8.8, we see that the performance of source clusters is virtually unaffected in terms of bridge throughput and packet blocking in both the source cluster and at the bridge.

However, increased GTS allocation because of the presence of two bridges mean that the CAP in the sink cluster is substantially reduced. As a result, the performance of the sink cluster itself begins to suffer, as can be seen from Figures 8.12(e) and 8.12(f). Although the traffic from ordinary nodes is decoupled from that of the bridges, their own collision rate increases, which causes an increase in the average duration of random backoff countdowns and leads to reduced throughput, as can be seen in Figure 8.12(f).

Non-acknowledged transfer, CSMA-CA bridges. The use of non-acknowledged transfer extends the non-saturation operating range and allows graceful degradation of performance, as can be seen from Figure 8.13. Packet blocking probability changes much less abruptly at all points of observation. Despite increased congestion, esp. at the beginning of the superframe in the sink cluster, packets lost due to collision are not re-transmitted and transmission of other packets may proceed with shorter average value of random backoff countdowns. Overall, this particular setup appears to be the most scalable one, provided non-acknowledged transfer is acceptable from the viewpoint of the application(s) executing in the network.

Non-acknowledged transfer, GTS bridges. Finally, Figure 8.14 presents results for the network with two source clusters connected to the sink cluster via separate GTS bridges. In order to keep the throughput of local traffic in the sink cluster within acceptable limits, each bridge is allocated only a single uplink and a single downlink GTS, rather than three as in the case of acknowledged traffic. As can be seen, the losses in the source cluster are higher than with the CSMA-CA bridges since the throughput of the bridges is limited. However, the packet flows from the bridges are completely decoupled from one another, and both are decoupled from the local traffic in the sink cluster; this is confirmed by fact that the maximum throughput in the sink cluster, Figure 8.14(f), is significantly higher than in the network with CSMA-CA bridges, Figure 8.13(f).

The results presented above allow the following conclusions to be made.

- In general, non-acknowledged transfer is preferred to the acknowledged one, since it improves the throughput and delays the onset of saturation. Note that this decision is dependent on the requirements of the application(s) executing in the network.

- Under small cluster sizes and low load in all clusters, the most efficient interconnection strategy is to use CSMA-CA bridges.

- Under moderate to high loads and/or large cluster sizes, the use of GTS bridges will allow higher throughput and widen the operating range without saturation.

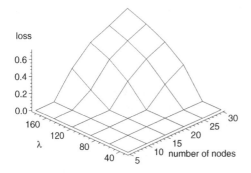

(a) Blocking at an ordinary node in one of the source clusters.

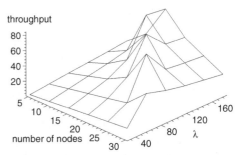

(b) Throughput from one of the source clusters to its bridge.

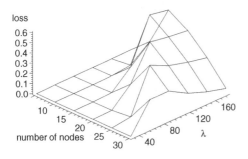

(c) Blocking at one of the bridges in its source cluster.

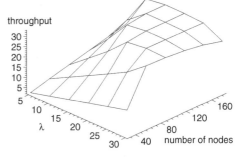

(d) Throughput from one of the bridges (i.e., source clusters) to the sink.

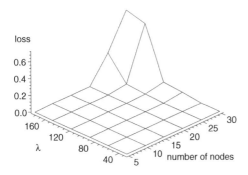

(e) Blocking at an ordinary node in the sink cluster.

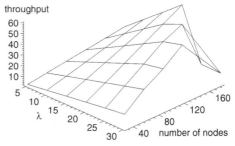

(f) Throughput from the sink cluster to the sink.

Figure 8.12 Blocking probability and throughput under acknowledged transfer in the network with two source clusters, each with a GTS bridge. Adapted from J. Mišić and C. J. Fung, 'The impact of master-slave bridge access mode on the performance of multi-cluster 802.15.4 network,' *Computer Networks* **51**: 2411–2449, © 2007 Elsevier B.V.

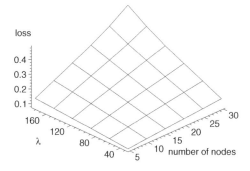

(a) Blocking at an ordinary node in one of the source clusters.

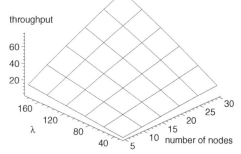

(b) Traffic from one of the source clusters to its bridge.

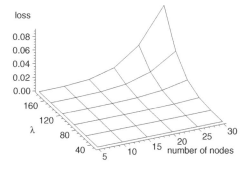

(c) Blocking at the bridge in one of the source clusters.

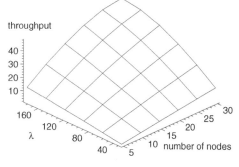

(d) Throughput from one of the bridges (i.e., source clusters) to the sink.

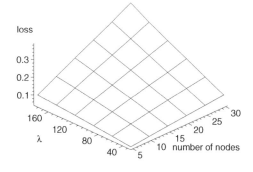

(e) Blocking at an ordinary node in the sink cluster.

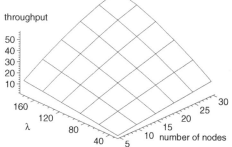

(f) Throughput from the sink cluster to the sink.

Figure 8.13 Blocking probability and throughput under non-acknowledged transfer in the network with two source clusters, each with a CSMA-CA bridge. Adapted from J. Mišić and C. J. Fung, 'The impact of master-slave bridge access mode on the performance of multi-cluster 802.15.4 network,' *Computer Networks* **51**: 2411–2449, © 2007 Elsevier B.V.

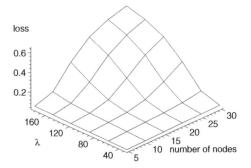

(a) Blocking at an ordinary node in the source cluster.

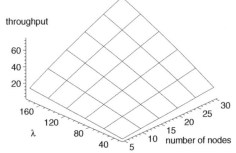

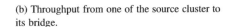

(b) Throughput from one of the source cluster to its bridge.

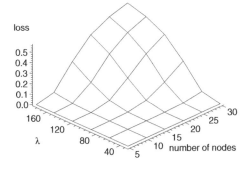

(c) Blocking at the bridge in one of the source clusters.

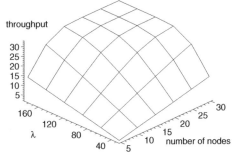

(d) Throughput from one of the bridges (i.e., source clusters) to the sink.

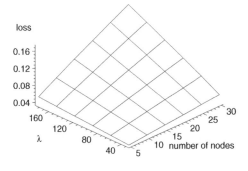

(e) Blocking at an ordinary node in the sink cluster.

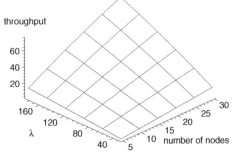

(f) Throughput from the sink cluster to the sink.

Figure 8.14 Blocking probability and throughput under non-acknowledged transfer in the network with two source clusters, each with a GTS bridge. Adapted from J. Mišić and C. J. Fung, 'The impact of master-slave bridge access mode on the performance of multi-cluster 802.15.4 network,' *Computer Networks* **51**: 2411–2449, © 2007 Elsevier B.V.

- However, if multiple child clusters are used, the use of GTS bridges decouples the traffic from those clusters at the expense of local traffic in the sink cluster. In such cases, reducing the number of ordinary nodes in the sink cluster would probably eliminate this source of inefficiency; in extreme cases, the sink cluster might contain only bridges but no ordinary nodes.

- The main design parameter for the network that uses GTS bridges is the selection of the optimum operating point with respect to the size of the superframe and the number of GTSs (i.e., packet lanes) to allocate to the bridges.

8.6 Modeling the Transmission Medium and Packet Service Times

In this section, which is effectively an appendix to the current chapter, we model the behavior of the transmission medium and packet service time for a network which consists of κ source clusters and one sink cluster, each of which contains n ordinary nodes.

Our analysis will make use of the following. From the viewpoint of a source cluster node, at any moment, q out of the remaining $n-1$ nodes have no packets deferred to the next superframe due to insufficient time in the current one, while $n-1-q$ nodes have such packets. The numbers q and $n-1-q$ follow a binomial distribution with the probability

$$P_q = \binom{n-1}{q}(1-P_d)^q P_d^{n-1-q} \tag{8.34}$$

Within the group of nodes without a deferred transmission, exactly r nodes have received packets during the inactive part of the superframe; the probability of this happening is

$$P_r = \binom{q}{r}\left(P_{0,src}P_{sync}\lambda(1-P_{src}^B)\right)^r\left(1-P_{0,src}P_{sync}\lambda(1-P_{src}^B)\right)^{q-r} \tag{8.35}$$

where $P_{0,src}$ denotes the probability that the node buffer is empty at an arbitrary time that was derived in Equation 8.15. Note also that the value of $\overline{D_d}$ (which was calculated in Section 8.2) depends on whether acknowledged or non-acknowledged transfer is used.

Analogous expressions hold in the sink cluster, provided the subscript src is replaced with snk.

8.6.1 Probability of success at first CCA

The first CCA succeeds only if no other node is currently transmitting a packet, or receiving the acknowledgment for a previously transmitted packet.

An ordinary node in a source cluster. The probability of success at first CCA, α, is obtained by simply dividing the mean number of non-busy backoff periods in the superframe by the total number of backoff periods in which the first CCA may occur.

The probability that any packet transmission will take place at the very beginning of the superframe is

$$n_{1,src} = 1 - (1-\tau_{1,src})^{n-1-q}(1-\tau_{2,src})^{q-r}(1-\tau_{3,1})^r \tag{8.36}$$

and the number of busy backoff periods due to these transmissions is $n_{1,src}(G'_p(1) + G'_a(1))$ and $n_{1,src}G'_p(1)$, for acknowledged and non-acknowledged transfer, respectively.

The probability of a transmission attempt after the third backoff period within the period of $\overline{D_d}$ backoff periods is

$$n_{2,src} = 1 - ((1 - \tau_{2,src})^{q-r}(1 - \tau_{3,2,src})^r)^{\overline{D_d}} \tag{8.37}$$

The probability of a transmission in the remaining part of the superframe within $\overline{D_d}$ backoff periods is

$$n_{3,src} = 1 - ((1 - \tau_{2,src})^q)^{\overline{D_d}} \tag{8.38}$$

The occupancy of the medium after the first transmission time can be found by dividing the superframe into chunks of $\overline{D_d}$ backoff periods and calculating the probability of transmission within each chunk.

On the other hand, first CCA can take place at any time from the beginning of the superframe (if the packet was deferred in the previous superframe, or the random countdown value was zero) up to the time $SM = SD - \overline{D_d} + 1$ (expressed in backoff periods), after which the transmission has to be deferred. Therefore, the probability α that the medium is idle at the first CCA is

$$
\begin{aligned}
\alpha_{src} = \sum_{q=0}^{n-1}\sum_{r=0}^{q} P_q P_r \cdot \\
\cdot \left(1 - \frac{(n_{1,src} + n_{2,src})\overline{D_d} + n_{3,src}(SD - 3\overline{D_d} + 1)}{SM} \cdot \frac{G'_p(1) + G'_a(1)}{\overline{D_d}} \right)
\end{aligned}
\tag{8.39}
$$

for acknowledged transfer; the expression for non-acknowledged transfer is the same, except that the term $G'_a(1)$ is absent.

CSMA-CA bridge in the sink cluster.   In the sink cluster each bridge has to compete for access with $\kappa - 1$ other bridges and with n ordinary nodes residing in the sink cluster. The expression for α_{bri} may be derived similarly to the previous one, but taking into account that the CSMA-CA bridge, while residing in the sink cluster, must compete for access to the medium with $\kappa - 1$ other bridges and n ordinary nodes from that cluster. Therefore,

$$
\begin{aligned}
n_{1,bri} &= 1 - (1 - \tau_{1,snk})^{n-q}(1 - \tau_{2,snk})^{q-r}(1 - \tau_{3,1,snk})^r \cdot \\
&\quad \cdot (1 - \tau_{1,bri} - \tau_{2,bri} - \tau_{3,1,bri})^{\kappa-1} \\
n_{2,bri} &= 1 - ((1 - \tau_{2,snk})^{q-r}(1 - \tau_{3,2,snk})^r(1 - \tau_{2,bri} - \tau_{3,2,bri})^{\kappa-1})^{\overline{D_d}} \\
n_{3,bri} &= 1 - ((1 - \tau_{2,snk})^q(1 - \tau_{2,bri})^{\kappa-1})^{\overline{D_d}}
\end{aligned}
\tag{8.40}
$$

after which α_{bri} can be expressed by Equation (8.39), provided that the summation has to be done over all n ordinary nodes, and that arguments of the sums correspond to access probabilities from the sink cluster.

An ordinary node in the sink cluster with CSMA-CA bridges.   Each ordinary node in the sink cluster must compete for medium access against κ bridges and $n - 1$ remaining nodes.

Therefore, the component probabilities are

$$
\begin{aligned}
n_{1,snk} &= 1 - (1 - \tau_{1,snk})^{n-1-q}(1 - \tau_{2,snk})^{q-r}(1 - \tau_{3,1,snk})^r \cdot \\
&\quad \cdot (1 - \tau_{1,bri} - \tau_{2,bri} - \tau_{3,1,bri})^\kappa \\
n_{2,snk} &= 1 - ((1 - \tau_{2,snk})^{q-r}(1 - \tau_{3,2,snk})^r (1 - \tau_{2,bri} - \tau_{3,2,bri})^\kappa)^{\overline{D_d}} \\
n_{3,snk} &= 1 - ((1 - \tau_{2,snk})^q (1 - \tau_{2,bri})^\kappa)^{\overline{D_d}}
\end{aligned}
\tag{8.41}
$$

and α_{snk} can be derived using expression (8.39).

The sink cluster with GTS bridges. In this case, the bridges need not do any CCA check, but the ordinary nodes see the following component probabilities:

$$
\begin{aligned}
n_{1,snk} &= 1 - (1 - \tau_{1,snk})^{n-1-q}(1 - \tau_{2,snk})^{q-r}(1 - \tau_{3,1,snk})^r \\
n_{2,snk} &= 1 - (1 - \tau_{2,snk})^{q-r}(1 - \tau_{3,2,snk})^r \\
n_{3,snk} &= 1 - (1 - \tau_{2,snk})^q
\end{aligned}
\tag{8.42}
$$

Also, since the effective superframe length is shorter on account of GTSs allocated to the bridges, it follows that

$$
\begin{aligned}
SD_{gts} &= SD - \kappa B(G'_p(1) + \Delta) \\
SM_{gts} &= SD_{gts} - \overline{D_d} + 1
\end{aligned}
$$

where B is the number of GTS packet lanes per bridge, while Δ presents the space needed for the transfer of acknowledgment and separation between the GTS lanes; these values have to be substituted in Equation (8.39).

8.6.2 Probability of success at second CCA

As explained earlier, the medium that will be idle on the second CCA is neither the cluster coordinator nor one of the remaining nodes that have started a transmission in that particular backoff period. The second CCA can be performed between the second backoff period in the superframe, up to the period in which there is no more time for packet transmission, which amounts to SM.

Ordinary node in one of the source clusters. Since each of the source clusters has n ordinary nodes and we assume there are no downlink transmissions, the probability of success at second CCA is

$$
\begin{aligned}
\beta_{src} = \sum_{q=0}^{n-1}\sum_{r=0}^{q} P_q P_r &\left(\frac{1}{SM} + \frac{(1 - \tau_{1,src})^{n-1-q}(1 - \tau_{2,src})^{q-r}(1 - \tau_{3,1,src})^r}{SM} \right. \\
&\left. + \frac{\overline{D_d} - 1}{SM}(1 - \tau_{2,src})^{q-r}(1 - \tau_{3,2,src})^r + \frac{SM - \overline{D_d}}{SM}(1 - \tau_{2,src})^q \right)
\end{aligned}
\tag{8.43}
$$

CSMA-CA bridge in the sink cluster. A CSMA-CA bridge must compete against all n nodes in the sink cluster, as well as the remaining $\kappa - 1$ bridges. Therefore,

$$
\begin{aligned}
\beta_{bri} = \sum_{q=0}^{n} \sum_{r=0}^{q} P_q P_r \bigg(\frac{1}{SM} + \\
+ \frac{(1 - \tau_{1,snk})^{n-1-q}(1 - \tau_{2,snk})^{q-r}(1 - \tau_{3,1,snk})^{r}(1 - \tau_{1,bri} - \tau_{2,bri} - \tau_{3,1,bri})^{\kappa-1}}{SM} \\
+ \frac{\overline{D_d} - 1}{SM}(1 - \tau_{2,snk})^{q-r}(1 - \tau_{3,2,snk})^{r}(1 - \tau_{2,bri} - \tau_{3,2,bri})^{\kappa-1} \\
+ \frac{SM - \overline{D_d}}{SM}(1 - \tau_{2,snk})^{q}(1 - \tau_{2,bri} - \tau_{3,2,bri})^{\kappa-1} \bigg)
\end{aligned}
\tag{8.44}
$$

An ordinary node in the sink cluster. An ordinary node in the sink cluster competes against the remaining $n - 1$ ordinary nodes and all κ CSMA-CA bridges. Therefore,

$$
\begin{aligned}
\beta_{snk} = \sum_{q=0}^{n-1} \sum_{r=0}^{q} P_q P_r \bigg(\frac{1}{SM} + \\
+ \frac{(1 - \tau_{1,snk})^{n-1-q}(1 - \tau_{2,snk})^{q-r}(1 - \tau_{3,1,snk})^{r}(1 - \tau_{1,bri} - \tau_{2,bri} - \tau_{3,1,bri})^{\kappa}}{SM} \\
+ \frac{\overline{D_d} - 1}{SM}(1 - \tau_{2,snk})^{q-r}(1 - \tau_{3,2,snk})^{r}(1 - \tau_{2,bri} - \tau_{3,2,bri})^{\kappa} \\
+ \frac{SM - \overline{D_d}}{SM}(1 - \tau_{2,snk})^{q}(1 - \tau_{2,bri} - \tau_{3,2,bri})^{\kappa} \bigg)
\end{aligned}
\tag{8.45}
$$

The sink cluster with GTS bridges. In this case, the bridges need not perform the second CCA, while ordinary nodes do not see any transmission from any bridge. Therefore,

$$
\begin{aligned}
\beta_{snk} = \sum_{q=0}^{n-1} \sum_{r=0}^{q} P_q P_r \bigg(\frac{1}{SM_{gts}} + \frac{(1 - \tau_{1,snk})^{n-1-q}(1 - \tau_{2,snk})^{q-r}(1 - \tau_{3,1,snk})^{r}}{SM_{gts}} \\
+ \frac{\overline{D_d} - 1}{SM_{gts}}(1 - \tau_{2,snk})^{q-r}(1 - \tau_{3,2,snk})^{r} + \frac{SM_{gts} - \overline{D_d}}{SM_{gts}}(1 - \tau_{2,snk})^{q} \bigg)
\end{aligned}
\tag{8.46}
$$

In fact, this model can easily accommodate any combination of GTS and CSMA-CA bridges, provided proper exponents that correspond to the numbers of nodes in each cluster, CSMA-CA bridges, and/or GTS bridges, are used.

8.6.3 Probability of successful transmission

Finally, we need the probability that a packet will not collide with other packet(s) that have undergone successful first and second CCAs. In general, this probability can be calculated as the probability that there are no accesses to the medium by the other nodes or the coordinator during the entire packet transmission time, including the optional acknowledgment. Note that a collision can happen in SM consecutive backoff periods starting from the third backoff period in the superframe.

An ordinary node in one of the source clusters. In this case, the success probability is

$$
\gamma_{src} = \sum_{q=0}^{n-1} \sum_{r=0}^{q} P_q P_r \left(\frac{((1 - \tau_{1,src})^{n-1-q}(1 - \tau_{2,src})^{q-r}(1 - \tau_{3,1,src})^r)^{\overline{D_d}}}{SM} \right.
$$
$$
\left. + \frac{\overline{D_d} - 1}{SM}((1 - \tau_{2,src})^{q-r}(1 - \tau_{3,2,src})^r)^{\overline{D_d}} + \frac{SM - \overline{D_d}}{SM}((1 - \tau_{2,src})^q)^{\overline{D_d}} \right)
$$

(8.47)

A CSMA-CA bridge in the sink cluster. The probability of successful transmission by a CSMA-CA bridge in the sink cluster may be calculated through an expression similar to Equation (8.47), but after taking into account the contention caused by the n ordinary nodes and $\kappa - 1$ remaining bridges:

$$
\gamma_{bri} = \sum_{q=0}^{n} \sum_{r=0}^{q} P_q P_r \left(\frac{\overline{D_d} - 1}{SM}((1 - \tau_{2,snk})^{q-r}(1 - \tau_{3,2,snk})^r(1 - \tau_{2,bri} - \tau_{3,2,bri})^{\kappa-1})^{\overline{D_d}} \right.
$$
$$
+ \frac{SM - \overline{D_d}}{SM}((1 - \tau_{2,snk})^{q-r}(1 - \tau_{3,2,snk})^r(1 - \tau_{2,bri})^{\kappa-1})^{\overline{D_d}}
$$
$$
\left. + \frac{((1 - \tau_{1,snk})^{n-1-q}(1 - \tau_{2,snk})^{q-r}(1 - \tau_{3,1,snk})^r(1 - \tau_{1,bri} - \tau_{2,bri} - \tau_{3,1,bri})^{\kappa-1})^{\overline{D_d}}}{SM} \right)
$$

(8.48)

An ordinary node in the sink cluster. Similar to the previous case, the probability of successful transmission for an ordinary node in the sink cluster is

$$
\gamma_{snk} = \sum_{q=0}^{n-1} \sum_{r=0}^{q} P_q P_r \left(\frac{SM - \overline{D_d}}{SM}((1 - \tau_{2,snk})^{q-r}(1 - \tau_{3,2,snk})^r(1 - \tau_{2,bri})^\kappa)^{\overline{D_d}} \right.
$$
$$
+ \frac{((1 - \tau_{1,snk})^{n-1-q}(1 - \tau_{2,snk})^{q-r}(1 - \tau_{3,1,snk})^r(1 - \tau_{1,bri} - \tau_{2,bri} - \tau_{3,1,bri})^\kappa)^{\overline{D_d}}}{SM}
$$
$$
\left. + \frac{\overline{D_d} - 1}{SM}((1 - \tau_{2,snk})^{q-r}(1 - \tau_{3,2,snk})^r(1 - \tau_{2,bri} - \tau_{3,2,bri})^\kappa)^{\overline{D_d}} \right)
$$

(8.49)

The case of the sink cluster with GTS bridges. Since the actions of ordinary nodes and bridge are separated, the probability γ_{snk} becomes

$$
\gamma_{snk} = \sum_{q=0}^{n-1} \sum_{r=0}^{q} P_q P_r \left(\frac{((1 - \tau_{1,snk})^{n-1-q}(1 - \tau_{2,snk})^{q-r}(1 - \tau_{3,1,snk})^r)^{\overline{D_d}}}{SM_{gts}} \right.
$$
$$
+ \frac{\overline{D_d} - 1}{SM_{gts}}((1 - \tau_{2,snk})^{q-r}(1 - \tau_{3,2,snk})^r
$$
$$
\left. + \frac{SM_{gts} - \overline{D_d}}{SM_{gts}}((1 - \tau_{2,snk})^{q-r}(1 - \tau_{3,2,snk})^r)^{\overline{D_d}} \right)
$$

(8.50)

8.6.4   Probability distribution for the packet service time

In order to derive the probability distribution for the packet service time, we need to model the initial waiting time between the packet arrival to the buffer during the inactive portion of the superframe, until the subsequent beacon frame after which the node will commence the random backoff countdown procedure. Since the packet arrival process is oblivious to the superframe timing, new packets can arrive at any time during the inactive portion of the superframe with the same probability. As a result, this waiting time has a uniform probability distribution. In case of an ordinary node in one of the source clusters, the PGF of this waiting time is

$$T_{sync}(z) = \pi_{0,src} P_{sync} \frac{z^{BI-SD} - 1}{(BI - SD)(z - 1)} + (1 - \pi_{0,src} P_{sync})z^0 \qquad (8.51)$$

A similar expression holds for the PGF of the waiting time in an ordinary node in the sink cluster, or the CSMA-CA bridge, provided appropriate values of π_0 are used.

We also need to model the freezing of backoff counter during the inactive portion of the superframe. The probability that a given backoff period is the last one within the active portion of the superframe is

$$P_{last} = \frac{1}{SD}$$
$$P_{last} = \frac{1}{SD_{gts}}$$

for any node (regardless of its role or the cluster in which it resides) when CSMA-CA bridges are used, and for the sink cluster with GTS bridges, respectively. The PGF for the effective duration of the backoff countdown interval, including the duration of the beacon frame, is

$$B_{off}(z) = (1 - P_{last})z + P_{last} z^{BI-SD+1} B_{ea}(z) \qquad (8.52)$$

The PGF for the duration of i-the backoff attempt is

$$B_i(z) = \sum_{k=0}^{W_i-1} \frac{1}{W_i} B_{off}^k(z) = \frac{B_{off}^{W_i}(z) - 1}{W_i(B_{off}(z) - 1)} \qquad (8.53)$$

As noted above, the transmission procedure will not start unless it can be finished within the current superframe. The number of backoff periods thus wasted can be described with the PGF of

$$B_p(z) = \frac{1}{D_d} \sum_{k=0}^{D_d-1} z^k. \qquad (8.54)$$

In case of acknowledged transfer, the PGF of the data packet transmission time for deferred and non-deferred transmissions, respectively, is

$$\begin{aligned} T_{d1}(z) &= B_p(z) z^{BI-SD} B_{ea}(z) G_p(z) t_{ack}(z) G_a(z) \\ T_{d2}(z) &= G_p(z) t_{ack}(z) G_a(z) \end{aligned} \qquad (8.55)$$

whereas the corresponding PGFs, in case of non-acknowledged transfer, are

$$\begin{aligned} T_{d1}(z) &= B_p(z) z^{BI-SD} B_{ea}(z) G_p(z) \\ T_{d2}(z) &= G_p(z) \end{aligned} \qquad (8.56)$$

A single transmission attempt. Let us denote the probability that a backoff attempt will be unsuccessful with $R_{ud} = 1 - P_d - (1 - P_d)\alpha\beta$. The function that describes the time needed for the backoff countdown and the transmission attempt itself can be written as

$$\mathcal{A}(z) = \sum_{i=0}^{m} \prod_{j=0}^{i} \left(B_j(z)R_{ud}\right) z^{2(i+1)} \left(P_d T_{d1}(z) + (1 - P_d)\alpha\beta T_{d2}(z)\right)$$
$$+ R_{ud}^{m+1} \prod_{j=0}^{m} B_j(z) z^{2(m+1)} \mathcal{A}(z) \tag{8.57}$$

where R_{ud}^{m+1} denotes the probability that $m + 1$ backoff attempts with non-decreasing backoff windows were not successful and the sequence of backoff windows has to be repeated, starting from the smallest backoff window. If we substitute $z = 1$ into Equation (8.57) we will obtain $\mathcal{A}(1) = 1$ which is a necessary condition for a function to be a valid PGF. From Equation (8.57), we obtain

$$\mathcal{A}(z) = \frac{\sum_{i=0}^{m} \prod_{j=0}^{i} \left(B_j(z)R_{ud}\right) z^{2(i+1)} \left(P_d T_{d1}(z) + (1 - P_d)\alpha\beta T_{d2}(z)\right)}{1 - R_{ud}^{m+1} \prod_{j=0}^{m} B_j(z) z^{2(m+1)}} \tag{8.58}$$

The case of non-acknowledged transfer. In this case, the transmission time calculated in Equation (8.56) has to be substituted in Equation (8.58). Note that this PGF depends on the context (i.e., different values correspond to ordinary nodes in source and sink cluster, and the bridge), and its final form is obtained by substituting appropriate values for different variables into Equation (8.58). Therefore, under non-acknowledged transfer, the PGF for the packet service time is

$$T_{t,src}(z) = \mathcal{A}_{src}(z)$$
$$T_{t,snk}(z) = \mathcal{A}_{snk}(z) \tag{8.59}$$
$$T_{t,bri}(z) = \mathcal{A}_{bri}(z)$$

for an ordinary node in one of the source clusters, an ordinary node in the sink cluster, and a CSMA-CA bridge in the sink cluster, respectively. There is no need for the corresponding PGF in case of a GTS bridge, as its transmissions are always successful.

The case of acknowledged, partially reliable transfer. In this case, the MAC layer will re-attempt transmission for up to a times. The probability that the transmission is successful in a attempts, P_a, was defined earlier in Section 8.2; for an ordinary node in one of the source clusters, it has the value of

$$P_{a,src} = \sum_{i=0}^{a-1} (1 - \gamma_{src}\delta_{src}(1 - P_{bri}^B))^i \gamma_{src}\delta_{src}(1 - P_{bri}^B)$$
$$= 1 - (1 - \gamma_{src}\delta_{src}(1 - P_{bri}^B))^a \tag{8.60}$$

where γ_{src} is the probability of no collisions in the source cluster; δ_{src} is the probability that a data packet or the subsequent acknowledgment packet will not be corrupted by the noise; and $1 - P_{bri}^B$ is the probability that packet is accepted by the bridge.

For an ordinary node in the sink cluster,

$$
\begin{aligned}
P_{a,snk} &= \sum_{i=0}^{a-1}(1 - \gamma_{snk}\delta_{snk})^i \gamma_{snk}\delta_{snk} \\
&= 1 - (1 - \gamma_{snk}\delta_{snk})^a
\end{aligned}
\tag{8.61}
$$

The probability $P_{a,bri}$ for the CSMA-CA bridge can simply be obtained by substituting the values with the subscript bri in lieu of those with the subscript snk.

With all these variables in place, the PGF for packet service time without the beacon synchronization time for the ordinary node in source cluster becomes:

$$
\begin{aligned}
T_{t,src}(z) =\ & \gamma_{src}\delta_{src}(1 - P_{bri}^B)\mathcal{A}_{src}(z) \\
& +\gamma_{src}\delta_{src}(1 - P_{bri}^B)(1 - \gamma_{src}\delta_{src}(1 - P_{bri}^B))\mathcal{A}_{src}(z)^2 \\
& \cdots \\
& +\gamma_{src}\delta_{src}(1 - P_{bri}^B)(1 - \gamma_{src}\delta_{src}(1 - P_{bri}^B))^{a-1}\mathcal{A}_{src}(z)^a \\
& + \left(1 - (1 - \gamma_{src}\delta_{src}(1 - P_{bri}^B))^a\right)z^0 \\
=\ & \frac{1 - \mathcal{A}_{src}(z)^a(1 - \gamma_{src}\delta_{src}(1 - P_{bri}^B))^a}{1 - \mathcal{A}_{src}(z)(1 - \gamma_{src}\delta_{src}(1 - P_{bri}^B))} \\
& + \left(1 - (1 - \gamma_{src}\delta_{src}(1 - P_{bri}^B))^a\right)z^0
\end{aligned}
\tag{8.62}
$$

Finally, by adding the beacon synchronization time for the nodes in the source and sink clusters, the PGFs for the packet transmission time become

$$
\begin{aligned}
T_{t,src}(z) =\ & T_{sync}(z)\frac{1 - \mathcal{A}_{src}(z)^a(1 - \gamma_{src}\delta_{src}(1 - P_{bri}^B))^a}{1 - \mathcal{A}_{src}(z)(1 - \gamma_{src}\delta_{src}(1 - P_{bri}^B))} \\
& + \left(1 - (1 - \gamma_{src}\delta_{src}(1 - P_{bri}^B))^a\right)z^0 \\
T_{t,snk}(z) =\ & T_{sync}(z)\frac{1 - \mathcal{A}_{snk}(z)^a(1 - \gamma_{snk}\delta_{snk})^a}{1 - \mathcal{A}_{snk}(z)(1 - \gamma_{snk}\delta_{snk})} \\
& + \left(1 - (1 - \gamma_{snk}\delta_{snk})^a\right)z^0 \\
T_{t,bri}(z) =\ & \frac{1 - \mathcal{A}_{bri}(z)^a(1 - \gamma_{bri}\delta_{bri})^a}{1 - \mathcal{A}_{bri}(z)(1 - \gamma_{bri}\delta_{bri})} \\
& + \left[1 - (1 - \gamma_{bri}\delta_{bri})^a\right]z^0
\end{aligned}
\tag{8.63}
$$

where the last equation holds only for a CSMA-CA bridge.

It can be verified that $T_t(1) = 1$ in all cases.

9

Equalization of Cluster Lifetimes

Timing of individual cluster operation and scheduling of bridge residence in different clusters are not the only factors that affect the performance of a multi-cluster network. In multi-hop and multi-level networks, the traffic volume in different clusters often differs, which leads to different power consumption and, by extension, different cluster lifetimes. We have seen that in a two cluster network from Section 8.4, the cluster where the sink node resides carries both its own traffic and the traffic from the entire source cluster; in this case, the cluster which contains the network sink will exhaust its power source earlier than the other cluster which is farther away from the sink. As the total throughput at the sink must be contributed by all the clusters in the network, the failure of one cluster means that the entire network effectively ceases to function, even though the total energy level may still suffice to support operation.

Since the desired goal of many such networks is to maintain the prescribed data rate at the sink for the maximum possible time, it becomes necessary that all clusters exhaust their power sources at about the same time. This goal can be accomplished through the use of redundant nodes coupled with activity management, similar to the approach described earlier in the context of single cluster networks. The reader will remember that, in Chapter 6, we developed an exact model for probability distribution for the node lifetime in a single cluster. This model can be coupled with the bridging model from Chapter 8 to give an accurate description of network operation and performance. However, the combined model is computationally complex and does not scale well. In this chapter we will describe a simplified model that provides sufficient accuracy whilst achieving good scalability.

9.1 Modeling the Clusters

Consider the network with the topology shown in Figure 9.1(a). We assume that all clusters operate in beacon enabled, slotted CSMA-CA mode under the control of their respective cluster coordinators. The coordinator of the top cluster acts as the network sink, while the coordinators of the middle and bottom clusters act as bridges which employ the CSMA-CA access mode to deliver their data.

Wireless Personal Area Networks Jelena Mišić and Vojislav B. Mišić
© 2008 John Wiley & Sons, Ltd

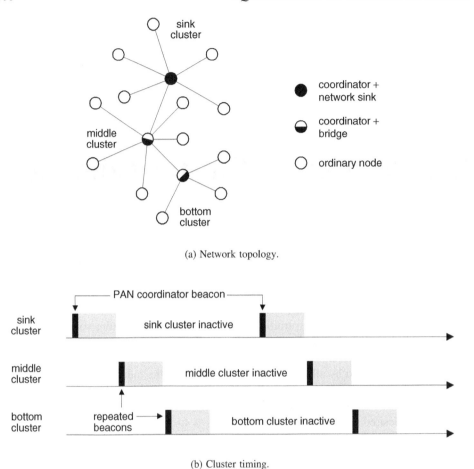

(a) Network topology.

(b) Cluster timing.

Figure 9.1 A three-cluster network.

All clusters contain redundant nodes, which allows the duty cycle of individual nodes to be reduced, and the network lifetime to be extended, through appropriate activity management (Mišić et al. 2006c). The goal is to maximize the lifetime of the entire network while maintaining the prescribed throughput at the sink, with each cluster expected to contribute exactly one-third of that amount. Ordinary nodes are battery operated, and their battery capacity is finite. However, the coordinators/bridges have to work continuously, therefore they do not sleep and their power budget is assumed to be infinite.

Bridge switching is schematically presented in Figure 9.1(b). All three clusters operate using the same RF channel, and use the same values for the superframe duration and beacon interval. As a result, the time between successive bridge visits to the 'upper' cluster is therefore the same as the period between successive beacons in its own, 'lower' cluster. In general, the beacon interval in the sink cluster must be long enough to accommodate superframes from all lower level clusters; for the case of three clusters as in Figure 9.1, this translates to the condition that the beacon order exceeds the superframe order by at least two, i.e., $BO \geq SO + 2$. (However, if the top and bottom clusters are beyond the transmission range of each other, this condition may be somewhat relaxed; the calculations are trivial.) Alternatively, different clusters may use different RF channels, and maybe even different PHY options, provided the node hardware provides the necessary capabilities.

As in the case of a single-cluster network, ordinary nodes spend most of their time sleeping, i.e., with their radio subsystem switched off. When a node wakes up and finds that there is a packet to transmit, it will transmit it, wait for the subsequent acknowledgment, and then go back to sleep. As explained in Chapter 6, this amounts to a 1-limited service policy (Takagi 1991). The actual sleep time is determined autonomously by each node, using the algorithm explained below.

Since the node needs to hear a beacon frame in order to synchronize with the cluster, and the cluster load is assumed to be light enough so that the cluster operates far below the saturation condition (Sections 3.7.2 and 8.4), it is safe to assume that each node that has a packet to transmit will hear one beacon frame per sleep cycle. As the minimum beacon frame size is two backoff periods, we assume that an additional backoff period would suffice to convey the required information about the number of live nodes and requested throughput to all the nodes that are awake at that time.

Let us denote energy consumption per backoff period during sleep, receiving, and transmitting with ω_s, ω_r, and ω_t, respectively. The exact numbers are, of course, implementation dependent; they can be obtained by consulting the manufacturer's datasheets for the hardware in question. As an example, the current and energy consumption for the ultra low power tmote_sky mote (wireless sensor node) operating in the ISM band (Moteiv Corporation 2006) are shown in Table 9.1. For reference, transmission power of 0 dBm allows the transmission range of about 50 meters indoors and up to 125 meters outdoors, depending on terrain conditions. Interestingly enough, receiving can consume more energy than transmitting, esp. when the transmission power level is reduced.

The energy consumption values in Table 9.1 are calculated for the nominal supply voltage of 2.85 V; according to the specification, the tmote_sky requires operating voltage between 2.1 and 3.6 V, which can be supplied by standard 1.5 V batteries. Battery capacity depends on the implementation: typical values are 400 to 900 mAh (milli-Amp-hours) for

Table 9.1 Current and energy consumption for the tmote_sky mote

Operating mode of the radio subsystem	Parameter	Current consumption	Energy consumption at 2.85 V
transmitting at 0 dBm	ω_t	17.4 mA	15.8 μJ
transmitting at −1 dBm	ω_t	16.5 mA	15.0 μJ
transmitting at −3 dBm	ω_t	15.2 mA	13.8 μJ
transmitting at −5 dBm	ω_t	13.9 mA	12.6 μJ
receiving	ω_r	19.7 mA	17.9 μJ
switched off (idle)	ω_s	20 μA	18.2 nJ

Note: Values of energy consumption calculated per backoff period.

zinc-carbon batteries, 1000 to 1500 mAh for zinc-chloride batteries and 800 to 1500 mAh for rechargeable nickel-cadmium batteries.

9.2 Distributed Activity Management

Under the activity management algorithm, coordinator broadcasts the required throughput (event sensing reliability) (Sankarasubramaniam et al. 2003) and the number of nodes which are currently alive. Ordinary nodes use this information to calculate the average time period between the packet transmissions. Note that this time period is much larger than the packet service time, since the total number of nodes in each cluster is larger than the minimum number needed to achieve the desired cluster throughput (i.e., redundant nodes are used). This time period is also used as the average value of the sleep period in which the node turns off its radio subsystem to reduce energy consumption. In order to avoid synchronized wake-up of nodes, sleep times are randomized using the geometric probability distribution with the same average value as the average period between packet transmissions. The parameter of the geometric distribution will be denoted as P_{sleep}, and the average value of sleep time is equal to $1/P_{sleep}$ backoff periods.

The detailed queueing model of two bridged clusters with power managed nodes is shown in Figure 9.2. If the node buffer is empty upon returning from sleep, the node will immediately start a new sleep period. If the node buffer is not empty, the node will transmit a single packet and go to sleep again. Input buffers are assumed to operate in push-out regime, which means that the buffer will always contain the most recent packets received from the higher layers of the protocol stack. The cluster coordinator transmits the required event sensing reliability per cluster $\mathcal{R}$ and the estimated number of live nodes necessary to achieve it. Nodes use this information to determine the mean of the geometric distribution and, subsequently, the duration of each sleep period.

The performance of the network operating in this regime may be analyzed using the approach described by Mišić et al. (2006d), but it requires solving a rather sizable system of equations that describes all clusters and bridges simultaneously. Instead, we propose a computationally lightweight technique in which node populations are calculated one cluster at a time, starting from the cluster which is farthest away from the sink. Namely, when redundant nodes are used, it is reasonable to assume that the required per-cluster throughput of $\mathcal{R}$ is much lower than the capacity of the cluster, which means that at any given time only a small fraction of nodes is active. The mean number of packets sent by the bridge in a single superframe can be approximated with

$$N = \mathcal{R} \cdot SD \cdot t_{boff}, \tag{9.1}$$

where $\mathcal{R}$ is the desired throughput, the beacon interval BI includes both active and inactive portion of the superframe, and t_{tboff} denotes the unit backoff period; when the 2450 MHz PHY option is used, $t_{tboff} = 0.32\,\mu s$.

We assume that top and bottom clusters are far enough apart so that their active periods can overlap, hence $SO = 0$, $BO = 1$, and the superframe duration SD has the default value of 48 backoff periods. Assuming the required throughput of $\mathcal{R} = 10$ packets per second per cluster, the nodes from each cluster send, on the average, $N = 10 \cdot 48 \cdot 0.00032 \approx 0.153$ packets per superframe to their respective cluster coordinators. However, the middle cluster

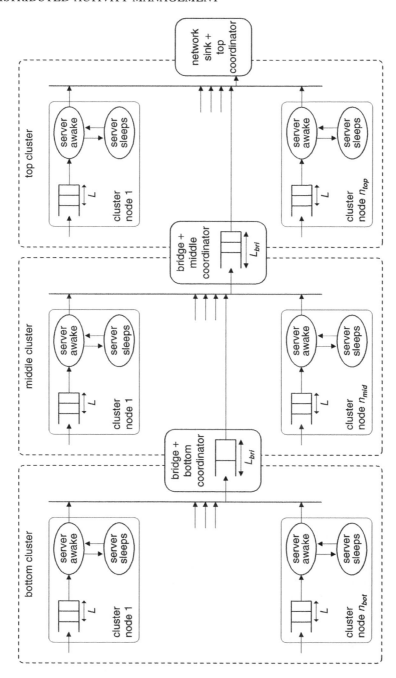

Figure 9.2 Simplified queueing model of network operation. Adapted from J. Mišić, 'Algorithm for equalization of cluster lifetimes in a multi-level beacon enabled 802.15.4 sensor network,' *Computer Networks*, **51**: 3252–3264, © 2007 Elsevier B. V.

coordinator receives twice that amount of traffic, while the network sink (which is the top cluster coordinator) receives three times this amount. (Note that even this last number amounts to less than one packet per two superframes.)

Under these assumptions, the probability that the bridge will block packets from the nodes in its cluster is negligible. Furthermore, the probability that the bridge will succeed in delivering all of its packets to the upper cluster coordinator during a single visit to that cluster is close to one; as a result, the bridge buffer will be empty after returning to its cluster. These notions allow for a simplified analysis, as will be seen from the following.

Let the clusters contain n_{bot}, n_{mid}, and n_{top} ordinary nodes, respectively, with the packet arrival rate of λ per node. (References to specific clusters will use the subscripts bot, mid, and top, respectively.) All nodes have buffers of finite capacity, L packets for an ordinary node and L_{bri} packets for the two bridge/coordinators; the top cluster coordinator acts as the network sink and it does not block packets.

Mean duration of the period between node transmissions is determined by each node as

$$B_{bot} = \frac{n_{bot}}{t_{boff}\mathcal{R}}$$

$$B_{mid} = \frac{n_{mid}}{t_{boff}\mathcal{R}} \tag{9.2}$$

$$B_{top} = \frac{n_{top}}{t_{boff}\mathcal{R}}$$

In order to calculate the actual sleep probability for a node, we must find the probability Q_c that a new sleep will immediately follow the previous one without an intervening packet transmission. Let π_k and q_k denote the steady state probabilities that there are k packets in the node buffer immediately upon packet departure and after returning from sleep, respectively. Then, the conditional probability that the Markov point corresponds to a return from the vacation, at which moment the node buffer is empty, is

$$Q_c = \frac{q_0}{\displaystyle\sum_{i=0}^{L} q_i} \tag{9.3}$$

Also, let a_k denote the probability that k packets will arrive to the node buffer during the packet service time, and let f_k stand for the probability that k packets will arrive to the node buffer during one sleep period. Since the network is lightly loaded, the packet service time does not exceed a couple of dozens of backoff periods and therefore we can assume that $a_0 \approx 1$ and $a_k \approx 0$, for $k = 1 .. \infty$.

For the buffer capacity of L packets, the steady state equations for state transitions are

$$q_0 = (q_0 + \pi_0) f_0$$

$$q_k = (q_0 + \pi_0) f_k + \sum_{j=1}^{k} \pi_j f_{k-j}, \qquad 1 \leq k \leq L - 1$$

$$q_L = (q_0 + \pi_0) \sum_{k=L}^{\infty} f_k + \sum_{j=1}^{L-1} \pi_j \sum_{k=L-j}^{\infty} f_k \tag{9.4}$$

$$\pi_k = \sum_{j=1}^{k+1} q_j a_{k-j+1-l}, \qquad 0 \leq k \leq L-2$$

$$\pi_{L-1} = \sum_{j=1}^{L} q_j \sum_{k=L-j}^{\infty} a_{k-l} \tag{9.4}$$

where subscripts indicate the relevant cluster. These transition probabilities must add up to one:

$$\sum_{k=0}^{L} q_k + \sum_{k=0}^{L-1} \pi_k = 1 \tag{9.5}$$

Solving the system (9.4) gives Q_c as function of f_k. The probability f_k that there are exactly k packet arrivals during the sleep period can be found as follows. First, the moment generating function for the sleep period is

$$\begin{aligned} V^*(s) &= \sum_{k=1}^{\infty}(1 - P_{sleep})P_{sleep}^{k-1}e^{-sk} \\ &= \frac{(1 - P_{sleep})e^{-s}}{1 - e^{-s}P_{sleep}} \end{aligned} \tag{9.6}$$

Since the PGF for the number of packet arrivals to the buffer during the sleep period can be found as

$$F(z) = V^*(\lambda - z\lambda) \tag{9.7}$$

the desired probability is

$$f_k = \frac{1}{k!}\frac{d^k F(z)}{dz^k}\bigg|_{z=0} \tag{9.8}$$

The probability distribution for the total inactive time of the node has a geometric distribution with the parameter Q_c, applied at the moments when the node returns from sleep. The corresponding moment generating function is

$$\begin{aligned} I^*(s) &= \sum_{k=1}^{\infty}(1 - Q_c)Q_c^{k-1}V^*(s)^k \\ &= \frac{(1 - Q_c)V^*(s)}{1 - V^*(s)Q_c} \end{aligned} \tag{9.9}$$

and its mean value is

$$\bar{I} = \frac{1}{(1 - Q_c)(1 - P_{sleep})} \tag{9.10}$$

Finally, by equating the average period between the transmissions with the average inactive time, we obtain

$$B_i = \bar{I} \tag{9.11}$$

which can be solved for P_{sleep} using the number of nodes in the cluster, the required sensing reliability, and packet arrival rate per node as independent variables.

A node that wakes up has to wait for the beacon for synchronization; the waiting time s_1 is uniformly distributed over the beacon interval BI. As there may be more than one node in this mode, increased collisions may result for the packets sent immediately after the beacon, as explained in Section 5.2. To avoid this, we introduce an additional waiting time s_2, the duration of which is uniformly distributed in the range $0..7$ backoff periods (Mišić et al. 2005b). The PGF for these two synchronization periods is

$$S_1(z) = \frac{1}{BI} \sum_{i=0}^{BI} z^i$$

$$S_2(z) = \frac{1}{8} \sum_{i=0}^{7} z^i$$

(9.12)

9.3 Energy Consumption in Interconnected Clusters

While the activity management extends the lifetime of each cluster, individual cluster lifetimes are not equal, and the network lifetime is determined by the shortest cluster lifetime. In order to equalize cluster lifetimes, which maximizes the network lifetime, individual node utilizations must be the same in each cluster. As the traffic load differs from one cluster to another (the middle cluster carries twice the load of the bottom one, and the top cluster carries three times the load of the bottom one), node utilizations differ; their values can be equalized by assigning different node population to each cluster. The algorithm to calculate node population considers one cluster at a time in an iterative fashion, starting with the cluster which is farthest away from the sink.

Let us now present the general approach, and then instantiate the values for each cluster in turn, beginning from the bottom cluster.

As mentioned above, we assume that all transmissions are acknowledged; if the acknowledgment packet is not received within the time prescribed by the standard (IEEE 2006), the transmission will be repeated. Let the PGF of the time interval between the data and subsequent ACK packet be $t_{ack}(z) = z^2$; actually its value is between *aTurnaroundTime* and *aTurnaroundTime + aUnitBackoffPeriod* (IEEE 2006), but we round the exponent to the next higher integer for simplicity.

Each CSMA-CA transmission has to be preceded by the random backoff countdown and two CCAs at unit backoff period boundaries. Throughout this time, the radio subsystem is switched on in the receiving mode; it is switched to the transmitting mode only if both CCAs are successful. The standard allows up to $m = macMaxCSMABackoffs = 5$ backoff attempts, during which the backoff window takes values of $W_0 = 7$, $W_1 = 15$, and $W_2 = W_3 = W_4 = 31$ unit backoff periods. We assume that the battery saving mode is not turned on; however, under the sleep management regime, all transmissions will complete in one or two backoff attempts, and battery saving mode is not that important. The PGF for the duration of j-th backoff attempt prior to transmission is equal to:

$$B_j(z) = \sum_{k=0}^{W_j-1} \frac{1}{W_j} z^k = \frac{z^{W_i} - 1}{W_j(z-1)}$$

(9.13)

In order to find the energy consumption during the j-th backoff attempt, we need to switch to the LST by substituting $z = e^{-s\omega_r}$ into the corresponding PGF:

$$E^*_{B_j}(s) = \frac{e^{-s\omega_r W_i} - 1}{W_j(e^{-s\omega_r} - 1)} \tag{9.14}$$

The LST for energy consumption during the two CCAs is $e^{-s2\omega_r}$.

Let the PGF of the data packet length be $G_p(z) = z^k$, and let $G_a(z) = z$ stand for the PGF of the ACK packet duration. Then, the PGF for the total transmission time of the data packet will be $D_d(z) = z^2 G_p(z) t_{ack}(z) G_a(z)$, and the mean value of the transmission time is $\overline{D_d} = 2 + G'_p(1) + t'_{ack}(1) + G'_a(1)$. The LST for the energy consumption during pure packet transmission time is $e^{-sk\omega_t}$.

The LST for energy consumption during waiting for and receiving the acknowledgment is $e^{-s3\omega_r}$; this value also describes the energy consumption during the reception of a beacon frame which lasts for three unit backoff periods.

Then, the PGF for the time needed for a single transmission attempt, including backoff attempts, is

$$A(z) = \frac{\displaystyle\sum_{i=0}^{m} \left(\prod_{j=0}^{i} B_j(z) \right) (1 - \alpha\beta)^i z^{2(i+1)} \left(\alpha\beta G_p(z) t_{ack}(z) G_a(z) \right)}{\displaystyle\sum_{i=0}^{m} (1 - \alpha\beta)^i \alpha\beta} \tag{9.15}$$

and the LST for the corresponding energy consumption is

$$\mathcal{E}^*_A(s) = \frac{\displaystyle\sum_{i=0}^{m} \left(\prod_{j=0}^{i} E^*_{B_j}(z) \right) (1 - \alpha\beta)^i e^{-s2\omega_r(i+1)} \left(\alpha\beta e^{-sk\omega_t} e^{-s3\omega_r} \right)}{\displaystyle\sum_{i=0}^{m} (1 - \alpha\beta)^i \alpha\beta} \tag{9.16}$$

By taking packet collisions into account, the probability distribution of the packet service time follows the geometric distribution, and its PGF becomes:

$$T(z) = \sum_{k=0}^{\infty} (A(z)(1 - \gamma))^k A(z)\gamma = \frac{\gamma A(z)}{1 - A(z) + \gamma A(z)} \tag{9.17}$$

In this case, mean packet service time can simply be written as

$$\overline{T} = T'(1) = \frac{A'(1)}{\gamma} \tag{9.18}$$

The LST for the energy spent during the packet service time is

$$E^*_T(s) = \frac{\gamma \mathcal{E}^*_A(s)}{1 - \mathcal{E}^*_A(s) + \gamma \mathcal{E}^*_A(s)} \tag{9.19}$$

We are now ready to apply those values for each cluster in turn, beginning with the bottom cluster, in order to find α, β, and γ, which denote the success probabilities for first CCA, second CCA, and packet transmission, respectively. The process is iterative, as will be seen.

Bottom cluster. The access probability for a node in the bottom cluster can be approximated with

$$\tau_{bot}^{(1)} = 1/\overline{I_{bot}} \tag{9.20}$$

where the exponent in parentheses indicates the index of the current iteration. Since $\tau_{bot}^{(1)}$ is very small and the number of nodes is large, we may estimate the per-cluster arrival rate of medium access events as

$$\lambda_{c,bot}^{(1)} = (n_{bot} - 1)\tau_{bot}^{(1)} \frac{SD}{16} \tag{9.21}$$

The probability that the medium is idle at the first CCA may be approximated with

$$\alpha_{bot}^{(1)} = \frac{1}{16} \sum_{i=0}^{15} e^{-i\lambda_{c,bot}^{(1)}} \tag{9.22}$$

The probability that the medium is idle on the second CCA is equal to the probability that neither one of the remaining $n_{bot} - 1$ nodes has started a transmission in that backoff period, which amounts to

$$\beta_{bot}^{(1)} = e^{-\lambda_{c,bot}^{(1)}} \tag{9.23}$$

By the same token, the overall success probability of a transmission attempt is

$$\gamma_{bot}^{(1)} = (\beta_{bot}^{(1)})^{\overline{D_d}} \tag{9.24}$$

This value of γ is used to revise the access probability to

$$\tau_{bot}^{(2)} = \frac{1}{\overline{I_{bot}}\gamma_{bot}^{(1)}} \tag{9.25}$$

which begins a new iteration cycle.

As the success probability under low traffic load is close to one, only a few iterations will suffice to achieve convergence of success probabilities $\alpha_{bot}^{(i)}$, $\beta_{bot}^{(i)}$, and $\gamma_{bot}^{(i)}$ to their limiting values α_{bot}, β_{bot}, and γ_{bot}, respectively.

The PGF of the time needed to conduct one transmission attempt is then obtained by substituting the limiting values of success probabilities in Equation (9.17). The LST for the energy spent in packet service is obtained by substituting those values in Equation (9.19). The average energy consumed during packet service is obtained as

$$\overline{E_{T,bot}} = -\frac{d}{ds} E_{T,bot}^*(s)\Big|_{s=0} \tag{9.26}$$

Finally, the average energy consumption per backoff period can be found as

$$u_{bot} = \frac{\overline{S_1}\omega_r + \overline{S_2}\omega_r + 3\omega_r + \overline{E_{T,bot}} + \overline{I_{bot}}\omega_s}{\overline{S_1} + \overline{S_2} + 3 + \overline{T_{bot}} + \overline{I_{bot}}}. \tag{9.27}$$

Given the battery budget of b, the average number of transmission/sleep cycles in the bottom cluster can be found as

$$n_{c,bot} = \left\lceil \frac{b}{\overline{S_1}\omega_r + 3\omega_r + \overline{S_2}\omega_r + \overline{E_{T,bot}} + \overline{I_{bot}}\omega_s} \right\rceil \tag{9.28}$$

By applying the law of large numbers (Grimmett and Stirzaker 1992), the PGF for total lifetime of the node in bottom cluster becomes

$$L_{bot}(z) = (S_1(z)S_2(z)T_{bot}(z)I_{bot}(z))^{n_{c,bot}} \tag{9.29}$$

By differentiating the respective PGFs we can obtain the standard deviation of the node lifetime, as well as the coefficient of skewness, μ, which measures the deviation of a distribution from symmetry (Pebbles, Jr. 1993).

Middle cluster. The iterative procedure is then applied to the middle cluster, where the presence of the bridge/coordinator from the bottom cluster must be accounted for. We begin by solving Equation (9.11) for the sleep probability in the upper cluster, while keeping n_{mid} as a parameter. The initial node access probability in the upper cluster is estimated as

$$\tau_{mid}^{(1)} = \frac{1}{\overline{I_{mid}}} \tag{9.30}$$

The impact of the bridge CSMA access in the middle cluster is modeled as

$$\tau_{bri,mid}^{(1)} = n_{bot}\tau_{bot}\frac{SD}{16} \tag{9.31}$$

The success probability for bridge transmissions depends on all the nodes in the cluster as

$$\gamma_{bri,mid}^{(1)} = (1 - \tau_{mid}^{(1)})^{\overline{D_d}n_{mid}}. \tag{9.32}$$

The process is then repeated to obtain the revised value for access probability for the bridge as

$$\tau_{bri,mid}^{(2)} = \frac{\tau_{bri,mid}^{(1)}}{\gamma_{bri,mid}^{(1)}} \tag{9.33}$$

Again, only a few iterations are needed to reach satisfactory accuracy. Note that the medium access event rate node must account for both the ordinary nodes and the bridge, hence

$$\lambda_{c,mid}^{(1)} = (n_{mid} - 1)\tau_{mid}^{(1)}\frac{SD}{16} + \tau_{bri,mid}^{(2)} \tag{9.34}$$

Success probabilities for ordinary nodes $\alpha_{mid}^{(1)}$, $\beta_{mid}^{(1)}$ and $\gamma_{mid}^{(1)}$ are calculated in a similar way to their bottom cluster counterparts:

$$\begin{aligned} \alpha_{mid}^{(1)} &= \frac{1}{16}\sum_{i=0}^{15} e^{-i\lambda_{c,mid}^{(1)}} \\ \beta_{mid}^{(1)} &= e^{-\lambda_{c,mid}^{(1)}} \\ \gamma_{mid}^{(1)} &= e^{-\lambda_{c,mid}^{(1)}\overline{D_d}} \end{aligned} \tag{9.35}$$

The PGFs for a single transmission attempt and for the overall packet transmission time can be calculated as $\mathcal{A}_{mid}(z)$ and $T_{mid}(z)$, respectively. Both PGFs depend on the number of nodes n_{mid} as the parameter. Finally, the average energy consumption per backoff period is calculated as

$$u_{mid} = \frac{\overline{S_1}\omega_r + \overline{S_2}\omega_r + 3\omega_r + \overline{E_{T,mid}} + \overline{I_{mid}}\omega_s}{\overline{S_1} + \overline{S_2} + 3 + \overline{T_{mid}} + \overline{I_{mid}}}. \tag{9.36}$$

Now, if the lifetime of the middle cluster is to be the same as that of the bottom cluster, the average energy consumed by a node per backoff period must be the same in both bottom and middle clusters:

$$u_{mid} = u_{bot} \tag{9.37}$$

from which we can obtain the initial population of the middle cluster n_{mid}.

Given the battery budget of b backoff periods, the average number of transmission/sleep cycles in bottom cluster can be found as

$$n_{c,mid} = \left\lceil \frac{b}{\overline{S_1}\omega_r + 3\omega_r + \overline{S_2}\omega_r + \overline{E_{T,mid}} + \overline{I_{mid}}\omega_s} \right\rceil \tag{9.38}$$

The PGF for total lifetime of the node in bottom cluster becomes

$$L_{mid}(z) = (S_1(z)S_2(z)T_{mid}(z)I_{mid}(z))^{n_{c,mid}} \tag{9.39}$$

Top (sink) cluster. The procedure is then repeated for the top cluster, starting from

$$\tau_{bri,top}^{(1)} = (n_{bot}\tau_{bot} + n_{mid}\tau_{mid})\frac{SD}{16} \tag{9.40}$$

It is worth noting that this algorithm can easily be scaled to networks with several clusters and/or several levels, as long as the clusters are not operating in the saturation condition.

9.4   Performance of Activity Management

In order to verify the algorithms for distributed calculation of the sleep interval, finding the cluster population, and determining the lifetime of the network, we have implemented the algorithms described above using Maple 10 from Waterloo Maple, Inc. (2005). The network is assumed to operate with the 2450 MHz PHY option. Packet arrivals to each node follow a Poisson distribution with the arrival rate of $\lambda = 1$ packet per second. The packet size is fixed at 30 bytes, including all PHY and MAC layer headers. All other parameters are set to default values prescribed in the 802.15.4 standard (IEEE 2006).

Ordinary nodes have buffers that can hold $L = 2$ packets each, while the capacity of the bridge buffer is $L_{bri} = 6$ packets. The required throughput was set to $\mathcal{R} = 10$ packets per second per cluster, respectively; note that the traffic load in middle and top clusters was $2\mathcal{R}$ and $3\mathcal{R}$ packets per second, respectively. Each ordinary node is assumed to be powered by two low-cost AA batteries with the capacity of 712.5 mAh which gives the total power budget of $b = 5130J$. Each node maintains the counter of the remaining backoff periods which was decremented after each backoff period when the radio subsystem was turned on. The coordinators/bridges and the top cluster coordinator/sink were assumed to have an infinite power supply, as explained above.

Table 9.2 Calculated network parameters for uniform population in each cluster

Parameter	Top	Middle	Bottom
number of nodes	100	100	100
inactive period (seconds)	10.00	10.00	10.00
success probability γ	0.7872	0.8649	0.9420
node utilization	0.00236	0.00226	0.00218
lifetime (days)	314.05	325.35	336.78
std. deviation	0.06%	3.09%	2.46%
skewness μ	1.33E-9	0.912E-14	1.68E-14

Uniform node population. In our first experiment, we have set the number of nodes in each of the clusters to 100, and then calculated the relevant network parameters which are given in Table 9.2. As can be seen, the standard deviation of the node lifetime is small and the skewness μ is close to zero. The obvious conclusion is, then, that all nodes in a given cluster will die within a short interval centered around the mean lifetime value for that cluster. At the same time, the success probability γ differs from cluster to cluster, as does the average node utilization and the ensuing cluster lifetime. The bottom cluster will live more than ten days longer than the middle one, which will live more than ten days longer than the top one, but the network will cease to provide the required throughput of $3\mathcal{R} = 30$ packets per second as soon as the top cluster dies.

These results are further verified through the diagrams in Figure 9.3, which show cluster lifetimes for $\mathcal{R} = 10$ and 12 packets per second. We notice that the lifetime decreases with

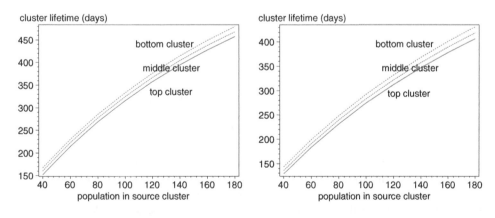

(a) $\mathcal{R} = 10$ packets per second per cluster. (b) $\mathcal{R} = 12$ packets per second per cluster.

Figure 9.3 Average lifetime in days when each cluster has 100 nodes. Adapted from J. Mišić, 'Algorithm for equalization of cluster lifetimes in a multi-level beacon enabled 802.15.4 sensor network,' *Computer Networks*, **51**: 3252–3264, © 2007 Elsevier B. V.

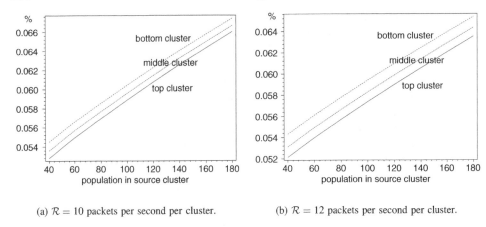

(a) $\mathcal{R} = 10$ packets per second per cluster.   (b) $\mathcal{R} = 12$ packets per second per cluster.

Figure 9.4 Ratio of standard deviation and mean of cluster lifetime, when each cluster has 100 nodes. Adapted from J. Mišić, 'Algorithm for equalization of cluster lifetimes in a multi-level beacon enabled 802.15.4 sensor network,' *Computer Networks*, **51**: 3252–3264, © 2007 Elsevier B. V.

increased remote traffic load in the cluster, as expected. Furthermore, the difference in cluster lifetimes are more pronounced when the required throughput is higher.

The diagrams in Figure 9.4 show the ratio of standard deviation and the mean of individual node lifetime, again for $\mathcal{R} = 10$ and 12 packets per second per cluster. The range of the values shown is well below 1 percent which shows that, despite the randomization of the sleep time, all the nodes will die within a very short time period. As expected, this ratio decreases when the required throughput increases, which is due to the decrease in the total number of transmission/sleep cycles.

Equalized cluster lifetimes. From the analysis and the numerical results presented above, there is a definite need for equalization of cluster lifetimes, which may be accomplished by adjusting the number of nodes in each cluster so as to make the individual node utilization uniform across all clusters. Let us assume that the required throughput is $\mathcal{R} = 10$ packets per second per cluster, and that the bottom cluster has $n_{bot} = 100$ nodes. Then, we have solved Equation (9.37) to obtain the population of $n_{mid} = 104$ nodes in the middle cluster, and $n_{top} = 109$ nodes in the top cluster. The relevant network parameters, in this case, are shown in Table 9.3, with node counts rounded to the next highest integer.

As can be seen, the node utilization is about the same in all clusters and, consequently, the curves for cluster lifetimes in Figures 9.5(a) and 9.5(b) are virtually indistinguishable. Although the increase in the number of nodes is only 4.3% (the total node count is 313 nodes, instead of the original 300), the lifetime of the three-cluster network has been extended to 333.12 days, which represents an increase of more than 6% over the previous value of 314.05 days.

Also, while the skewness values are somewhat different, all three of them are well below 0.1%, and it is the vicinity to zero that counts.

Table 9.3 Calculated network parameters for equalized cluster lifetimes

Parameter	Top	Middle	Bottom
number of nodes	109	104	100
inactive period (seconds)	10.799	10.399	10.00
success probability γ	0.7872	0.8649	0.9420
utilization	0.00218	0.00217	0.00218
lifetime (days)	333.12	333.86	336.78
std. deviation	0.061%	0.059%	2.465%
skewness μ	1.38E-9	1.3E-9	1.68E-14

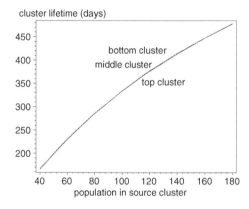

(a) Average lifetime in days for $R = 10$.

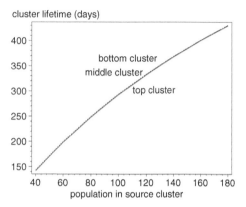

(b) Average lifetime in days for $R = 12$.

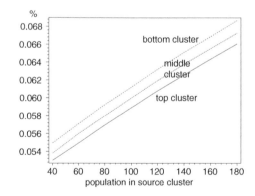

(c) Ratio of standard deviation and mean of lifetime for $R = 10$.

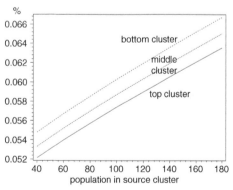

(d) Ratio of standard deviation and mean of lifetime for $R = 12$.

Figure 9.5 Cluster performance with initial node population adjusted to achieve equalized cluster lifetimes. Adapted from J. Mišić, 'Algorithm for equalization of cluster lifetimes in a multi-level beacon enabled 802.15.4 sensor network,' *Computer Networks*, **51**: 3252–3264, © 2007 Elsevier B. V.

The cluster lifetime and the ratio of standard deviation and mean node lifetime, for required per-cluster throughput of $R = 10$ and 12 packets per second, are shown in Figure 9.5. Notice that the ratio of standard deviation and mean node lifetime is even lower than in the case where cluster populations were uniform, Figure 9.4, which shows that all the nodes are operational almost up to the end, and then die in a short interval.

10

Cluster Interconnection with Slave-Slave Bridges

Another possible approach to implementing a multi-cluster 802.15.4 network is to employ a different kind of bridges: those that are just ordinary nodes, but not coordinators, in each cluster they visit. Such bridges will be referred to as slave-slave or SS bridges, by analogy with Bluetooth networks (Mišić and Mišić 2005, Chapter 10). An example topology of a network with SS bridges is shown in Figure 10.1. For clarity, Figure 10.1 shows only coordinators and bridges, but not ordinary nodes; furthermore, the transmission ranges of different cluster coordinators are outlined with semicircular arcs (sink cluster and cluster 3) and semicircles with different fill patterns (clusters 1 and 2).

The operation of an 802.15.4 network that uses SS bridges differs in certain important aspects from those in a comparable network that uses MS bridges. Some of the differences can be seen from Figure 10.1, while other require a more detailed analysis of bridge and cluster operation and timing.

Topology-wise, the use of an SS bridge allows the clusters to be spaced farther apart, as the cluster coordinators are not required to be located within the transmission range of each other. Of course, the bridge must be within the transmission range of the coordinators of both such clusters, and vice versa.

If the coordinators are within the transmission range of each other, as is the case with the coordinators of the sink cluster and cluster 1 in Figure 10.1(a), their respective clusters may form a multi-cluster tree. In that case, the superframes will remain in perfect synchronization because the coordinator of cluster 1 simply repeats the beacon received from the sink cluster, Figure 10.1(b). This is the preferred setup for clusters interconnected with an MS bridge, as discussed in Chapter 8.

However, the clusters that the bridge visits may operate independently, in which case their coordinators may or may not be within the transmission range of each other. Independent operation means there is no synchronization between the superframes, even if the cluster coordinators use the same superframe parameters, which they may or may not do.

Wireless Personal Area Networks   Jelena Mišić and Vojislav B. Mišić
© 2008 John Wiley & Sons, Ltd

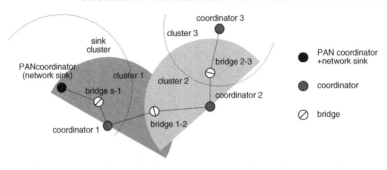

(a) Network topology with SS bridges.

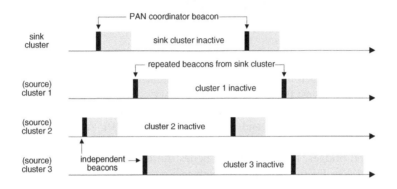

(b) Superframe timing in different clusters.

Figure 10.1 Pertaining to the operation of SS bridges.

For example, clusters 1 and 2 in Figure 10.1 use the same parameters for both beacon order *BO* and superframe order *SO*, while clusters 2 and 3 use the same value for the beacon order but different values for the superframe order.

If the clusters operate independently but use the same RF channel, their transmission ranges should overlap as little as possible, otherwise increased interference may result.

The clusters may also use different RF channels (Section 2.1) in which case no interference occurs, regardless of the relative position of the cluster coordinators. This setup allows the cluster coordinators to set their superframe timing independently, but a multi-cluster tree structure cannot be used.

While both SS and MS bridges can, in theory, visit more than two clusters to perform their function, the MS bridge is somewhat limited in that respect because it must monitor and control the operation of the cluster of which it is the coordinator. To that end, it must not be absent from that cluster for prolonged periods of time, lest the ordinary nodes in the cluster conclude that the cluster is not operational any more (and try to associate with some other cluster, or to form a cluster of their own). The SS bridge has no such limitations.

Finally, we note that a given cluster may host one or more bridges, be they of SS or MS type, but the performance of a cluster may suffer because of the increase in traffic load supplied by the bridges, as shown in Section 8.5 of the previous chapter.

10.1 Operation of the SS Bridge

Let us now discuss and, subsequently, analyze the operation of the bridge in more detail. For simplicity, we assume that the network consists of two clusters with n ordinary nodes each, interconnected with a single SS bridge, as shown in Figure 10.2; the operation of the bridge in this network is schematically shown in Figure 10.3.

We assume that clusters operate independently, using the same set of superframe parameters BO and SO, with their superframes overlapped in time. We also assume that the transmissions from one cluster do not interfere with those from the other. Note that all these assumptions do not incur a loss in generality, and our analysis could easily be extended to correspond to networks with more complex topology and superframe timing.

In this network, each ordinary node in either cluster receives packets from the upper layers of the network protocol stack and queues them for delivery to the cluster coordinator. We assume that the packet arrival process follows a Poisson distribution with the mean packet arrival rate of λ at each node. Since the buffers at each node have a finite size of L packets, some packets that arrive may be blocked, with the probability P_{src}^B or P_{snk}^B, for nodes in source and sink cluster, respectively.

Packets queued at each ordinary node are sent in the uplink direction to the corresponding cluster coordinator. We assume that all such transmissions use the slotted CSMA-CA medium access mechanism. The coordinator of the source cluster has a finite buffer for downlink packets that can accommodate up to L_d packets; if this buffer is full, the source cluster coordinator will not accept any new packets. Packets that are received at that time will be dropped; the probability of this happening will be denoted with P_d^B. The coordinator of the sink cluster acts as the network sink and is assumed capable of accepting any number of packets without blocking.

The source cluster coordinator sends the queued data packets to the bridge which stores them for subsequent delivery to the sink. Downlink transmissions follow the CSMA-CA medium access mechanism explained in Section 2.5. In this procedure, the coordinator

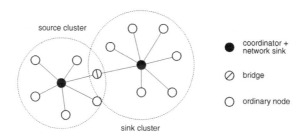

source cluster

coordinator + network sink

bridge

ordinary node

sink cluster

Figure 10.2 Two clusters interconnected with an SS bridge.

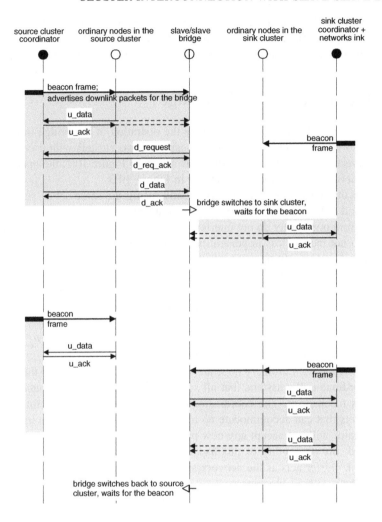

Figure 10.3 SS bridge switching between the clusters. Shaded areas denote time intervals where the bridge is listening to communications in a cluster. Solid arrows show packet transmissions, dashed arrows show transmissions that may be overheard by the bridge and, thus, affect its behavior. Adapted from J. Mišić, 'Analytical model for slave-slave bridging in 802.15.4 beacon enabled networks,' submitted to *IEEE Trans. Vehicular Technology*, 2008.

must first announce the presence of pending packets in the beacon frame; the bridge explicitly requests the downlink transmission with a data request packet, which must be acknowledged; finally, the coordinator sends the actual data packet to the bridge using the CSMA-CA medium access mechanism.

Since the coordinator uses CSMA-CA medium access mechanism for sending the queued data to the bridge, any data packets sent in the uplink direction during the random backoff countdown will simply be ignored by the coordinator, as explained in Section 5.4. The label P_c^B denotes the probability of this type of blocking.

The bridge might also use a dedicated GTS for its communication with the source cluster coordinator. However, the bridge will not be present in the source cluster in every superframe, for reasons that should be obvious from Figure 10.3, and some of the packets sent during its absence may be lost. The loss might be alleviated by allocating a separate GTS for subsequent acknowledgments; but the bridge may be absent for more than one source cluster superframe and delivery may still not be guaranteed while the blocking probability at the coordinator will increase. Furthermore, prolonged absence might even lead the coordinator to conclude that the bridge has left the source cluster for good, which further complicates the task of network management. For all these reasons, we have chosen to limit our analysis to the case where the communication between the source cluster coordinator and the bridge are conducted using only CSMA-CA access mode.

If there is more than one packet pending for delivery to the bridge, the coordinator will set the Frame Pending field in the MAC frame header (Table 2.4, p. 36) of the first data packet sent in the downlink. (For simplicity, we assume that data packets are not aggregated by either the source cluster coordinator or the bridge.) The bridge will then have to send a separate data request packet for each of the pending packets.

The bridge buffer has a finite size of L_{bri}; when this buffer is full, the bridge will ignore the announcement about a pending packet, or packets, and simply refuse to send the data request packet. The probability of this type of blocking will be denoted with P_f^B, so as to distinguish it from the other type of blocking encountered in MS bridges, Chapter 8, as well as in ordinary nodes.

When the bridge switches to the sink cluster, it must first synchronize with the beacon. If both clusters use identical values for the beacon order BO and superframe order SO, their beacon interval, BI, and superframe duration, SD, will be the same, but they need not be aligned. The average synchronization time is, then, a random value uniformly distributed between 0 and BI.

Once synchronized with the sink cluster superframe, the bridge sends queued data packets in the uplink direction to the coordinator of the sink cluster. To that end, the SS bridge may use CSMA-CA or GTS access mode, just like an MS one. In the former case, the bridge must compete against all n ordinary nodes in the sink cluster; in the latter, the transmissions from the bridge are decoupled from those of ordinary nodes, and vice versa. The price paid by the ordinary nodes is the shortening of the available CAP time with respect to the case where the bridge uses CSMA-CA access.

The bridge remains in the sink cluster for one superframe only; once the superframe ends, the bridge returns to the source cluster where it again has to synchronize with the beacon. The bridge operating in the CSMA-CA mode may encounter the situation in which it has a packet to transmit to the sink, but is unable to do so because the remaining time in the active portion of the sink cluster superframe is insufficient for the two CCAs, data packet transmission, and (optional) acknowledgment packet. In this case, we assume that the bridge will freeze its backoff counter and leave the sink cluster; the backoff countdown will be resumed upon returning to the sink cluster. (The same assumption was made in

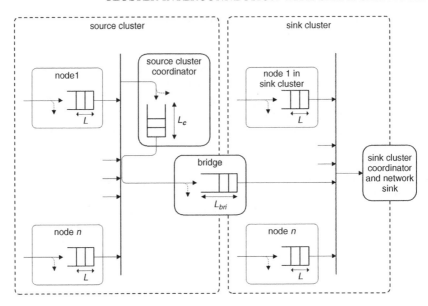

Figure 10.4 Queueing model for the bridging process between source and sink clusters interconnected with an SS bridge (only the data paths are shown). Adapted from J. Mišić, 'Analytical model for slave-slave bridging in 802.15.4 beacon enabled networks,' submitted to *IEEE Trans. Vehicular Technology*, 2008.

Section 8.1, when discussing the behavior of an MS bridge.) Obviously, this procedure requires a certain level of functionality beyond the one prescribed by the standard; but so does the bridging operation itself.

A simplified queueing model of the network is shown in Figure 10.4.

As before, clusters may use acknowledged or non-acknowledged transfer. For simplicity, we will assume that both clusters operate in the same manner, but our analysis may easily be extended to the alternative case in which one of cluster uses acknowledgments while the other does not, although it seems to be of little value in practice.

Given that the number of nodes n is relatively large, and that the events of packet corruption, collision in case of simultaneous transmission, and blocking at different queues, are non-correlated, we may safely assume that packet arrival processes to the coordinator and the bridge can be approximated with Poisson processes with mean arrival rates of λ_c and λ_{bri}, respectively. In this case, depending on whether acknowledgments are used or not, the arrival rates to the queues in Figure 10.4 can take the following values.

Non-acknowledged transfer. The traffic admitted to the source cluster (and, subsequently, sent to the source cluster coordinator) is $n\lambda(1 - P_{src}^B)$, while the total packet arrival rate offered to the source cluster coordinator is

$$\lambda_c = n\lambda(1 - P_{src}^B)\gamma_{src/o}\delta_{src}(1 - P_c^B), \qquad (10.1)$$

where $\gamma_{src/o}$ denotes the probability that no collision has occurred for a particular uplink data packet in source cluster, and δ_{src} denotes the probability that such a packet is not corrupted by noise and interference. The latter can be calculated from the given bit error rate, *BER*.

The bridge will request a packet from the downlink queue only when there is empty space in its buffer. While the data request packet has to be acknowledged and, if necessary, repeated, no acknowledgment is expected (or, indeed, sent) after a downlink packet transmission. Given the blocking probability of the downlink queue at the coordinator, P_d^B, and the probability that the bridge buffer is full, P_f^B, the offered load to the bridge is

$$\lambda_{bri} = \lambda_c (1 - P_d^B)(1 - P_f^B) \tag{10.2}$$

Acknowledged transfer. In this case, the traffic blocked by the coordinator or the bridge will 'remain' in the source cluster until acknowledged, and thus increase channel utilization as well as the collision rate. The total arrival rate offered to the downlink queue at the coordinator, λ_c, satisfies the relation

$$n\lambda(1 - P_{src}^B) = \lambda_c(1 - P_d^B)(1 - P_c^B), \tag{10.3}$$

and the total arrival rate offered to the bridge satisfies the equality

$$n\lambda(1 - P_{src}^B) = \lambda_{bri}(1 - P_d^B)(1 - P_c^B)(1 - P_f^B), \tag{10.4}$$

From these expressions, the offered packet arrival rate toward the bridge may be obtained as

$$\lambda_{bri} = \frac{n\lambda(1 - P_{src}^B)}{(1 - P_f^B)(1 - P_c^B)(1 - P_d^B)} \tag{10.5}$$

10.1.1 Choice of Markov points for queueing analysis

Blocking probabilities P_{src}^B and P_f^B may be found by analyzing the probability distribution for the number of packets in corresponding buffers at certain discrete moments. These moments, shown together with the appropriate probabilities in Figure 10.5, can be determined as follows.

An ordinary node in source cluster transmits packets during the active part of the source cluster superframe. During the inactive part of the superframe, packets arrive at the node buffer but are not transmitted. Therefore, from the queueing theoretic viewpoint, an ordinary node can be considered as the server which serves a random number of packets, as determined by the packet service time at the MAC and PHY layers, in exhaustive mode during the active part of the superframe, and takes a vacation during the inactive part. The important moments to use in our analysis are, thus, the moments of packet departure, and the end of the inactive part of the superframe, as shown in Figure 10.5.

Similar reasoning holds for ordinary nodes in the sink cluster as well.

The source cluster coordinator receives data packets from ordinary nodes and sends them to the bridge, but only during the active portion of the superframe; there is no activity during the inactive portion. Therefore, it can be considered as a M/G/1/K system without vacations, and its queue should be analyzed at the times of downlink packet departures.

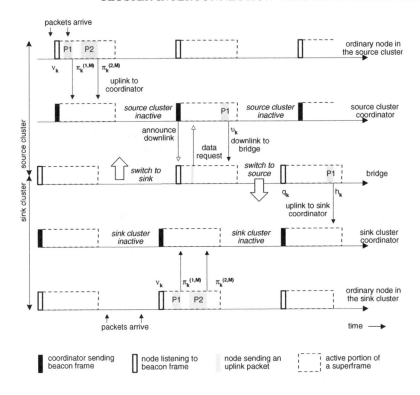

Figure 10.5 Pertaining to the choice of Markov points for queueing analysis of the network with an SS bridge. Adapted from J. Mišić, 'Analytical model for slave-slave bridging in 802.15.4 beacon enabled networks,' submitted to *IEEE Trans. Vehicular Technology*, 2008.

The bridge acts in the following manner. During the active part of the source cluster superframe, it receives packets from the nodes in the source cluster via the coordinator, but does not transmit anything. Once it switches to the sink cluster, the bridge transmits the packets from its buffer at a certain rate, but does not receive any new packets.

Therefore, when present in the source cluster, the bridge may be considered as a server which is on vacation; when present in the sink cluster, the bridge may be considered to be a gated server which begins service with a certain number of packets in its buffer; this number is the same as the number of packets that were present in the bridge buffer at the end of its residence in the source cluster. The service rate of a CSMA-CA bridge while in the sink cluster depends on the volume of traffic in the sink cluster; the service rate of a GTS bridge is constant determined by the number of GTSs allocated to the bridge.

10.1.2   Case 1: ordinary nodes in either cluster

Limited buffer capacity of L packets in an ordinary node makes it necessary to determine the extent of blocking at this point. Let us first note that the Laplace-Stieltjes Transform (LST) of the packet service times $T^*_{t,src}(s)$ and $T^*_{t,snk}(s)$ can be obtained by substituting the variable z with e^{-s} in the corresponding PGFs

$$
\begin{aligned}
T_{t,src}(z) &= \sum_{k=0}^{\infty} p_{t,src}(k) z^k \\
T_{t,snk}(z) &= \sum_{k=0}^{\infty} p_{t,snk}(k) z^k
\end{aligned}
\tag{10.6}
$$

The derivation of these PGFs is somewhat involved, which is why it will be separately presented in Section 10.7 at the end of this chapter.

The PGFs for duration of two, three, etc. packet service times are

$$
\begin{aligned}
T_{t,src}^n(z) &= \sum_{k=0}^{\infty} p_{nt,src}(k) z^k \quad n = 2, 3, \ldots \\
T_{t,snk}^n(z) &= \sum_{k=0}^{\infty} p_{nt,snk}(k) z^k \quad n = 2, 3, \ldots
\end{aligned}
\tag{10.7}
$$

where the required mass probabilities can be found by equating the coefficients $p_{nt,src}(k)$ or $p_{nt,snk}$, for $n = 2, 3, \ldots$, with matching terms in $T_{t,src}^n(z)$ or $T_{t,snk}^n(z)$, respectively, expressed in polynomial form. The corresponding pdfs will be denoted with $t_{dt,src}(x)$ and $t_{dt,snk}(x)$, respectively.

We also need the probability distribution of the number of packets that can be served during the active part of a superframe. In the source cluster, this discrete probability distribution has only a few mass probabilities, since the packet service time is bounded from below (i.e., has a minimum value) dictated by the MAC layer. Therefore, the maximum number of packets that are served in one service period (i.e., during a single active period of the superframe) is

$$
\widehat{M} = \left\lfloor \frac{SD}{T_{min,src}} \right\rfloor,
\tag{10.8}
$$

where SD is the superframe duration and $T_{min,src}$ is the minimum packet service time. This time includes the minimum duration of the backoff countdown, two backoff periods for the CCAs, the actual packet transmission time, and the time to receive the acknowledgment, if required.

The mass probabilities for the number of packet transmissions in one service period are

$$
pm(0) = \sum_{k=SD+1}^{\infty} p_{t,src}(k)
$$

$$
pm(1) = (1 - pm(0)) \sum_{k=SD+1}^{\infty} p_{2t,src}(k)
$$

$$
pm(2) = (1 - pm(0) - pm(1)) \sum_{k=SD+1}^{\infty} p_{3t,src}(k)
$$

$$
\cdots
$$

$$
pm(\widehat{M}) = 1 - \sum_{i=0}^{\widehat{M}-1} pm(i)
$$

$$
\tag{10.9}
$$

Note that it is possible to have a situation when no packets will be served in one service period (although the device's buffer is not empty), i.e., to have $pm(0) > 0$. This happens under high loads in the cluster when a packet experiences many collisions and backoff attempts before it is successfully transmitted.

The probability of k packet arrivals to a given node during a single packet service time is

$$a_k = \int_0^\infty \frac{(\lambda x)^k}{k!} e^{-\lambda x} t_{dt,src}(x) dx \qquad (10.10)$$

and the corresponding PGF, according to Takagi (1991), is

$$A_{src}(z) = \sum_{k=0}^\infty a_k z^k = \int_0^\infty e^{-x\lambda(1-z)} t_{dt,src}(x) dx = T_{t,src}^*(\lambda - z\lambda) \qquad (10.11)$$

When the LST of the packet service time is known, the probability a_k can be obtained as

$$a_k = \frac{1}{k!} \cdot \frac{d^k A_{src}(z)}{dz^k}\bigg|_{z=0} \qquad (10.12)$$

Since the duration of inactive superframe part has the LST of

$$V_c^*(s) = e^{-s(BI-SD)}, \qquad (10.13)$$

the PGF for the number of packet arrivals to an ordinary node in the source cluster during a single inactive part of the superframe is

$$F_c(z) = V_c^*(\lambda - z\lambda) \qquad (10.14)$$

and the probability of k packet arrivals to the node buffer during that same time (i.e., the server vacation) is

$$f_k = \frac{1}{k!} \cdot \frac{d^k F_c(z)}{dz^k}\bigg|_{z=0}. \qquad (10.15)$$

Let $\pi_k^{(m,M)}$ denote the steady state probability that there are k packets in the node buffer immediately after m-th packet serviced in the service period in which a total of M packets are served. As noted before, M is a random variable which depends on the traffic intensity in the cluster. Also, let v_k denote the steady state probability that there are exactly k packets in the node buffer at the end of the server vacation. Sample moments when $\pi_k^{(m,M)}$ and v_k are evaluated for one of the ordinary nodes in each cluster are shown in Figure 10.5.

Then, the state of the node buffer in the source cluster at the beginning of the active portion of the superframe, and after packet departures from the buffer, can be described with the following equations:

$$v_k = pm(0) \sum_{j=1}^k v_j f_{k-j} + \left(v_0 + \sum_{M=2}^{\widehat{M}} \sum_{m=1}^{M-1} \pi_0^{(m,M)} \right) f_k$$

$$+ \sum_{M=1}^{\widehat{M}} \sum_{j=0}^k \pi_j^{(M,M)} f_{k-j}, \qquad 0 \le k \le L-1 \qquad (10.16)$$

$$v_L = pm(0) \sum_{j=1}^{L} v_j \sum_{k=L-j}^{\infty} f_k + \left(v_0 + \sum_{M=2}^{\widehat{M}} \sum_{m=1}^{M-1} \pi_0^{(m,M)} \right) \sum_{k=L}^{\infty} f_k$$

$$+ \sum_{M=1}^{\widehat{M}} \sum_{j=0}^{L-1} \pi_j^{(M,M)} \sum_{k=L-j}^{\infty} f_k$$

$$\pi_k^{(1,M)} = pm(M) \sum_{j=1}^{k+1} v_j a_{k-j+1} \qquad\qquad 0 \le k \le L-2; \; 1 \le M \le \widehat{M}$$

$$\pi_{L-1}^{(1,M)} = pm(M) \sum_{j=1}^{L} v_j \sum_{k=L-j}^{\infty} a_k, \qquad\qquad 1 \le M \le \widehat{M}$$

$$\pi_k^{(m,M)} = \sum_{j=1}^{k+1} \pi_j^{(m-1,M)} a_{k-j+1}, \qquad 0 \le k \le L-2; \; 2 \le m \le M; \; 1 \le M \le \widehat{M}$$

$$\pi_{L-1}^{(m,M)} = \sum_{j=1}^{L-1} \pi_j^{(m-1,M)} \sum_{k=L-j}^{\infty} a_k$$

$$\tag{10.16}$$

The sum of all probabilities has to be equal to 1:

$$\sum_{k=0}^{L} v_k + \sum_{M=1}^{\widehat{M}} \sum_{m=1}^{M} \sum_{k=0}^{L-1} \pi_k^{(m,M)} = 1 \tag{10.17}$$

The probability distribution of the device queue length at the time of packet departure can be found by solving the system (10.16) together with the normalization Equation (10.17).

In order to find the buffer blocking probability at arbitrary time, we need to find the average time period between Markov points. Since the probability of a vacation period starting after an arbitrary Markov point is equal to

$$v_{src} = \sum_{k=0}^{L} v_k,$$

the average distance between two consecutive Markov points is

$$\eta_{src} = v_{src}(BI - SD) + (1 - v_{src})T'_{t,src}(1) \tag{10.18}$$

where $T'_{t,src}(1)$ denotes the mean packet service time for an ordinary node in the source cluster. The carried load for an ordinary node in the source cluster can be determined as

$$\rho' = \frac{(1 - v_{src})T'_{t,src}(1)}{\eta_{src}} \tag{10.19}$$

Given that the offered load for the node is $\rho = \lambda T'_{t,src}(1)$, the blocking probability for the buffer at an ordinary node is

$$P_{src}^B = 1 - \frac{\rho'}{\rho} = 1 - \frac{(1 - v_{src})}{\lambda \eta_{src}}, \tag{10.20}$$

The probability that the node buffer is empty at the end of a service period is

$$\pi_{0,src} = \sum_{M=1}^{\hat{M}} \sum_{m=1}^{M} \pi_0^{(m,M)} \tag{10.21}$$

However, the probability that the node buffer is empty at arbitrary time, including the inactive part of the superframe, can be obtained by multiplying the probability $\pi_{0,src}$ with the ratio of packet inter-arrival time $1/\lambda$ and the average distance between Markov points η_{src}:

$$P_{0,src} = \frac{\pi_{0,src}}{\lambda \eta_{src}}. \tag{10.22}$$

Note that the impact of bridge node does not appear explicitly in the Equations (10.16), (10.17), and (10.20), although it is actually contained in the probability distribution of the duration of the packet service time which will be derived in Section 10.7 at the end of this chapter.

Similar considerations apply to the probability of packet blocking at the device buffers in the sink cluster, which is why the labels for Markov points for ordinary nodes are the same in both clusters, Figure 10.5. However, the model of the packet service time for an ordinary node in the sink cluster will include the impact of the bridge activity, as will be shown in Section 10.7.

10.1.3   Case 2: source cluster coordinator

The downlink queue at the source cluster coordinator is assumed to have a finite capacity of L_c packets. The relevant Markov points, in this case, correspond to the downlink packet departure times; these are denoted with υ_k in Figure 10.5. However, each downlink packet transmission is preceded by a successful transmission of the data request packet from the bridge, hence the effective service time for downlink packets includes the service time for data request packets. Let us assume that LST for overall downlink packet transmission time, $T_{t,dtot}^*(s)$, is known (its derivation will be presented in Section 10.7 at the end of this chapter), while $t_{dtot}(x)$ denotes the corresponding pdf. Then, the probability of k packet arrivals to the coordinator queue during the downlink packet service time is

$$d_k = \int_0^\infty \frac{(\lambda_c x)^k}{k!} e^{-\lambda_c x} t_{dt}(x) dx \tag{10.23}$$

Let υ_k denote the steady state probability that there are k packets in the downlink buffer immediately after a packet departure; the steady state values of these probabilities are

$$\upsilon_k = \upsilon_0 d_k + \sum_{j=1}^{k+1} \upsilon_j \, d_{k-j+1}, \qquad 0 \le k \le L_c - 2$$

$$\upsilon_{L-1} = \upsilon_0 \sum_{k=L_c-1}^{\infty} d_k + \sum_{j=1}^{L_c-1} \upsilon_j \sum_{k=L_c-j}^{\infty} d_k \tag{10.24}$$

The probability distribution of the node buffer length at the time of packet departure can be found by solving the system in a recursive manner. If we introduce the substitution

$$v'_k = \frac{v_k}{v_0},$$

as outlined by Takagi (1993), we obtain

$$v'_0 = 1$$

$$v'_{k+1} = \frac{1}{d_0} \left(v'_k - \sum_{j=1}^{k} v'_j d_{k-j+1} - d_k \right), \quad 0 \le k \le L_c - 2 \tag{10.25}$$

The third equation describes the devices with finite queues with

$$v_0 = \frac{1}{\displaystyle\sum_{k=0}^{L-1} v'_k} \tag{10.26}$$

However, new packets arrive at the downlink buffer at arbitrary time and we need to find the corresponding packet blocking probability. To that end, let us define the total offered load toward the downlink queue as $\rho_{dtot} = \lambda_c T'_{t,dtot}(1)$. Then, from the equality

$$(1 - P_d^B)\rho_{dtot} = 1 - (1 - P_d^B)v_0,$$

we obtain

$$P_d^B = 1 - \frac{1}{v_0 + \rho_{dtot}} \tag{10.27}$$

Then, the carried load from the downlink buffer becomes

$$\rho'_{dtot} = (1 - P_d^B)\rho_{dtot},$$

which is equal to the probability that the downlink queue is not empty.

10.1.4 Case 3: the SS bridge

The Markov points relevant for the bridge buffer include the time when the bridge switches to the sink cluster, which corresponds to the end of the bridge vacation in the source cluster, and the moments after each service period during the bridge residence in the sink cluster, as shown in Figure 10.5. It is unnecessary to consider the moments of every packet departure, as there are no packet arrivals while the bridge is in the sink cluster.

The probability distribution for the packet service time of the bridge in the sink cluster, assuming the bridge operates in the CSMA-CA access mode, may be described with the PGF of

$$T_{t,bri}(z) = \sum_{k=0}^{\infty} p_{t,bri}(k)z^k \tag{10.28}$$

Detailed derivation of this PGF will be presented in Section 10.7 at the end of this chapter. The PGFs for duration of two, three, etc. bridge packet service times are

$$T^n_{t,bri}(z) = \sum_{k=0}^{\infty} p_{nt,bri}(k) z^k \quad n = 2, 3, \ldots \tag{10.29}$$

where, as before, the required mass probabilities can be found by coefficient matching.

If the bridge is operating in the GTS access mode, the number of packets delivered to the sink cluster is constant in every superframe, as it is determined by the number of GTSs allocated to the bridge.

We will also need the probability distribution of the number of packets which the bridge can deliver in the active part of the sink superframe; the derivation is similar to the case of ordinary node in the source cluster presented in Section 10.1.2. The maximum number of packets that are served in one service period (i.e., during the active part of the sink cluster superframe) is finite and equal to the ratio of the active superframe size in sink cluster and minimum packet service time:

$$\widehat{B} = \left\lfloor \frac{BI - SD}{T'_{t,bri}(1)} \right\rfloor \tag{10.30}$$

Actually, the mean service time for packets sent to the sink by the bridge operating in the CSMA-CA access mode is longer than the packet service time for packets sent by ordinary nodes in the source cluster, since the bridge must compete for medium access against the ordinary nodes in the sink cluster. The mass probabilities for the number of bridge packet transmissions B during the active part of the superframe in the sink cluster is

$$pb(0) = \sum_{k=SD+1}^{\infty} p_{t,bri}(k)$$

$$pb(1) = (1 - pb(0)) \sum_{k=SD+1}^{\infty} p_{2t,bri}(k)$$

$$pb(2) = (1 - pb(0) - pb(1)) \sum_{k=SD+1}^{\infty} p_{3t,bri}(k) \tag{10.31}$$

$$\ldots$$

$$pb(\widehat{B}) = 1 - \sum_{i=0}^{\widehat{B}-1} pb(i)$$

Since packets can arrive to the bridge only when the bridge is present in the source cluster, and the bridge residence in the source cluster lasts for SD backoff periods, the probability of k packet arrivals to the bridge is

$$u_k = \frac{(\lambda_{bri} SD)^k}{k!} e^{-\lambda_{bri} SD} \tag{10.32}$$

Let q_k denote the probability that the bridge buffer contains k packets when it switches to the sink cluster, and let h_k stand for the probability that the bridge buffer contains k packets at the end of a service period. If B is the number of packets that can be transmitted in one service period, the state of the bridge buffer can be described by

$$h_0 = \sum_{B=1}^{L_{bri}} pb(B) \sum_{k=1}^{B} q_k$$

$$h_k = \sum_{B=1}^{L_{bri}-k} pb(B)q_{k+B}, \qquad\qquad 1 \le k \le L_{bri} - 1$$

$$q_k = q_0 u_k + pb(0) \sum_{j=1}^{k} q_j u_{k-j} + \sum_{j=0}^{k} h_j u_{k-j}, \qquad 0 \le k < L_{bri} - B \qquad (10.33)$$

$$q_k = q_0 u_k + pb(0) \sum_{j=1}^{k} q_j u_{k-j} + \sum_{j=0}^{L_{bri}-B} h_j u_{k-j}, \quad L_{bri} - B + 1 \le k < L_{bri}$$

$$q_{L_{bri}} = q_0 \sum_{l=L_{bri}}^{\infty} u_l + pb(0) \sum_{j=0}^{L_{bri}} q_j \sum_{l=L_{bri}-j}^{\infty} u_l + \sum_{j=0}^{L_{bri}-B} h_j \sum_{l=L_{bri}-j}^{\infty} u_l$$

The buffer occupancies at all Markov points should add up to one:

$$\sum_{k=0}^{L_{bri}} q_k + \sum_{k=0}^{L_{bri}-B} h_k = 1 \qquad\qquad (10.34)$$

By solving the system (10.33) together with the normalization Equation (10.34), we obtain the probability distribution of bridge buffer occupancy at the end of visit to the source cluster (end of vacation) and at the end of service period in the sink cluster.

The probability that the bridge buffer is full is obtained as

$$P_f^B = \frac{q_{Lbri}}{\displaystyle\sum_{i=0}^{L_{bri}} q_i} \qquad\qquad (10.35)$$

Under acknowledged transfer, this parameter is critical for the performance of the source cluster since packets blocked by the bridge have to be retransmitted until acknowledged.

10.2 Markov Chain Model for the SS Bridge

In order to arrive at the Markov chain model for the clusters interconnected with an SS bridge, let us note that different nodes in the network use the same slotted CSMA-CA medium access mechanism described in Section 2.3. This is the case with ordinary nodes in the source cluster, ordinary nodes in the sink cluster, and also with the SS bridge when present in the sink cluster, provided it uses the CSMA-CA access mode.

However, when the bridge is present in the source cluster, the CSMA-CA medium access mechanism will be used in two related, yet distinct operations: when the bridge sends data request packets to the coordinator, and when the coordinator sends the data packets to the bridge. In this case, the generic CSMA-CA access mechanism may be conveniently described with a generic Markov sub-chain or block, shown in Figure 10.6.

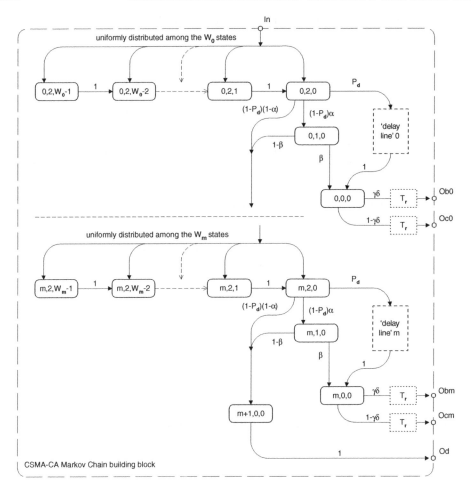

Figure 10.6 Markov sub-chain for the CSMA-CA access mechanism. Adapted from J. Mišić, 'Analytical model for slave-slave bridging in 802.15.4 beacon enabled networks,' submitted to *IEEE Trans. Vehicular Technology*, 2008.

In Figure 10.6, the two boxes represent the time taken by packet transmission (T_r) and the additional time needed when a packet is deferred to the next superframe because of insufficient time in the current one (the 'delay line' box); the probability of latter event is $P_d = \overline{D_d}/SD$. Both of them are shown in more detail in Figure 10.7.

We assume that the Markov sub-chain has a stationary distribution. At the boundaries of unit backoff periods, this distribution can be described by the process $\{i, c, k, d\}$, where

- $i \in (0 .. m)$ is the index of the current backoff attempt, where m is a constant with the default value of 4.

- $c \in (0, 1, 2)$ denotes the index of the current CCA.

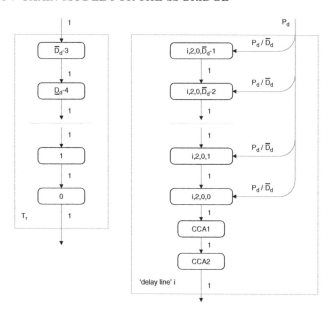

Figure 10.7 Delay lines for the Markov sub-chain block in Figure 10.6. Adapted from J. Mišić, 'Analytical model for slave-slave bridging in 802.15.4 beacon enabled networks,' submitted to *IEEE Trans. Vehicular Technology*, 2008.

- $k \in (0 .. W_i - 1)$ is the current value of the backoff counter during the random backoff countdown. The label $W_i = W_0 2^{min(i, 5-macMinBE)}$ denotes the size of the backoff window in i-th backoff attempt. The minimum window size is $W_0 = 2^{macMinBE}$; by default, $macMinBE = 3$.

- $d \in (0 .. \overline{D_d} - 1)$ denotes the index of the state within the delay line mentioned above; in order to reduce notational complexity, it will be shown only within the delay line and omitted elsewhere.

The high level discrete time Markov chains that describe the operation of the SS bridge in the source cluster under non-acknowledged and acknowledged transfer are presented in Figures 10.8 and 10.9, respectively. The basic building block on these figures is the Markov sub-chain for the CSMA-CA access, presented in Figure 10.6, which models all states related to the backoff procedure, CCAs and packet transmissions. The main difference between the non-acknowledged and acknowledged transfer is that, in the former case, unsuccessful downlink transmissions need not be repeated, whereas in the latter, a downlink transmission that has not been successful will be re-attempted from the very beginning (i.e., from the data request packet). In either case, the data request packet has to be acknowledged before the actual downlink transmission of the data packet.

Let us now analyze the operation of this model in more detail. As before, we will denote the PGF of the data packet length with $G_p(z) = z^k$, where the packet size is equal

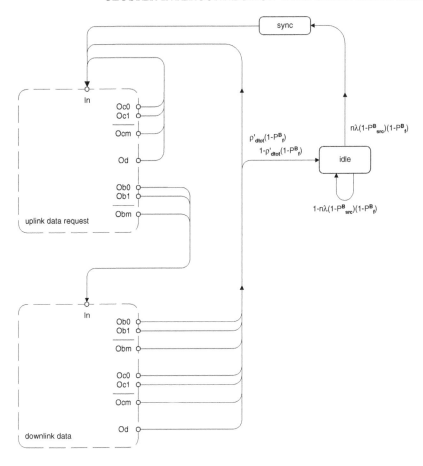

Figure 10.8 Markov chain for the SS bridge under non-acknowledged transfer. Adapted from J. Mišić, 'Analytical model for slave-slave bridging in 802.15.4 beacon enabled networks,' submitted to *IEEE Trans. Vehicular Technology*, 2008.

to k backoff periods. Data request packets have a length of two backoff periods, so the corresponding PGF is $G_r(z) = z^2$.

In the case of non-acknowledged transfer, the PGF for the total transmission time of a data packet case is $D_d(z) = z^2 G_p(z)$, where z^2 stands for two backoff periods that are needed to conduct the two CCAs. Under acknowledged transfer, the PGF for the total transmission time of a data packet will be denoted with $D_d(z) = z^2 G_p(z) t_{ack}(z) G_a(z)$, while its mean value is $\overline{D_d} = 2 + G'_p(1) + t'_{ack}(1) + G'_a(1)$. In this case, the PGF of the time interval between the packet transmission and subsequent acknowledgment is $t_{ack}(z) = z^2$, while $G_a(z) = z$ stands for the PGF of the acknowledgment packet duration.

Also, the PGF of the duration of the beacon frame is denoted as $B_{ea}(z) = z^2$.

The time needed to send a data request packet and receive the subsequent acknowledgment is $\overline{D_r} = 2 + G'_r(1) + t'_{ack}(1) + G'_a(1)$.

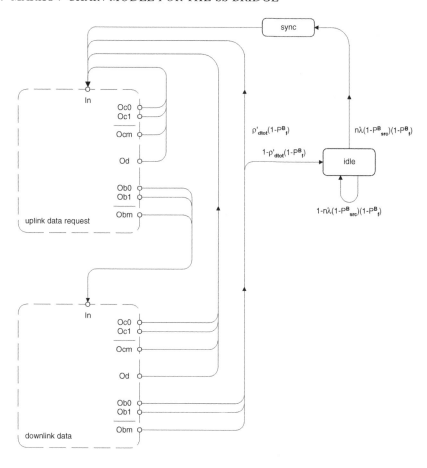

Figure 10.9 Markov chain for the SS bridge under acknowledged transfer. Adapted from J. Mišić, 'Analytical model for slave-slave bridging in 802.15.4 beacon enabled networks,' submitted to *IEEE Trans. Vehicular Technology*, 2008.

The probability that the medium is idle at the first and second CCA will be denoted with α and β, respectively; the probability that a given packet did not suffer a collision will be denoted with γ; finally, the probability that the given packet was not corrupted by noise will be denoted with δ. The last two probabilities have already been used in the discussions pertaining to the queueing model in Figure 10.4.

We note that different nodes in the network, i.e., the source cluster coordinator, the bridge, and the ordinary nodes in either cluster, will have different views of the medium, depending on the activity of other nodes in the cluster. Thus an ordinary node in the source cluster will experience the activity of the remaining $n - 1$ ordinary nodes, but also the activity of the bridge which generates both uplink and downlink traffic. The variables 'seen' from an ordinary node will be denoted with the subscript src/o. On the other hand, the source cluster coordinator and the bridge, when it resides in the source cluster, will experience the activity of the n ordinary nodes. The relevant variables 'seen' from these nodes will be denoted using the subscript src/b.

Similar distinction must be observed in the sink cluster, where an ordinary node will experience the activity of the remaining $n-1$ ordinary nodes and the bridge (but the latter will generate only uplink traffic), while the bridge will experience the activity of all n ordinary nodes. Where appropriate, the subscripts snk/o and snk/b will refer to the relevant variable as 'seen' from the viewpoint of an ordinary node and the bridge, respectively.

Let us denote the input probability to the basic MAC block as Prob(In), which means that the probability of arriving to the finishing state after the first backoff will have the value of $x_{0,2,0} = \text{Prob}(In)$. Let us also represent the state after the last unsuccessful backoff phase as $x_{m+1,0,0}$; this facilitates the presentation although it has no physical meaning. Using the transition probabilities indicated in Figure 10.6 and 10.7, we can derive the relationships between the state probabilities and solve the Markov chain. For brevity, we will omit l whenever it is zero, and introduce the auxiliary variables C_1, C_2, C_3, and C_4 using the equations

$$x_{0,1,0} = \text{Prob}(In)(1 - P_d)\alpha = \text{Prob}(In)C_1$$

$$x_{1,2,0} = \text{Prob}(In)(1 - P_d)(1 - \alpha\beta) = \text{Prob}(In)C_2$$

$$x_{0,0,0} = \text{Prob}(In)((1 - P_d)\alpha\beta + P_d) = \text{Prob}(In)C_3 \tag{10.36}$$

$$C_4 = \frac{1 - C_2^{m+1}}{1 - C_2}$$

Using these values, we obtain the equations that describe the Markov sub-chain:

$$x_{i,0,0} = \text{Prob}(In)C_3 C_2^i, \qquad\qquad\qquad i = 0 .. m$$

$$x_{i,2,k} = \text{Prob}(In)\frac{W_i - k}{W_i} \cdot C_2^i, \qquad i = 1 .. m; k = 0 .. W_i - 1$$

$$x_{i,1,0} = \text{Prob}(In)C_1 C_2^i \qquad\qquad\qquad i = 0 .. m$$

$$x_{0,2,k} = \text{Prob}(In)\frac{W_0 - k}{W_0} \qquad\qquad k = 1 .. W_0 - 1 \tag{10.37}$$

$$x_{m+1,0,0} = \text{Prob}(In)C_2^{m+1}$$

$$\sum_{l=0}^{\overline{D_d}-1} x_{i,2,0,l} = \frac{\text{Prob}(In)C_2^i P_d(\overline{D_d} - 1)}{2}$$

The sum of probabilities for sub-chain with transmission of uplink request (having superscript r) is

$$s^r = \sum_{i=0}^{m}\sum_{k=0}^{W_i-1} x_{i,2,k}^r + (\overline{D_r} - 2)\sum_{i=0}^{m} x_{i,0,0}^r + \sum_{i=0}^{m} x_{i,1,0}^r$$

$$+ x_{m+1,0,0}^r + \sum_{i=0}^{m}\sum_{l=0}^{\overline{D_r}-1} x_{i,2,0,l}^r \tag{10.38}$$

which can be simplified to

$$s^r = (\text{Prob}(In))^r \left(C_4 \left(C_3(\overline{D_r} - 2) + C_1 + \frac{P_d(\overline{D_r} - 1)}{2} \right) + \left(\sum_{i=0}^{m} \frac{C_2^i(W_i + 1)}{2} + C_2^{m+1} \right) \right)$$

(10.39)

By the same token, the sum of the states for the downlink transmission block is

$$s^d = (\text{Prob}(In))^d \left(C_4 \left(C_3(\overline{D_r} - 2) + C_1 + \frac{P_d(\overline{D_r} - 1)}{2} \right) + \left(\sum_{i=0}^{m} \frac{C_2^i(W_i + 1)}{2} + C_2^{m+1} \right) \right)$$

(10.40)

where the input probability to the downlink transmission block is

$$(\text{Prob}(In))^d = \gamma_{src/b} \delta \sum_{i=0}^{m} x_{i,0,0}^d = \gamma_{src/b} C_3 C_4 (\text{Prob}(In))^r$$

(10.41)

The synchronization block has a uniform probability distribution between 0 and $SD - 1$ backoff periods, with the average value of $SD/2$.

Under acknowledged transfer, the following equation holds:

$$\begin{aligned}
(\text{Prob}(In))^r = \quad & P_z n \lambda (1 - P_{src}^B)(1 - P_f^B) + C_3 C_4 (1 - \gamma_{src/b}\delta)(\text{Prob}(In))^r \\
& + (\text{Prob}(In))^r C_2^{(m+1)} + (\text{Prob}(In))^d C_3 C_4 (1 - \gamma_{src/b}\delta) \\
& + (\text{Prob}(In))^d C_3 C_4 \gamma_{src/b}\delta \rho'_{dtot}(1 - P_f^B)
\end{aligned}$$

(10.42)

from which the input probability to the request block may be obtained as

$$(\text{Prob}(In))^r = $$
$$\frac{P_z n \lambda (1 - P_{src}^B)(1 - P_f^B)}{1 - C_3 C_4 (1 - \gamma_{src/b}\delta) - C_2^{(m+1)} + (C_3 C_4)^2 \gamma_{src/b}(1 - \gamma_{src/b}\delta + \gamma_{src/b}\delta \rho'_{dtot}(1 - P_f^B))}$$

(10.43)

In the above expressions, $\text{Prob}(idle) = P_z$ denotes the probability of being in the idle state.

If downlink transmissions towards the bridge are not acknowledged, the following holds:

$$\begin{aligned}
(\text{Prob}(In))^r = \quad & P_z n \lambda (1 - P_{src}^B)(1 - P_f^B) + C_3 C_4 (1 - \gamma_{src/b}\delta)(\text{Prob}(In))^r \\
& + (\text{Prob}(In))^r C_2^{(m+1)} + (\text{Prob}(In))^d C_3 C_4 \rho'_{dtot}(1 - P_f^B)
\end{aligned}$$

(10.44)

which gives the input probability as

$$(\text{Prob}(In))^r = \frac{P_z n \lambda (1 - P_{src}^B)(1 - P_f^B)}{1 - C_3 C_4 (1 - \gamma_{src/b}\delta) - C_2^{(m+1)} + \rho'_{dtot}(1 - P_f^B)}$$

(10.45)

The sum of probabilities within the beacon synchronization line is

$$s^b = P_z n \lambda (1 - P_{src}^B)(1 - P_f^B) \sum_{i=0}^{SD-1} \frac{i}{SD} = \frac{SD}{2} \tag{10.46}$$

Then, the normalization condition for this Markov chain becomes

$$s^r + s^d + s^b + P_z = 1 \tag{10.47}$$

By solving Equation 10.47, the value of P_z is obtained as a function of different cluster parameters.

Furthermore, the probabilities to access the medium during the transmission of an uplink data request and the subsequent downlink data packet are

$$\begin{aligned} \tau_r &= C_3 C_4 (\text{Prob}(In))^r \\ \tau_d &= \gamma_{src/b}(C_3 C_4)^2 (\text{Prob}(In))^r \end{aligned} \tag{10.48}$$

while the probability that the coordinator is involved in downlink backoff and, thus, unable to accept any uplink data packet transmission is

$$P_c^B = s^d - \tau_d(\overline{D_d} - 2) \tag{10.49}$$

Finally, the packet arrival rate towards the bridge becomes

$$\lambda_{bri} = \tau_d \gamma_{src/b} \tag{10.50}$$

10.3   Markov Chain for Non-Bridge Nodes

As mentioned above, ordinary nodes in both the source and sink clusters, as well as the bridge node in the sink cluster, use the same CSMA-CA algorithm with active and inactive periods. The corresponding discrete-time Markov chain is shown in Figure 10.10, with boxes representing packet transmission time and additional delay incurred when a data packet is deferred to the next superframe presented in more detail in Figure 10.7.

In this Markov chain, the probability that a new packet arrives to the node which is in the idle state in the active and inactive part of the superframe are denoted as ϕ_a and ϕ_i, respectively; θ_0 is the probability that the node buffer is empty after a successful packet transmission. This chain is general in the sense that the transition probabilities are generally labeled; actual values have to be substituted in order to model a particular node in the network. For clarity, we will first solve the general model and then substitute the specific values to obtain the solutions for an ordinary node in the source and sink cluster, and for the bridge in the sink cluster.

We note that the first backoff phase in the Markov chain actually has two parts. The part which is connected to the idle state with the probability ϕ_i represents the situation when a new packet arrives to the empty buffer during the inactive portion of the superframe. In that case, the first backoff countdown will start immediately after the beacon and the value of backoff counter will be in the range $0 .. W_0 - 1$. Those states will be denoted as $x_{i,c,k}^s$. On the other hand, if the packet arrives to a node during the active portion of the

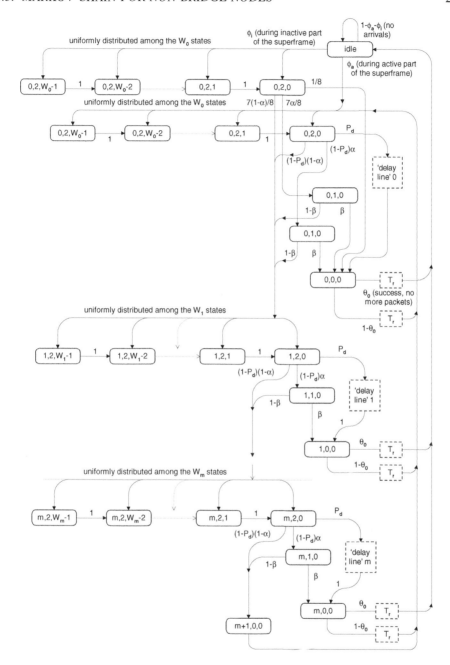

Figure 10.10 Markov chain model of the CSMA-CA algorithm used by ordinary nodes in the source and sink cluster, and by the bridge in the sink cluster. Adapted from J. Mišić, 'Analytical model for slave-slave bridging in 802.15.4 beacon enabled networks,' submitted to *IEEE Trans. Vehicular Technology*, 2008.

superframe, the backoff countdown will start at a random position within the superframe, and those states will be denoted as $x_{i,c,k}$. As these two cases will have different effect on the behavior of the medium, they have to be modeled separately.

From the Markov chain, we define the probability to access the medium as in uplink data transmission

$$\tau_u = \sum_{i=0}^{m} x_{i,0,0} \tag{10.51}$$

Then the probability of switching into the idle state is equal to $\tau_u \theta_0$. After setting the balance equation for the idle state, we obtain the probability of the node being in the idle state as

$$\text{Prob(idle)} = P_z = \tau_u \frac{\theta_0}{\phi_a + \phi_i} \tag{10.52}$$

Furthermore, if we consider the output from the idle state and set the balance equations for the first backoff phase after the idle state started during the inactive part of the superframe, we obtain that

$$x^s_{0,2,W_0-k} = k P_z \frac{\phi_i}{W_0}, \quad 1 \le k < W_0 \tag{10.53}$$

First backoff phase in the active superframe part is started after the packet arrival during the idle state, after packet transmission (regardless of the transmission success), or after last unsuccessful backoff phase. The state probabilities of the first backoff phase started in the active superframe part are represented as $x_{0,2,k}$, for $0 \le k < W_0$. The input probability for that set of states is

$$\begin{aligned} U_a &= P_z \phi_a + \tau(1 - \theta_0) + x_{m+1,0,0} \\ &= \tau_u \left(1 - \theta_0 \frac{\phi_i}{\phi_i + \phi_a}\right) + x_{m+1,0,0} \end{aligned} \tag{10.54}$$

By setting balance equations we obtain

$$x_{0,2,W_0-k} = k \frac{U_a}{W_0}, \quad 1 \le k < W_0 \tag{10.55}$$

The values of $x_{i,2,k}$ for subsequent backoff attempts, $i = 2 \dots m$, can be similarly obtained.

Using the transition probabilities indicated in Figures 10.10 and 10.7 we can derive the relationships between the state probabilities and solve the Markov chain. For brevity, we will omit index d whenever it is zero, and introduce the auxiliary variables, C^s_1, C^s_2, and C^s_3, defined via the following equations:

$$\begin{aligned} x_{0,1,0} &= x_{0,2,0}(1 - P_d)\alpha + x^s_{0,2,0}\frac{7\alpha}{8} \\ &= x_{0,2,0}C_1 + x^s_{0,2,0}C^s_1 \\ x_{1,2,0} &= x_{0,2,0}(1 - P_d)(1 - \alpha\beta) + x^s_{0,2,0}\frac{7}{8}(1 - \alpha\beta) \\ &= x_{0,2,0}C_2 + x^s_{0,2,0}C^s_2 \\ &= \tau\left(\theta_0\frac{\phi_i}{\phi_i + \phi_a}C^s_2 + \left(1 - \theta_0\frac{\phi_i}{\phi_i + \phi_a}\right)C_2\right) + C_2 x_{m+1,0,0} \\ x_{0,0,0} &= x_{0,2,0}((1 - P_d)\alpha\beta + P_d) + x^s_{0,2,0}(\frac{7}{8}\alpha\beta + \frac{1}{8}) \\ &= x_{0,2,0}C_3 + x^s_{0,2,0}C^s_3 \end{aligned} \tag{10.56}$$

From the expressions that describe the Markov chain we obtain

$$
\begin{aligned}
x_{i,1,0} &= C_1 C_2^{i-1} x_{1,2,0}, & i &= 1 .. m \\
x_{i,2,0} &= C_2^{i-1} x_{1,2,0}, & i &= 1 .. m \\
x_{i,0,0} &= C_3 C_2^{i-1} x_{1,2,0}, & i &= 1 .. m+1 \\
\sum_{d=0}^{\overline{D_d}-1} x_{i,2,0,d} &= x_{i,2,0} C_2^i \frac{\overline{D_d}-1}{2}
\end{aligned}
\tag{10.57}
$$

Of course, the sum of all probabilities in the Markov chain must be equal to one:

$$
\begin{aligned}
P_z + \sum_{k=0}^{W_0-1} x_{0,2,k}^s + x_{0,1,0}^s + \sum_{i=0}^{m} \sum_{k=0}^{W_i-1} x_{i,2,k} + \sum_{i=0}^{m} x_{i,0,0}(\overline{D_d}-2) \; + \\
+ \sum_{i=0}^{m} x_{i,1,0} + x_{m+1,0,0} + \sum_{i=0}^{m} \sum_{d=0}^{\overline{D_d}-1} x_{i,2,0,d} \; = 1
\end{aligned}
\tag{10.58}
$$

which has to be solved for the access probability τ_u. Note that the value for τ_u obtained in this manner is just the average value during the active portion of the superframe. However, access to the medium is prohibited in the first two and the last $\overline{D_d}-1$ backoff periods in the superframe. Therefore, it becomes necessary to identify the parts of the superframe where some types of access can occur and scale the access probabilities accordingly. To that end, let $SM = SD - \overline{D_d} + 1$ denote the duration of the part of the superframe where either CCAs or actual access can occur. As before, we will separately consider different time intervals where access can occur, and then combine them to obtain the final expression.

Considering the packets that arrive to an idle node during the inactive portion of the superframe, medium access is allowed from the third backoff period after the beacon frame, up to the $W_0 + 2$-th backoff period of the superframe. The solution of the Markov chain gives us $x_{0,2,0}^s$ as the probability that this access is possible over the entire active part of the superframe. Since the initial value for the random backoff countdown is chosen between 0 and $W_0 - 1$, the probability of access in the third backoff period after the beacon frame is

$$
\tau_{u,3,1} = x_{0,2,0}^s \frac{1}{W_0} \cdot \frac{SM-2}{W_0}
\tag{10.59}
$$

The scaling factor is the total number of backoff periods where any access is allowed, $SM-2$, divided by the number of backoff periods W_0 where access can actually take place.

The probability that access will occur in some other (fourth to $W_0 + 2$-th) backoff period after the beacon is

$$
\tau_{u,3,2} = x_{0,2,0}^s \frac{W_0-1}{W_0} \cdot \alpha\beta \frac{SM-2}{W_0-1}
\tag{10.60}
$$

The reason for separating $\tau_{u,3,1}$ from $\tau_{u,3,2}$ is that former one overlaps with transmissions delayed from the previous superframe due to insufficient time. The probability to access the medium in this case is $SM-2$ times higher than the value averaged over the whole superframe – since it can happen only within the third backoff period after the beacon. Therefore,

$$\tau_{u,1} = (SM - 2)\frac{P_d(\tau_u - x_{0,2,0}C_3^s)}{C_3} \tag{10.61}$$

while the probability to have non-delayed access is

$$\tau_{u,2} = \left(1 - \frac{P_d}{C_3}\right)(\tau_u - x_{0,2,0}C_3^s) \tag{10.62}$$

We are now equipped to derive the relevant probability distributions for nodes in the specific locations in the network.

10.3.1 An ordinary node in the source cluster

Non-acknowledged transfer. The idle state of the Markov chain is reached when the node buffer is empty after a packet transmission, regardless of whether the packet suffered a collision or it was rejected by the bridge. As derived in Equation (10.21) above, $\theta_{0,src} = \pi_{0,src}$. Since the packet arrival rate to a node in the source cluster is small, the probability of zero Poisson arrivals during the unit backoff period can be approximated with a Taylor series, i.e., $e^{-\lambda} \approx 1 - \lambda$. The probability of non-zero packet arrivals, i.e., the probability of leaving the idle state, is λ, which further gives

$$\begin{aligned} \phi_{i,src} &= P_{sync}\lambda \\ \phi_{a,src} &= (1 - P_{sync})\lambda \end{aligned} \tag{10.63}$$

where $P_{sync} = 1 - 2^{SO-BO}$ denotes the conditional probability that the packet arrives at an idle node during the inactive portion of the superframe.

Acknowledged transfer. In this case, the idle state is reached only if the node buffer is empty after a packet transmission and the transmission was successful and the packet was accepted by the bridge. Therefore,

$$\theta_{0,src} = \gamma_{src/o}\delta_{src}(1 - P_f^B)(1 - P_c^B)(1 - P_d^B)\pi_{0,src} \tag{10.64}$$

Using the same Taylor series approximation, we obtain

$$\begin{aligned} \phi_{i,src} &= P_{sync}\lambda \\ \phi_{a,src} &= (1 - P_{sync})\lambda \end{aligned} \tag{10.65}$$

10.3.2 CSMA-CA bridge in the sink cluster

Let us recall that the probability that bridge buffer becomes empty during active part of the sink cluster superframe is h_0, as derived in Equation (10.33).

Non-acknowledged transfer. The probability of entering the idle state of the Markov chain is

$$\theta_{0,bri} = \frac{h_0}{\sum\limits_{i=0}^{L_{bri}-B} h_i} \tag{10.66}$$

The packets can arrive at the bridge only during its residence in the source cluster; since the bridge does not send any data packets at that time, this residence period may be considered as its inactive period. The packet arrival rate at the bridge queue during this period is

$$\lambda_{bri} = n\lambda(1 - P^B_{src})\gamma_{src/o}\delta_{src}(1 - P^B_c)(1 - P^B_f)(1 - P^B_d) \qquad (10.67)$$

from which we obtain

$$\begin{aligned}\phi_{a,bri} &= 0 \\ \phi_{i,bri} &= \lambda_{bri}\end{aligned} \qquad (10.68)$$

Acknowledged transfer. In this case, the probability of entering the idle state of the Markov chain is equal to

$$\theta_{0,bri} = \frac{h_0}{\sum\limits_{i=0}^{L_{bri}-B} h_i}\gamma_{snk/b}\delta_{bri} \qquad (10.69)$$

where δ_{bri} denotes the probability that a packet transmitted by the bridge is not corrupted by noise. The packet arrival rate at the bridge queue is

$$\lambda_{bri} = \frac{\tau_d \gamma_{src/b}}{1 - P^B_{bri}} \qquad (10.70)$$

from which we obtain

$$\begin{aligned}\phi_{i,bri} &= \lambda_{bri} \\ \phi_{a,bri} &= 0\end{aligned} \qquad (10.71)$$

10.3.3 An ordinary node in the sink cluster

Non-acknowledged transfer. In this case, the idle state of the Markov chain is reached when the node buffer becomes empty, regardless of the transmission success, and

$$\theta_{0,snk} = \pi_{0,snk} \qquad (10.72)$$

as well as

$$\begin{aligned}\phi_{i,snk} &= P_{sync}\lambda \\ \phi_{a,snk} &= (1 - P_{sync})\lambda\end{aligned} \qquad (10.73)$$

where P_{sync}, as above, denotes the conditional probability that the packet arrives at an idle node during the inactive portion of the superframe.

Acknowledged transfer. In this case, an ordinary node in the sink cluster can reach the idle state only if its buffer remains empty and the transmission was successful, so that

$$\theta_{0,snk} = \gamma_{snk/o}\delta_{snk}\pi_{0,snk} \qquad (10.74)$$

Note that δ_{snk} should be equal to δ_{bri}, as the probability that the packet is corrupted by noise does not depend on the actual node that sent it.

The idle state is left upon the arrival of a new packet, i.e., with

$$\begin{aligned}\phi_{i,snk} &= P_{sync}\lambda \\ \phi_{a,snk} &= (1 - P_{sync})\lambda\end{aligned} \qquad (10.75)$$

10.3.4   On the bridge access mode in the sink cluster

When the bridge is in the sink cluster, it may operate in the CSMA-CA access mode, in which it competes against all ordinary nodes in that cluster, or in the GTS mode, in which it enjoys contention-free access through GTSs allocated by the sink. At the same time, ordinary nodes in the sink cluster operate in the CSMA-CA mode regardless of the bridge access mode, and the Markov chain for both access modes is virtually the same. As a result, when the bridge operates in the CSMA-CA access mode, the ordinary nodes experience increased contention and their performance suffers. When the bridge operates in the GTS mode, its traffic is entirely decoupled from that generated by the ordinary nodes and vice versa; however, the duration of the CAP in the sink cluster superframe is shortened due to the presence of the CFP, which increases contention among ordinary nodes. In this respect, the performance of the sink cluster is similar to that in the network with an MS bridge, analyzed in detail in Chapter 8.

10.4   Performance Evaluation

In the next two sections we present performance results for the network with one source and one sink cluster, each of which has n ordinary nodes, interconnected with an SS bridge, as shown in Figure 10.2. We assume that the network uses the 2450 MHz PHY option, in which case the raw data rate is 250kbps, *aUnitBackoffPeriod* has 10 bytes, and *aBaseSlotDuration* has 30 bytes. The superframe size in both clusters was controlled with $SO = 0$, $BO = 1$, but no attempt has been made to synchronize the clusters. As the value of *aNumSuperframeSlots* is 16, the *aBaseSuperframeDuration* is exactly 480 bytes for both clusters. The minimum and maximum values of the backoff exponent, *macMinBE* and *aMaxBE*, were set to three and five, respectively, while the maximum number of backoff attempts was five. Other parameter values were set as shown in Table 10.1. When present in the source cluster, the bridge operated using CSMA-CA access mode; in the sink cluster, the bridge could use either CSMA-CA or GTS access mode, for reasons explained in Section 10.1. In the latter mode, the bridge was allocated a single GTS with three backoff periods, which suffices for a single packet of the chosen size.

Table 10.1   Parameters used to model the behavior of the network. Adapted from J. Mišić, 'Analytical model for slave-slave bridging in 802.15.4 beacon enabled networks,' submitted to *IEEE Trans. Vehicular Technology*, 2008.

number of nodes in each cluster, n	5–30
packet arrival rate, λ	30–240 packets per minute per node
packet size	3 backoff periods
buffer size at an ordinary node, L	2 packets
buffer size at the coordinator, L_c	20 packets
buffer size at the bridge, L_{bri}	6 packets
superframe size, SD	480 bytes
maximum number of re-transmissions, a	5
bit error rate, BER	10^{-4}

Each component of the model (i.e., source cluster, sink cluster and bridge) in different operating modes was described by dedicated set of equations, as follows.

Regardless of the bridge access mode in the sink cluster,

- behavior of ordinary nodes in the source cluster is modeled with the system formed by Equations (10.20), (10.21), (10.22), (10.58), (10.81), (10.90), and (10.96);

- behavior of source cluster coordinator/bridge is modeled with Equations (10.26) and (10.27);

- behavior of the bridge in the source cluster may be modeled using Equations (10.38), (10.40), (10.47), (10.48), (10.49).

When the bridge uses the CSMA-CA access mode in the sink cluster,

- behavior of ordinary nodes in the sink cluster is modeled using Equations (10.20), (10.21), (10.22), (10.58), (10.81), (10.94), and (10.100);

- behavior of the bridge in the sink cluster is modeled through Equations (10.5), (10.31), (10.33), (10.34), (10.35), (10.58), (10.81), (10.93), and (10.99).

Alternatively, when the bridge uses the GTS access mode in the sink cluster,

- behavior of ordinary nodes in the sink cluster is modeled through Equations (10.20), (10.21), (10.22), (10.58), (10.81), (10.95), and (10.101), where SD_{gts} and SM_{gts} are used;

- behavior of bridge in the sink cluster is modeled through Equations (10.5), (10.33), (10.34) and (10.35), since the number of packets delivered to the sink cluster is constant in every service period.

The resulting systems of equations were numerically solved under varying network size and packet arrival rates. Analytical processing was done using Maple 10 from Waterloo Maple, Inc. (2005). The results can be summarized as follows.

10.5   To Acknowledge or Not To Acknowledge: The CSMA-CA Bridge

Figure 10.11 presents access probabilities for different nodes in the source cluster, assuming that the bridge uses the CSMA-CA access mode in the sink cluster, for both acknowledged and non-acknowledged transfer. Although the vertical scales differ, there is not much difference between the two types of transfer when the cluster size is small. However, the discrepancy increases with traffic load. At higher packet arrival rate and/or larger cluster size, access probability under acknowledged traffic begins to increase due to the increased retransmission rate caused by collisions. Beyond a certain point, however, the access probability for data request packets and downlink packets show a sharp drop. Namely, the increased collision rate, together with blocking of data requests at the coordinator, lead to increased packet blocking at all points. As more packets are blocked, more retransmissions are needed, and the overall service rate deteriorates rapidly – in effect, the source cluster saturates.

Under non-acknowledged transfer, this cumulative effect does not occur, at least not in the range of values of n and λ shown in the diagrams in the right column of Figure 10.11. Since data packets are not acknowledged, packet loss does not lead to retransmission and the network still operates in unsaturated regime.

Similar behavior may be observed in access probabilities in the sink cluster, as shown in Figure 10.12. Under acknowledged transfer, the absence of downlink traffic allows the sink cluster to operate in non-saturated regime in a much wider range of packet arrival rates and/or cluster sizes than the source cluster. Under non-acknowledged transfer, the limits of non-saturated operation cannot be detected within the range of values shown.

These conclusions are further confirmed by the diagrams of aggregate throughput at various points in the network, which are shown in Figures 10.13 and 10.14 for acknowledged and non-acknowledged transfer, respectively. In both cases, there is a noticeable difference between the traffic volume sent to the nodes in the source cluster (i.e., offered load) and the traffic volume actually admitted by the bridge. This difference quickly increases with cluster size and packet arrival rate due to packet loss at different points along the way. In both cases, loss is caused by collisions and noise; note that the route from an ordinary node in the source cluster to the network sink includes three hops when an SS bridge is used, as opposed to only two when an MS bridge is used (Chapter 8). Under non-acknowledged transfer, additional loss can be attributed to packet blocking. The reader will recall that Equation 10.5 includes the effects of no less than four different mechanisms of blocking: data packet blocking may occur at ordinary nodes, at the co-ordinator, or at the bridge, while data request packets may experience blocking at the coordinator.

Another interesting observation can be made when comparing local and remote traffic in the sink cluster under non-acknowledged transfer. Namely, the volume of local traffic, i.e., the traffic generated by the ordinary nodes, that reaches the sink, Figure 10.13(e), is about 50% higher than the corresponding volume of the traffic originating in the source cluster, Figure 10.13(d). This should come as no surprise since the local traffic experiences only collisions from the bridge traffic, while the remote traffic suffers collisions in both clusters as well as blocking, as noted above.

Under acknowledged transfer, throughput is shown only before the ordinary nodes in the source cluster, Figure 10.14(a), and at the network sink, Figure 10.14(b), since the use of acknowledgments and up to a attempts at re-transmission significantly reduce or even eliminate packet losses. However, increase in traffic volume caused by packet re-transmissions and (to a lesser extent) by acknowledgment packets, leads to an abrupt onset of saturation in the source cluster, as discussed above in the context of access probabilities.

The effects of saturation are clear in both Figure 10.14(b) and Figure 10.14(c); however, the region in which the sink cluster operates in non-saturated regime is much wider than the corresponding region in the source cluster. Furthermore, Figures 10.14(c) and 10.14(b) show a clear difference between the volumes of local and remote traffic received by the sink, although it is much less conspicuous than in the case of non-acknowledged transfer because of saturation.

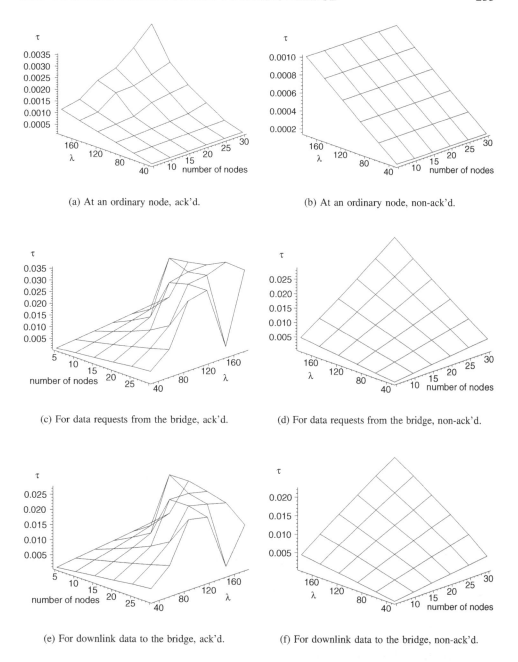

(a) At an ordinary node, ack'd.

(b) At an ordinary node, non-ack'd.

(c) For data requests from the bridge, ack'd.

(d) For data requests from the bridge, non-ack'd.

(e) For downlink data to the bridge, ack'd.

(f) For downlink data to the bridge, non-ack'd.

Figure 10.11 Access probabilities in the source cluster, CSMA-CA bridge. Acknowledged transfer on the left, non-acknowledged transfer on the right. Adapted from J. Mišić, 'Analytical model for slave-slave bridging in 802.15.4 beacon enabled networks,' submitted to *IEEE Trans. Vehicular Technology*, 2008.

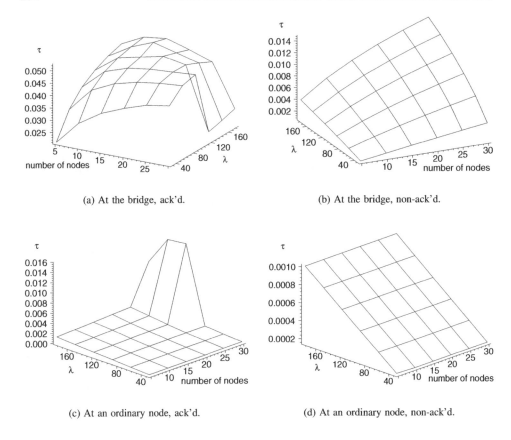

(a) At the bridge, ack'd.   (b) At the bridge, non-ack'd.

(c) At an ordinary node, ack'd.   (d) At an ordinary node, non-ack'd.

Figure 10.12 Access probabilities in the sink cluster, CSMA-CA bridge. Acknowledged transfer on the left, non-acknowledged transfer on the right. Adapted from J. Mišić, 'Analytical model for slave-slave bridging in 802.15.4 beacon enabled networks,' submitted to *IEEE Trans. Vehicular Technology*, 2008.

10.6 Thou Shalt Not Acknowledge: The GTS Bridge

The use of acknowledged transfer results in numerous retransmissions which increase the traffic volume and, by extension, limit the operating range. Since this increase is mainly confined to the source cluster, there is no reason to doubt that the network with a GTS bridge will not suffer from similar performance problems. At the same time, the losses in the sink cluster will be reduced due to the decoupling of bridge traffic from the local one. On account of this, we present performance results for non-acknowledged GTS access only.

We begin with access probabilities for different nodes in the network, which are shown in Figure 10.15; their values appear rather similar to those obtained, under non-acknowledged transfer, in the network with the CSMA-CA bridge (right-hand columns of Figures 10.11 and 10.12).

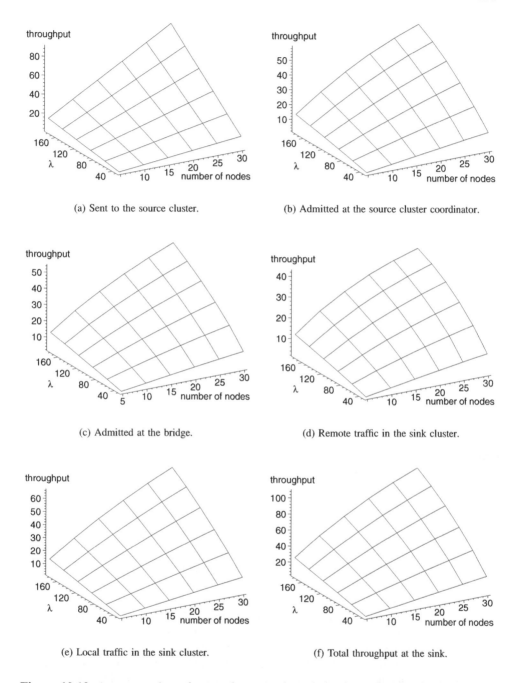

(a) Sent to the source cluster.

(b) Admitted at the source cluster coordinator.

(c) Admitted at the bridge.

(d) Remote traffic in the sink cluster.

(e) Local traffic in the sink cluster.

(f) Total throughput at the sink.

Figure 10.13 Aggregate throughput under non-acknowledged transfer, CSMA-CA bridge. Adapted from J. Mišić, 'Analytical model for slave-slave bridging in 802.15.4 beacon enabled networks,' submitted to *IEEE Trans. Vehicular Technology*, 2008.

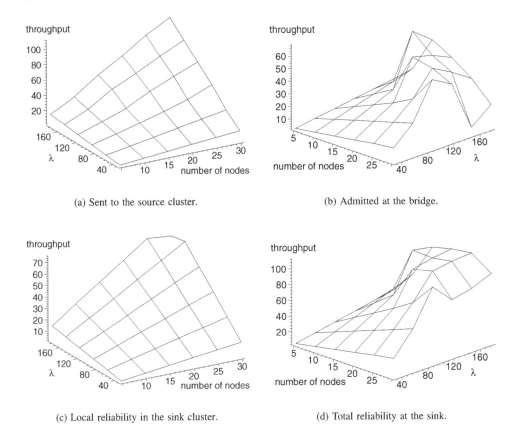

(a) Sent to the source cluster.

(b) Admitted at the bridge.

(c) Local reliability in the sink cluster.

(d) Total reliability at the sink.

Figure 10.14   Aggregate throughput under acknowledged transfer, CSMA-CA bridge. Adapted from J. Mišić, 'Analytical model for slave-slave bridging in 802.15.4 beacon enabled networks,' submitted to *IEEE Trans. Vehicular Technology*, 2008.

The values of throughput at different points in the network are shown in Figure 10.16. An interesting point is that the throughput admitted by the GTS bridge, Figure 10.16(c), begins to flatten under high traffic load, while the corresponding throughput admitted by the CSMA-CA bridge, Figure 10.13(c), continues to increase. As could be expected, flattening also affects the throughput of remote traffic in the sink cluster, shown in Figure 10.16(d), unlike its CSMA-CA bridge counterpart, Figure 10.13(d). Flattening results from the hard bandwidth limitation imposed by the fixed size of the GTS allocated to the bridge in the sink cluster, regardless of the amount of traffic that the bridge might receive from the source cluster coordinator. On the other hand, the CSMA-CA bridge competes for medium access with ordinary nodes in the sink cluster; consequently, its bandwidth limit is soft and, if it receives more traffic from the sink cluster, it can still manage to deliver most of it to the network sink. The same observation holds for the throughput of local traffic in the sink cluster, which shows no sign of saturation within the range of n and λ depicted in Figure 10.16(e).

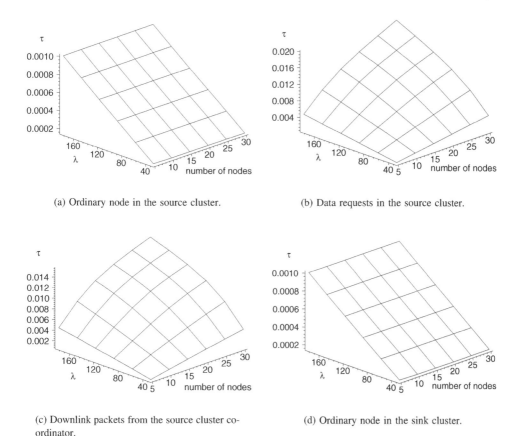

(a) Ordinary node in the source cluster.

(b) Data requests in the source cluster.

(c) Downlink packets from the source cluster co-ordinator.

(d) Ordinary node in the sink cluster.

Figure 10.15 Access probabilities under non-acknowledged transfer, GTS bridge. Adapted from J. Mišić, 'Analytical model for slave-slave bridging in 802.15.4 beacon enabled networks,' submitted to *IEEE Trans. Vehicular Technology*, 2008.

The GTS bandwidth limitation can be extended by allocating two or more GTSs to the bridge; however, our analysis of the operation of the sink cluster with two MS bridges in Section 8.5 indicates that the increase in the duration of the CFP incurs the risk of affecting the local traffic in the sink cluster.

Finally, packet loss probability for transmissions in the sink cluster, where it is due to collision and noise corruption, and for end-to-end transmissions is shown in Figure 10.17. As expected, the network with the GTS bridge outperforms the one with the CSMA-CA bridge with respect to packet losses in the sink cluster; the advantage is about 20% in the entire range of values for n and λ shown in the diagrams. End-to-end packet loss is also lower when the bridge uses GTS access mode, up to the traffic volume where the GTS bandwidth limitation begins to show; at the high end of the range of independent variables in the diagram, the CSMA-CA bridge offers about 15% lower loss than the GTS bridge.

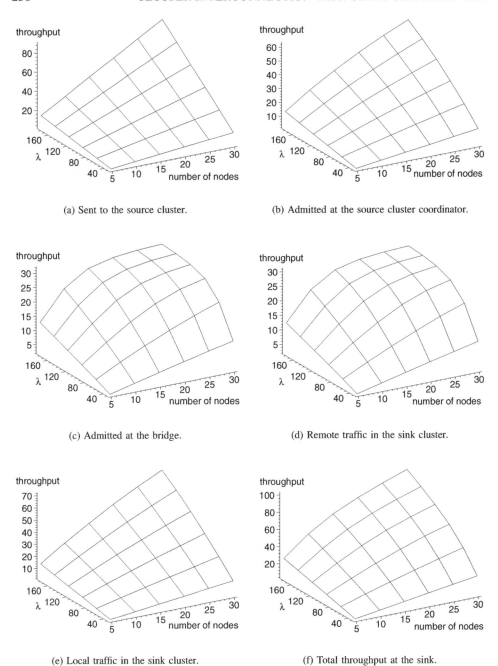

(a) Sent to the source cluster.

(b) Admitted at the source cluster coordinator.

(c) Admitted at the bridge.

(d) Remote traffic in the sink cluster.

(e) Local traffic in the sink cluster.

(f) Total throughput at the sink.

Figure 10.16 Aggregate throughput under non-acknowledged transfer, GTS bridge. Adapted from J. Mišić, 'Analytical model for slave-slave bridging in 802.15.4 beacon enabled networks,' submitted to *IEEE Trans. Vehicular Technology*, 2008.

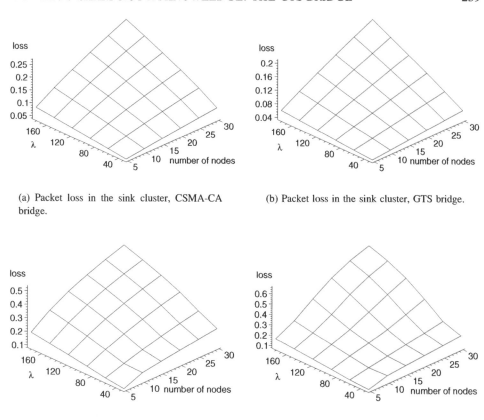

(a) Packet loss in the sink cluster, CSMA-CA bridge.

(b) Packet loss in the sink cluster, GTS bridge.

(c) End-to-end loss, CSMA-CA bridge.

(d) End-to-end loss, GTS bridge.

Figure 10.17 Probability of packet loss: a comparison of CSMA-CA and GTS bridge access modes. Adapted from J. Mišić, 'Analytical model for slave-slave bridging in 802.15.4 beacon enabled networks,' submitted to *IEEE Trans. Vehicular Technology*, 2008.

The results presented above allow the following conclusions to be made:

- In general, non-acknowledged transfer is preferred to the acknowledged one, since it improves the throughput and delays the onset of saturation. Furthermore, it offers graceful degradation since there is no abrupt increase due to increased retransmission rate. On the other hand, the use of acknowledged transfer results in reliable packet transmission. The decision on whether to use acknowledgments or not is, therefore, mostly dependent on the requirements of the application(s) executing in the network.

- Local packet losses in the sink cluster are almost independent of the activity of the bridge, except in the case of the CSMA-CA bridge, under acknowledged transfer, and at high traffic load. Reducing the number of ordinary nodes in the sink cluster would probably eliminate this source of inefficiency.

- Under small cluster sizes and low load in all clusters, the most efficient interconnection strategy is to use CSMA-CA bridges. Under moderate to high loads and/or large cluster sizes, the use of GTS bridges will allow higher throughput and widen the operating range without saturation.

In comparison with the topology that uses MS bridge (or bridges), the network with a SS bridge offers somewhat lower performance, mainly because of the increased number of hops (three, for the SS bridge, vs. two, for the MS bridge) that the non-local traffic has to go through in order to be delivered to the network sink. The extra hop is the downlink connection from a source cluster coordinator to the bridge; since downlink transfers are more complex than the uplink ones (Section 2.5), this hop is the most critical one with respect to performance.

However, the SS bridge could use a dedicated GTS to receive data from the source cluster, in which case two CSMA-CA transmissions would be replaced with a single contention-free transmission; as a result, the unpredictable delay and blocking in the downlink hop would be virtually eliminated.

Moreover, the use of the SS bridge mode allows the network to be set with two or more bridges per cluster interconnection, which might allow for some load balancing. A SS bridge is also able to visit more than two clusters, which may offer some advantages with respect to routing in applications that require a multi-hop topology.

10.7 Modeling the Transmission Medium and Packet Service Times

In this section, which is effectively an appendix to the current chapter, we model the behavior of the transmission medium and packet service time for the network which consists of one source cluster and one sink cluster, each of which contains exactly n ordinary nodes; an additional node acts as the SS bridge and switches periodically between the clusters in order to deliver the packets from the source cluster to the sink.

Our analysis will make use of the following. From the viewpoint of a node in the source cluster, at any moment q stations out of $n - 1$ have no packets deferred to the next superframe due to insufficient time in the current one, while $n - 1 - q$ nodes have such packets. The numbers q and $n - 1 - q$ follow a binomial distribution with the probability

$$P_q = \binom{n-1}{q}(1 - P_d)^q P_d^{n-1-q} \tag{10.76}$$

Within the group of nodes without a deferred transmission, exactly r nodes have received packets during the inactive part of the superframe; the probability of this happening is

$$P_r = \binom{q}{r}(P_{0,src}P_{sync}\lambda(1 - P_{src}^B))^r(1 - P_{0,src}P_{sync}\lambda(1 - P_{src}^B))^{q-r} \tag{10.77}$$

where $P_{0,src}$ denotes the probability that the node buffer is empty at an arbitrary time; this probability was derived in Equation 10.22. Note also that appropriate values of $\overline{D_d}$, which are calculated in Section 10.2, have to be used for non-acknowledged and acknowledged transfer respectively. Analogous expressions hold in the sink cluster, provided the subscript *src* is replaced with *snk*.

10.7.1 Probability of success at the first CCA

The first CCA succeeds only if no other node is currently transmitting a packet or receiving a previously transmitted packet. Note that an SS bridge is present in both clusters, albeit at different time periods; moreover, both uplink and downlink transmissions are allowed in the source cluster because of the bridge.

In order to calculate α_{src}, the probability that the medium is idle on the first CCA test, we have to find the mean number of busy backoff periods within the superframe. This number will be divided into the total number of backoff periods in the superframe in which the first CCA can occur. Note that the first CCA will not take place if the remaining time in the superframe is insufficient to complete the transaction, which amounts to $SM = SD - \overline{D_d} + 1$ backoff periods. We will evaluate views of the medium separately for ordinary nodes and for the bridge.

An ordinary node in the source cluster. An ordinary node must compete with the remaining $n - 1$ ordinary nodes and with the bridge which can have uplink request access or downlink data access. Then, we have the following component probabilities.

The probability that any packet transmission will take place at the very beginning of the superframe is

$$n_{1,src}^{(o)} = 1 - (1 - \tau_{1,src}^{(u)})^{n-1-q}(1 - \tau_{2,src}^{(u)})^{q-r}(1 - \tau_{3,1}^{(u)})^r(1 - \tau_r - \tau_d) \qquad (10.78)$$

and the number of busy backoff periods due to these transmissions is $n_{1,src}(G_p'(1) + G_a'(1))$ and $n_{1,src}G_p'(1)$ for acknowledged and non-acknowledged transfer, respectively.

The probability of a transmission attempt after the third backoff period within the period of $\overline{D_d}$ backoff periods is

$$n_{2,src}^{(o)} = 1 - ((1 - \tau_{2,src}^{(u)})^{q-r}(1 - \tau_{3,2,src}^{(u)})^r(1 - \tau_r - \tau_d))^{\overline{D_d}} \qquad (10.79)$$

The probability of a transmission in the remaining part of the superframe within $\overline{D_d}$ backoff periods is

$$n_{3,src}^{(o)} = 1 - ((1 - \tau_{2,src}^{(u)}(1 - \tau_r - \tau_d))^q)^{\overline{D_d}} \qquad (10.80)$$

The occupancy of the medium after the first transmission time can be found by dividing the superframe into chunks of $\overline{D_d}$ backoff periods and calculating the probability of transmission within each chunk.

The total number of unit backoff periods in which the first CCA can occur is $SM = SD - \overline{D_d} + 1$. Then, the probability that the medium is idle at the first CCA is

$$\alpha_{src/o} = \sum_{q=0}^{n-1}\sum_{r=0}^{q} P_q P_r \cdot$$

$$\cdot \left(1 - \left(\frac{n_{1,src}^{(o)}\overline{D_d}}{SM} + \frac{n_{2,src}^{(o)}\overline{D_d}}{SM} + \frac{n_{3,src}^{(o)}(SD - 3\overline{D_d} + 1)}{SM}\right)\frac{G_p'(1) + G_a'(1)}{\overline{D_d}}\right) \qquad (10.81)$$

in case acknowledged transfer is used; for non-acknowledged transfer, the expression is the same except for the term $G_a'(1)$ which is absent.

Bridge in the source cluster. The bridge must compete for medium access against all of the n ordinary nodes in the source cluster. As before, we begin by modeling the probabilities of packet transmissions at different parts of the superframe. The probability that any packet transmission will take place at the beginning of the superframe is

$$n_{1,src}^{(b)} = 1 - (1 - \tau_{1,src}^{(u)})^{n-q}(1 - \tau_{2,src}^{(u)})^{q-r}(1 - \tau_{3,1}^{(u)})^r \qquad (10.82)$$

and the number of busy backoff periods due to these transmissions is $n_{1,src}(G_p'(1) + G_a'(1))$ for acknowledged and $n_{1,src}G_p'(1)$ for non-acknowledged transfers.

The probability of a transmission attempt after the third backoff period within the period of $\overline{D_d}$ backoff periods is

$$n_{2,src}^{(b)} = 1 - ((1 - \tau_{2,src}^{(u)})^{q-r}(1 - \tau_{3,2,src}^{(u)})^r)^{\overline{D_d}} \qquad (10.83)$$

Finally, the probability of a transmission in the remaining part of the superframe within the period of $\overline{D_d}$ backoff periods is

$$n_{3,src}^{(b)} = 1 - ((1 - \tau_{2,src}^{(u)})^q)^{\overline{D_d}} \qquad (10.84)$$

The occupancy of the medium after the first transmission time can be found by dividing the superframe into chunks of $\overline{D_d}$ backoff periods and calculating the probability of transmission within each chunk.

On the other hand, the total number of backoff periods in which the first CCA can occur is $SM = SD - \overline{D_d} + 1$. Then, the probability that the medium is idle at the first CCA can be obtained as

$$\alpha_{src/b} = \sum_{q=0}^{n}\sum_{r=0}^{q} P_q P_r \cdot$$
$$\cdot \left(1 - \left(\frac{n_{1,src}^{(b)}\overline{D_d}}{SM} + \frac{n_{2,src}^{(b)}\overline{D_d}}{SM} + \frac{n_{3,src}^{(b)}(SD - 3\overline{D_d} + 1)}{SM}\right)\frac{G_p'(1) + G_a'(1)}{\overline{D_d}}\right) \qquad (10.85)$$

in case of acknowledged transfer; the expressions for non-acknowledged transfer is the same, except that the term $G_a'(1)$ is not present.

The average probability of success at first CCA in the source cluster has the value:

$$\alpha_{src} = \frac{n\tau_{u,src}}{n\tau_{u,src} + \tau_r + \tau_d}\alpha_{src/o} + \frac{\tau_d + \tau_r}{n\tau_{u,src} + \tau_r + \tau_d}\alpha_{src/b} \qquad (10.86)$$

CSMA-CA bridge in the sink cluster. When the bridge uses the CSMA-CA access mode, we need to calculate $\alpha_{snk/b}$, the probability that the first CCA performed by the bridge will be successful. The derivation is similar to the one conducted for the source cluster, except that the arguments of summation have to be calculated using the access probabilities from the sink cluster. Therefore,

$$\begin{aligned}
n_{1,bri} &= 1 - (1 - \tau_{u,1,snk})^{n-q}(1 - \tau_{u,2,snk})^{q-r}(1 - \tau_{u,3,1,snk})^r \\
n_{2,bri} &= 1 - ((1 - \tau_{u,2,snk})^{q-r}(1 - \tau_{u,3,2,snk})^r)^{\overline{D_d}} \\
n_{3,bri} &= 1 - ((1 - \tau_{u,2,snk})^q)^{\overline{D_d}}
\end{aligned} \qquad (10.87)$$

The value of α_{bri} can be derived using Equation (10.81), provided summation is done over all n ordinary nodes, instead of just $n - 1$.

An ordinary node in the sink cluster. In this case, node competes with the bridge and $n - 1$ other nodes. Therefore, the components which contribute to the calculation of $\alpha_{snk/o}$ are

$$
\begin{aligned}
n_{1,snk} =\ & 1 - (1 - \tau_{u,1,snk})^{n-1-q}(1 - \tau_{u,2,snk})^{q-r} \cdot \\
& \cdot (1 - \tau_{u,3,1,snk})^{r}(1 - \tau_{u,1,bri} - \tau_{u,2,bri} - \tau_{u,3,1,bri}) \\
n_{2,snk} =\ & 1 - ((1 - \tau_{u,2,snk})^{q-r}(1 - \tau_{u,3,2,snk})^{r}(1 - \tau_{u,2,bri} - \tau_{u,3,2,bri}))^{\overline{D_d}} \\
n_{3,snk} =\ & 1 - ((1 - \tau_{u,2,snk})^{q}(1 - \tau_{u,2,bri}))^{\overline{D_d}}
\end{aligned}
\tag{10.88}
$$

The value of $\alpha_{snk/o}$ can be derived using Equation (10.81).

GTS bridge in the sink cluster. In this case the bridge need not do any CCA checks, but the ordinary nodes observe the following probabilities:

$$
\begin{aligned}
n_{1,snk} =\ & 1 - (1 - \tau_{u,1,snk})^{n-1-q}(1 - \tau_{u,2,snk})^{q-r}(1 - \tau_{u,3,1,snk})^{r} \\
n_{2,snk} =\ & 1 - (1 - \tau_{u,2,snk})^{q-r}(1 - \tau_{u,3,2,snk})^{r} \\
n_{3,snk} =\ & 1 - (1 - \tau_{u,2,snk})^{q}
\end{aligned}
\tag{10.89}
$$

Also, since the effective superframe length is shorter because a part of the active portion of the superframe is used by the GTS, it follows that $SD_{gts} = SD - B(G'_p(1) + \Delta)$ and $SM_{gts} = SD_{gts} - D_d + 1$, where B is the number of GTSs ('packet lanes') allocated to the bridge, and Δ presents the space needed for the transfer of acknowledgment and separation between the GTSs lanes; these values can then be substituted into Equation (10.81) to obtain the desired probability of success at first CCA.

10.7.2 Probability of success on the second CCA

The probability that the medium is idle on the second CCA for a given node is equal to the probability that neither the coordinator nor the one of the remaining nodes have started a transmission in that backoff period. The second CCA can be performed in any backoff period from the second backoff period in the superframe, up to the period in which there is no more time for packet transmission, which amounts to SM.

Ordinary node in the source cluster. Again, ordinary nodes and bridge have different view of the medium. For ordinary nodes this probability for the source cluster is equal to:

$$
\begin{aligned}
\beta_{src/o} = \sum_{q=0}^{n-1} \sum_{r=0}^{q} P_q P_r \cdot \Bigg(& \frac{1}{SM} \\
& + \frac{(1 - \tau_{u,1,src})^{n-1-q}(1 - \tau_{u,2,src})^{q-r}(1 - \tau_{u,3,1,src})^{r}(1 - \tau_r - \tau_d)}{SM} \\
& + \frac{\overline{D_d} - 1}{SM}(1 - \tau_{u,2,src})^{q-r}(1 - \tau_{u,3,2,src})^{r}(1 - \tau_r - \tau_d) \\
& + \frac{SM - \overline{D_d}}{SM}(1 - \tau_{u,2,src})^{q}(1 - \tau_r - \tau_d) \Bigg)
\end{aligned}
\tag{10.90}
$$

Bridge in the source cluster. For a bridge which observes n nodes this probability has the value

$$
\begin{aligned}
\beta_{src/b} = \sum_{q=0}^{n}\sum_{r=0}^{q} P_q\, P_r & \left(\frac{1}{SM} + \frac{(1-\tau_{u,1,src})^{n-q}(1-\tau_{u,2,src})^{q-r}(1-\tau_{u,3,1,src})^{r}}{SM} \right. \\
& \left. + \frac{\overline{D_d}-1}{SM}(1-\tau_{u,2,src})^{q-r}(1-\tau_{u,3,2,src})^{r} + \frac{SM-\overline{D_d}}{SM}(1-\tau_{u,2,src})^{q} \right)
\end{aligned}
$$

$$(10.91)$$

Average value of β in the source cluster is:

$$
\beta_{src} = \frac{n\tau_{u,src}}{n\tau_{u,src}+\tau_r+\tau_d}\beta_{src/o} + \frac{(\tau_d+\tau_r)}{n\tau_{u,src}+\tau_r+\tau_d}\beta_{src/b} \tag{10.92}
$$

CSMA-CA bridge in the sink cluster. In order to calculate $\beta_{snk/b}$, we must account for all the n nodes and bridge in the sink cluster. Therefore,

$$
\begin{aligned}
\beta_{snk/b} = \sum_{q=0}^{n}\sum_{r=0}^{q} P_q\, P_r & \left(\frac{1}{SM} \right. \\
& + \frac{(1-\tau_{u,1,snk})^{n-1-q}(1-\tau_{u,2,snk})^{q-r}(1-\tau_{u,3,1,snk})^{r}(1-\tau_{u,1,bri}-\tau_{u,2,bri}-\tau_{u,3,1,bri})}{SM} \\
& + \frac{\overline{D_d}-1}{SM}(1-\tau_{u,2,snk})^{q-r}(1-\tau_{u,3,2,snk})^{r}(1-\tau_{u,2,bri}-\tau_{u,3,2,bri}) \\
& \left. + \frac{SM-\overline{D_d}}{SM}(1-\tau_{u,2,snk})^{q}(1-\tau_{u,2,bri}-\tau_{u,3,2,bri}) \right)
\end{aligned}
$$

$$(10.93)$$

Ordinary node in the sink cluster with the CSMA-CA bridge. In order to calculate $\beta_{snk/o}$, we must account for all the $n-1$ ordinary nodes and CSMA-CA bridge in the sink cluster. Therefore,

$$
\begin{aligned}
\beta_{snk/o} = \sum_{q=0}^{n-1}\sum_{r=0}^{q} P_q\, P_r & \left(\frac{1}{SM} \right. \\
& + \frac{(1-\tau_{u,1,snk})^{n-1-q}(1-\tau_{u,2,snk})^{q-r}(1-\tau_{u,3,1,snk})^{r}(1-\tau_{u,1,bri}-\tau_{u,2,bri}-\tau_{u,3,1,bri})}{SM} \\
& + \frac{\overline{D_d}-1}{SM}(1-\tau_{u,2,snk})^{q-r}(1-\tau_{u,3,2,snk})^{r}(1-\tau_{u,2,bri}-\tau_{u,3,2,bri}) \\
& \left. + \frac{SM-\overline{D_d}}{SM}(1-\tau_{u,2,snk})^{q}(1-\tau_{u,2,bri}-\tau_{u,3,2,bri}) \right)
\end{aligned}
$$

$$(10.94)$$

GTS bridge in the sink cluster. In this case, the bridge does not perform the second CCA and the ordinary nodes need not care for it. Therefore, the value of $\beta_{snk/o}$

becomes

$$
\beta_{snk/o} = \sum_{q=0}^{n-1}\sum_{r=0}^{q} P_q P_r \left(\frac{1}{SM_{gts}} \right.
$$

$$
+ \frac{(1 - \tau_{u,1,snk})^{n-1-q}(1 - \tau_{u,2,snk})^{q-r}(1 - \tau_{u,3,1,snk})^r}{SM_{gts}}
$$

$$
+ \frac{\overline{D_d} - 1}{SM_{gts}}(1 - \tau_{u,2,snk})^{q-r}(1 - \tau_{u,3,2,snk})^r
$$

$$
+ \left. \frac{SM_{gts} - \overline{D_d}}{SM_{gts}}(1 - \tau_{u,2,snk})^q \right)
$$

(10.95)

10.7.3 Probability of successful transmission

Finally, we need the probability that a packet will not collide with other packets that have undergone successful first and second CCAs. In the source cluster, this probability can be calculated as the probability that there are no accesses to the medium by the other nodes or the coordinator during the period of one complete packet transmission time. (Note that a collision can happen in SM consecutive backoff periods starting from the third backoff period in the superframe.)

Source cluster. In source cluster an ordinary node can collide with any of other $n - 1$ nodes and the bridge. Therefore the probability of no collision has the form:

$$
\gamma_{src/o} = \sum_{q=0}^{n-1}\sum_{r=0}^{q} P_q P_r \left(\frac{SM - \overline{D_d}}{SM}((1 - \tau_{u,2,src})^q (1 - \tau_r - \tau_d))^{\overline{D_d}} \right.
$$

$$
+ \frac{\overline{D_d} - 1}{SM}((1 - \tau_{u,2,src})^{q-r}(1 - \tau_{u,3,2,src})^r (1 - \tau_r - \tau_d))^{\overline{D_d}}
$$

(10.96)

$$
+ \left. \frac{((1-\tau_{u,1,src})^{n-1-q}(1-\tau_{u,2,src})^{q-r}(1-\tau_{u,3,1,src})^r (1-\tau_r - \tau_d))^{\overline{D_d}}}{SM} \right)
$$

For the bridge, the probability of no collision is:

$$
\gamma_{src/b} = \sum_{q=0}^{n}\sum_{r=0}^{q} P_q P_r \left(\frac{((1-\tau_{u,1,src})^{n-q}(1-\tau_{u,2,src})^{q-r}(1-\tau_{u,3,1,src})^r)^{\overline{D_d}}}{SM} \right.
$$

$$
+ \frac{\overline{D_d} - 1}{SM}((1 - \tau_{u,2,src})^{q-r}(1 - \tau_{u,3,2,src})^r)^{\overline{D_d}}
$$

(10.97)

$$
+ \left. \frac{SM - \overline{D_d}}{SM}((1 - \tau_{u,2,src})^q)^{\overline{D_d}} \right)
$$

Average value of γ_{src} has the form:

$$\gamma_{src} = \frac{n\tau_{u,src}}{n\tau_{u,src} + \tau_r + \tau_d}\gamma_{src/o} + \frac{\tau_d + \tau_r}{n\tau_{u,src} + \tau_r + \tau_d}\gamma_{src/b} \qquad (10.98)$$

CSMA-CA bridge in the sink cluster. Again, we can calculate $\gamma_{snk/b}$, the probability of successful transmission for the bridge, by considering the impact of all the n ordinary nodes in the sink cluster and CSMA-CA bridge. This can be accomplished by using the expression similar to Equation (10.96), namely:

$$\gamma_{snk/b} = \sum_{q=0}^{n}\sum_{r=0}^{q}P_qP_r$$

$$\left(\frac{((1-\tau_{u,1,snk})^{n-1-q}(1-\tau_{u,2,snk})^{q-r}(1-\tau_{u,3,1,snk})^r(1-\tau_{u,1,bri}-\tau_{u,2,bri}-\tau_{u,3,1,bri}))^{\overline{D_d}}}{SM}\right.$$

$$+\frac{\overline{D_d}-1}{SM}((1-\tau_{u,2,snk})^{q-r}(1-\tau_{u,3,2,snk})^r(1-\tau_{u,2,bri}-\tau_{u,3,2,bri}))^{\overline{D_d}}$$

$$\left.+\frac{SM-\overline{D_d}}{SM}((1-\tau_{u,2,snk})^{q-r}(1-\tau_{u,3,2,snk})^r(1-\tau_{u,2,bri}))^{\overline{D_d}}\right)$$

$$(10.99)$$

Ordinary node in the sink cluster with the CSMA-CA bridge. The probability of successful transmission for an ordinary node in sink cluster $\gamma_{snk/o}$ has to be calculated by considering the impact of CSMA-CA bridge and the other $n-1$ nodes:

$$\gamma_{snk/o} = \sum_{q=0}^{n-1}\sum_{r=0}^{q}P_qP_r\cdot$$

$$\left(\frac{((1-\tau_{u,1,snk})^{n-1-q}(1-\tau_{u,2,snk})^{q-r}(1-\tau_{u,3,1,snk})^r(1-\tau_{u,1,bri}-\tau_{u,2,bri}-\tau_{u,3,1,bri}))^{\overline{D_d}}}{SM}\right.$$

$$+\frac{\overline{D_d}-1}{SM}((1-\tau_{u,2,snk})^{q-r}(1-\tau_{u,3,2,snk})^r(1-\tau_{u,2,bri}-\tau_{u,3,2,bri}))^{\overline{D_d}}$$

$$\left.+\frac{SM-\overline{D_d}}{SM}((1-\tau_{u,2,snk})^{q-r}(1-\tau_{u,3,2,snk})^r(1-\tau_{u,2,bri}))^{\overline{D_d}}\right)$$

$$(10.100)$$

Ordinary node in the sink cluster with GTS bridge. Since the actions of ordinary nodes and bridge are separated, the probability $\gamma_{snk/o}$ becomes

$$\gamma_{snk/o} = \sum_{q=0}^{n-1}P_q\sum_{r=0}^{q}P_r\left(\frac{((1-\tau_{u,1,snk})^{n-1-q}(1-\tau_{u,2,snk})^{q-r}(1-\tau_{u,3,1,snk})^r)^{\overline{D_d}}}{SM_{gts}}\right.$$

$$+\frac{\overline{D_d}-1}{SM_{gts}}((1-\tau_{u,2,snk})^{q-r}(1-\tau_{u,3,2,snk})^r$$

$$\left.+\frac{SM_{gts}-\overline{D_d}}{SM_{gts}}((1-\tau_{u,2,snk})^{q-r}(1-\tau_{u,3,2,snk})^r)^{\overline{D_d}}\right)$$

$$(10.101)$$

10.7.4 Probability distribution for the packet service time

The packet service time is, in fact, the service time for the packet queue at the network node. Following all the derivations above, we can now derive the probability distribution of the packet service time at the MAC layer, which we have used in Sections 10.1.2 and 10.1.3. We will keep the derivation generic for both non-acknowledged and acknowledged modes as far as possible and branch to separate results at the moment when differences in derivation occur. Regarding the cases for source, sink and bridge node we will again try to model the most general case and then derive the particular cases as needed.

In order to derive the aforementioned distribution, we need to model the initial waiting time between the packet arrival to an empty buffer during the inactive portion of the superframe until the following beacon when node starts the random backoff countdown procedure. Since the packet arrival process is totally oblivious to the superframe timing, a new packet can arrive at any instant of time during inactive superframe part with the same probability, and therefore this waiting time has uniform probability distribution. For the transmission from a node in the source cluster, its PGF is

$$T_{sync}(z) = \pi_{0,src} P_{sync} \frac{z^{BI-SD} - 1}{(BI - SD)(z - 1)} + (1 - \pi_{0,src} P_{sync}) z^0 \tag{10.102}$$

A similar expression holds for the PGF of the transmissions from the node in the sink cluster or the bridge, provided the appropriate value of π_0 is used.

We also need to model the effect of freezing the backoff counter during the inactive portion of the superframe. The probability that the given backoff period is the last one within the active portion of the superframe is $P_{last} = \frac{1}{SD}$ for source cluster and sink cluster with CSMA bridge and $P_{last} = \frac{1}{SD_{gts}}$ for sink cluster with GTS bridge. Then, the PGF for the effective duration of the backoff countdown interval, including the duration of the beacon frame, is

$$B_{off} z = (1 - P_{last}) z + P_{last} z^{(BI-SD+1)} B_{ea} z \tag{10.103}$$

The PGF for the duration of i-the backoff attempt is

$$B_i(z) = \sum_{k=0}^{W_i-1} \frac{1}{W_i} B_{off}^k(z) = \frac{B_{off}^{W_i}(z) - 1}{W_i(B_{off}(z) - 1)} \tag{10.104}$$

As noted above, the transmission procedure will not start unless it can be finished within the current superframe. The number of backoff periods thus wasted can be described with the PGF of $B_p(z) = \frac{1}{D_d} \sum_{k=0}^{D_d-1} z^k$. Then, for the case of acknowledged transmission, the PGF of the data packet (uplink or downlink) transmission time for deferred and non-deferred transmissions, respectively, is

$$\begin{aligned} T_{d1} z &= B_p z^{(BI-SD)} B_{ea} z G_p z t_{ack} z G_a z \\ T_{d2}(z) &= G_p(z) t_{ack}(z) G_a(z) \end{aligned} \tag{10.105}$$

For the case of non-acknowledged transmission the equivalent PGFs are:

$$\begin{aligned} T_{d1} z &= B_p z^{(BI-SD)} B_{ea} z G_p z \\ T_{d2}(z) &= G_p(z) \end{aligned} \tag{10.106}$$

For the transmission of request packet, $G_p(z)$ has to be replaced with z^2.

PGF for a single transmission attempt. For simplicity, let us denote the probability that a backoff attempt will be unsuccessful as $R_{ud} = 1 - P_d - (1 - P_d)\alpha\beta$. The function that describes the time needed for the backoff countdown and the transmission attempt itself can be presented in the following equation:

$$P(z) = \sum_{i=0}^{m} \prod_{j=0}^{i} \left(B_j(z)R_{ud}\right) z^{2(i+1)} \left(P_d T_{d1}(z) + (1 - P_d)\alpha\beta T_{d2}(z)\right)$$

$$+ R_{ud}^{m+1} \prod_{j=0}^{m} B_j(z) z^{2(m+1)} P(z) \tag{10.107}$$

where R_{ud}^{m+1} denotes the probability that $m + 1$ backoff attempts with non-decreasing backoff windows were not successful and the sequence of backoff windows has to be repeated starting from the smallest backoff window. If we substitute $z = 1$ into Equation (10.107) we will obtain $P(1) = 1$ which is a necessary condition for a PGF. From Equation (10.107) we obtain:

$$P(z) = \frac{\displaystyle\sum_{i=0}^{m} \prod_{j=0}^{i} \left(B_j(z)R_{ud}\right) z^{2(i+1)} \left(P_d T_{d1}(z) + (1 - P_d)\alpha\beta T_{d2}(z)\right)}{1 - R_{ud}^{m+1} \displaystyle\prod_{j=0}^{m} B_j(z) z^{2(m+1)}} \tag{10.108}$$

Packet service time under non-acknowledged transfer. In this case, pure transmission time from expression (10.106) has to be substituted in (10.108). Note that this PGF depends on the environment (i.e., source or sink cluster or the bridge) and the final form is obtained by substituting appropriate values of $\alpha, \beta, \gamma, \ldots$ and δ in expression (10.108). Therefore for non-acknowledged transfer this function has the values:

$$T_{t,src}(z) = P_{src}(z)$$

$$T_{t,snk}(z) = P_{snk}(z)$$

$$T_{t,bri}(z) = P_{bri}(z)$$

for node in source cluster, sink cluster and bridge respectively. The PGF for the bridge transmission applies only to the CSMA-CA access mode, since the bridge transmission in the GTS mode is collision-free.

Packet service time under acknowledged transfer. For the acknowledged transfer we have to take into account that transmission will be attempted until packet is acknowledged.

Let us denote the probability that transmission attempt will be successful as

$$P_s = \gamma_{src}\delta_{src}(1 - P_f^B)(1 - P_c^B)(1 - P_c^B) \tag{10.109}$$

Then the PGF for packet service time without the beacon synchronization time for the ordinary node in source cluster becomes:

$$
\begin{aligned}
T_{t,src}(z) &= \sum_{k=0}^{\infty} P_s (1 - P_s)^k (P_{src}(z))^{k+1} \\
&= \frac{P_s P_{src}(z)}{1 - P_s P_{src}(z)}
\end{aligned}
\tag{10.110}
$$

The PGF for the service time of the request packet, $T_{t,req}(z)$, can be obtained by substituting z^2 for $G_p(z)$.

Finally, by adding the beacon synchronization time for the nodes in source and sink clusters, the PGFs for packet transmission time become

$$
\begin{aligned}
T_{t,src}(z) &= T_{sync}(z) \frac{1 - P_{src}(z)^a (1 - \gamma_{src}\delta_{src}(1 - P_{bri}^B))^a}{1 - P_{src}(z)(1 - \gamma_{src}\delta_{src}(1 - P_{bri}^B))} \\
&\quad + \left(1 - (1 - \gamma_{src}\delta_{src}(1 - P_{bri}^B))^a\right) z^0 \\
T_{t,snk}(z) &= T_{sync}(z) \frac{1 - P_{snk}(z)^a (1 - \gamma_{snk}\delta_{snk})^a}{1 - P_{snk}(z)(1 - \gamma_{snk}\delta_{snk})} \\
&\quad + \left(1 - (1 - \gamma_{snk}\delta_{snk})^a\right) z^0 \\
T_{t,bri}(z) &= \frac{1 - P_{bri}(z)^a (1 - \gamma_{bri}\delta_{bri})^a}{1 - P_{bri}(z)(1 - \gamma_{bri}\delta_{bri})} + \left(1 - (1 - \gamma_{bri}\delta_{bri})^a\right) z^0
\end{aligned}
\tag{10.111}
$$

It can be verified that $T_t(1) = 1$, which is a necessary condition that a given function is a PGF.

Of course, the equations above hold only in the case of an CSMA-CA bridge.

Downlink transmission. A downlink transmission has to be preceded by the acknowledged transmission of the data request packet. If the downlink transmission is not acknowledged itself, the PGF for the downlink packet service time is

$$
T_{t,dtot}(z) = T_{t,req}(z) P_{src}(z)
\tag{10.112}
$$

For acknowledged downlink transmission PGF for packet service time becomes:

$$
\begin{aligned}
T_{t,dtot}(z) &= \sum_{k=0}^{\infty} P_s (1 - P_s)^k (T_{t,req}(z) P_{src}(z))^{k+1} \\
&= \frac{P_s P_{src}(z)}{1 - P_s T_{t,req}(z) P_{src}(z)}
\end{aligned}
\tag{10.113}
$$

Part III Summary and Further Reading

In Part III, we have presented basic tenets of cluster interconnection in multi-cluster 802.15.4 networks. Cluster interconnection schemes are classified into master-slave and slave-slave, using the dichotomy that was proposed for Bluetooth networks (Bluetooth SIG 2004; Mišić and Mišić 2005). The former offer better performance as inter-cluster traffic has to traverse fewer hops; however, the latter offers more flexibility as several bridges may connect a given pair of clusters, and a single bridge can visit more than two clusters. In both cases, acknowledged transmissions place a serious burden on the performance of data traffic and should be avoided whenever possible. Furthermore, the CSMA-CA access mode is more adaptable to traffic conditions under low to moderate loads, but GTS access gives better performance under high traffic load. However, in cases where multiple bridges connect to a single sink cluster, GTS access by the bridges tends to give preference to inter-cluster (i.e., bridge) traffic at the expense of local traffic in the sink cluster. In this case, use of the CSMA-CA bridge access mode might offer a better balance between local and non-local traffic. If possible, slave-slave bridges should use GTS access mode in the source cluster, as this will reduce the overhead caused by the comparatively inefficient download mechanism prescribed by the 802.15.4 standard, Section 2.5. Some additional considerations regarding the impact of bridge residence times are discussed in (Mišić et al. 2006a), while Mišić and Udayshankar (to appear) provide an in-depth comparison of two types of bridges; it is worth noting that both papers are based exclusively on simulations.

Other authors have also investigated issues related to multi-cluster 802.15.4 networks; we will mention here only a few. Ha et al. (2005) have looked at the optimum setting of various MAC parameters in a multi-cluster tree. Kiri et al. (2006) compare the performance of several protocols, including 802.15.4, in a multi-cluster topology. Koubaa et al. (2006b) discuss delay bounds and other performance measures (albeit using mean value analysis only) for multi-cluster 802.15.4 topologies, and use those results to arrive at practical guidelines for designing and dimensioning such networks; a subsequent paper (Koubaa et al. 2007b) discusses beacon scheduling in 802.15.4/ZigBee environment. Some other peformance issues, including energy considerations, in a mixed environment wherein 802.15.4 networks providing transport support for ZigBee applications, are presented by Ding et al. (2006) and Kohvakka et al. (2006). Xing et al. (2006) modify the 802.15.4

MAC protocol in order to improve its performance; the modified protocol (referred to as IC-MAC) uses the CSMA-CA medium access at low to moderate load, but switches to scheduled transmissions at high loads. Also, Kim et al. (2006a) proposed the concept of virtual channels to efficiently schedule transmissions and improve performance in a multi-cluster environment.

Part IV

Security

11

Security in 802.15.4 Specification

Successful deployment of wireless personal area networks and sensor networks necessitates a thorough understanding of many facets of their design and operation, not the least important of which is their security (Chan and Perrig 2003; Shi and Perrig 2004; Wood and Stankovic 2002). However, a number of challenges exist that make this task difficult.

First, access to the wireless medium is open to all, including potential adversaries that don't even have to be physically close to the sensor field.

Second, the network may consist of many nodes which are expected to operate with little human intervention for prolonged periods of time. Furthermore, some or all of the nodes may be mobile. As a result, attacks and disruptions more difficult to detect.

Third, most of the nodes in a WPAN or a sensor network have limited energy at their disposal and the chips' computational power is limited; this restricts the choice of cryptographic techniques that can be applied to ensure that privacy and integrity are adequately supported.

Finally, actual nodes may be subject to damage, or even physical capture and subsequent subversion by a hostile adversary.

Consequently, sophisticated techniques are required to monitor and detect possible intrusions and, if necessary, launch appropriate countermeasures. From the networking perspective, security threats may occur at different layers of the ISO/OSI model (Chan and Perrig 2003):

- *Routing layer* attacks include spoofed, altered, or replayed routing information spread by an adversary, selective forwarding of packets, sinkhole attacks that attract traffic from a specific area to a compromised node (or nodes), Sybil attacks in which a compromised node assumes many identities, acknowledgment spoofing, injecting corrupted packets, neglecting routing information, or forward messages along wrong paths (Hu and Perrig 2004; Ren and Liang 2004).

- *MAC layer* attacks typically focus on disrupting channel access for regular nodes, thus disrupting the information flow both to and from the sensor node; this leads to a DoS condition at the MAC layer (Wood and Stankovic 2002). Security at the MAC layer has been mostly studied in the context of 802.11 MAC layer (Gupta et al.

Wireless Personal Area Networks   Jelena Mišić and Vojislav B. Mišić
© 2008 John Wiley & Sons, Ltd

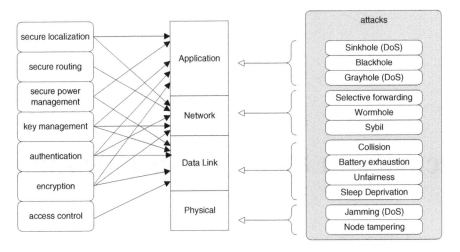

Figure 11.1 Security services and possible attacks with respect to layers of the network protocol stack.

2002; Karlof et al. 2004; Ren and Liang 2004; Zhou et al. 2004) but also in the more general context of different types of attacks (Bellardo and Savage 2003; Newsome et al. 2004; Wood and Stankovic 2002).

- Finally, *physical layer* (jamming) attacks consist of the attacker sending signals that disrupt the information flow through radio frequency interference. Jamming at the MAC level may be accomplished through sending large size packets with useless information.

Some of those attacks, together with the security services that address the threat they pose, are listed in Figure 11.1.

11.1 Security Services

Among the security services listed in Figure 11.1, the 802.15.4 standard (IEEE 2006) provides support for the following:

- data confidentiality, which ensures that the transmitted data is only available to their proper addressees and not to anyone else;

- data authenticity or integrity, which ensures that the data received was not modified in transit; and

- replay protection, which ensures that duplicated data transmissions are detected.

Communications are protected on a frame-by-frame basis, which allows the application to make a tradeoff between increased security and reduced performance due to security-related overhead. The use of data confidentiality is optional, and several alternatives to support data integrity are provided. Furthermore, whenever one or both of those two services are used, replay protection is added as well. It should be noted that security protection applies to beacon, data, and MAC command frames; acknowledgment frames are not protected.

To support these services, the standard employs cryptographic mechanisms based on symmetric-key cryptography that uses keys provided by the higher layers of the network protocol stack. The keys may be shared between two peers or among a group of devices; these are referred to as the link key and the group key, respectively. Keys and related information which is necessary for securing incoming and unsecuring outgoing data frames are kept in the so-called key table. A special variable, *macDefaultKeySource*, holds the information commonly shared between the originator and recipient, or recipients, of a secured frame; it is used to simplify the process of securing an outgoing frame or unsecuring an incoming one.

A separate table, known as the device table, holds addressing information and other security-related information related to specific devices. When protection is required for a specific frame, the information from this table is combined with that in the key table to provide all necessary ingredients to secure outgoing frames and unsecure incoming ones.

Another table is used in conjunction with the incoming data frames; it is referred to as the minimum security level table. Namely, when a frame is received, the node may expect a certain level of security based on the available information about the sender of the frame. The minimum security level also depends on the type of the frame and, in certain cases, on its content. If the incoming frame is not secured appropriately, security processing may fail or, in some cases, pass conditionally; in the latter case, the node may choose to accept the frame or not.

Finally, a frame counter is used to ensure replay protection for outgoing frames. The frame counter is incremented each time an outgoing frame is secured; its current value is included in each secured outgoing frame. It is also one of the elements needed to unsecure an incoming frame. When the counter reaches its maximum value, the associated keying material is deemed invalidated and needs to be replaced. In this manner, sequential freshness is supported to protect data communication from replay attacks.

11.2 Auxiliary Security Header

As noted in Section 2.9, the MAC packet header may include the Auxiliary Security Header, which contains information related to security. The Auxiliary Security Header is present only when the Security Enabled subfield of the Frame Control field is set to one. The Auxiliary Security Header consists of Security Control, Frame Counter, and Key Identifier fields.

Security Control field occupies one byte; it begins with the Security Level subfield, in bits 0 to 2, which specifies the security level used for the current frame. The meaning of specific values in this subfield is explained in Table 11.1, where ENC denotes encryption and MIC stands for Message Integrity Code. While MIC is often referred to as the Message Authentication Code, we have chosen a less common acronym in order to avoid confusion with the concept of Medium Access Control which is frequently used in this book.

Table 11.1   Values allowed in the Security Level subfield

Identifier	Attributes	Confidentiality	Integrity	M (in bytes)
00_{16}	None	–	–	none
01_{16}	MIC-32	–	yes	4
02_{16}	MIC-64	–	yes	8
03_{16}	MIC-128	–	yes	16
04_{16}	ENC	yes	–	none
05_{16}	ENC-MIC-32	yes	yes	4
06_{16}	ENC-MIC-64	yes	yes	8
07_{16}	ENC-MIC-128	yes	yes	16

Note: M denotes the length of the authentication tag, in bytes.

Table 11.2   Values allowed in the Key Identifier Mode subfield

Key Identifier Mode	Key determined from ...	Field Length (in bytes)
00_{16}	implicitly	0
01_{16}	Key Index with *macDefaultKeySource*	1
02_{16}	Key Index and 4-byte Key Source	5
03_{16}	Key Index and 8-byte Key Source	9

The next two bits provide information about the Key Identifier Mode, which determines the manner in which the key used to protect the frame is derived, as well as the representation of the Key Identifier field in the Auxiliary Security Header. Values allowed in these two bits are summarized in Table 11.2. If the value in this field is different from 00_{16}, an additional field known as the Key Identifier field, must also be present in the Auxiliary Security Header; this field contains the 4- or 8-byte Key Source subfield that holds the address of the originator of a group key, followed by a one-byte Key Index subfield that allows unique identification of different keys used by the same originator node.

Frame Counter is used to ensure sequential freshness and protect data and command transmissions against replay attacks, as mentioned above; it occupies four bytes (32 bits).

11.3   Securing and Unsecuring Frames

Using the information provided by the application that requests secure communication, security services listed above are implemented as follows. Note that the discussion here is but a brief overview; the reader should consult the official standard (IEEE 2006) for exact details.

The authors of the 802.15.4 standard have chosen to implement the security services listed above using CCM*, a generic approach that offers encryption, authentication, or combined encryption and authentication service. The input data for the CCM* forward transformation, which corresponds to the operation of securing an outgoing data frame, consists of:

- the key to be used for encryption, formed according to the specified Key Identifier mode (Table 11.2);

- a nonce formed by the extended source address, frame counter, and desired security level;

- the MAC layer header, auxiliary security header, and non-payload fields, designated as *a* data;

- finally, the unsecured payload fields, designated as *m* data.

Higher layers of the network protocol stack (i.e., the application that requests secure communication) provide the information about the desired Security Level, Key Identification Mode, Key Source, and Key Index parameters.

If the chosen security level was 00_{16}, nothing is done and both *a* and *m* data are simply copied to the outgoing frame.

If the chosen security level was 01_{16}, 02_{16}, 03_{16}, 05_{16}, 06_{16}, or 07_{16}, the authentication tag or MIC is generated. The length of the MIC corresponds to the chosen security level, as per Table 11.1. Authentication tag is added to the frame header, while the payload fields are simply copied to the outgoing frame.

If the chosen security level was 04_{16}, 05_{16}, 06_{16}, or 07_{16}, the *m* data is encrypted using the AES block cipher (FIPS 2001) to generate the so-called *c* data. At security levels 05_{16}, 06_{16}, and 07_{16}, encryption is applied to the authentication tag as well. The *c* data replaces the *m* data in the outgoing frame, while the encrypted authentication tag replaces the original authentication tag.

The frame formats that result from those security levels are illustrated in Figure 11.2, where lighter, striped shading denotes integrity protection while darker shading denotes confidentiality protection (i.e., encryption).

The procedure to generate the MIC and/or the encrypted data is described in the specification for the combined counter with CBC-MAC (cipher block chaining message authentication code) mode of operation known as CCM*. This is a minor modification of the well-known CCM approach (ANSI 2001) but augmented with the ability to separately apply encryption and integrity protection (which eliminates the need for separate mechanisms to achieve the two) and the ability to use a single key for all of its security functions (which simplifies key management).

An inverse procedure is used for unsecuring an incoming frame. If the incoming frame originates at an unknown node, or the authentication process shows that the message has been tampered with, higher layers of the network protocol stack should be informed that the authenticity verification failed and all relevant data should be destroyed.

In order to enhance the security of encrypted communication, the standard limits the total amount of data that is encrypted with a single key to at most 2^{61} block cipher encryption invocations. However, it does not specify any particular key management algorithm, and actual implementations are free to choose a suitable algorithm without restrictions. One such implementation, which is recommended as part of the ZigBee standard (ZigBee Alliance 2006), will be described in Appendix A.

(a) Security level 00_{16}: no security.

(b) Security levels 01_{16}, 02_{16}, and 03_{16}: message integrity only.

(c) Security level 04_{16}: encryption only.

| SYNC | PHY HDR | MAC HDR | Aux Sec HDR | encrypted MAC payload | MIC |

(d) Security levels 05_{16}, 06_{16}, and 07_{16}: message integrity and encryption.

Figure 11.2 Structure of secured frames (Security Enabled subfield set to one).

11.4 Attacks

We conclude this chapter with a brief overview of possible attacks on 802.15.4 networks operating in beacon enabled, slotted CSMA-CA mode. In this environment, attacks can be broadly classified in two categories, depending on whether the attacker follows the rules of the 802.15.4 MAC protocol, either fully or only to a certain extent, or not. While the attacks from the latter category are potentially more dangerous, defense against them is much more difficult, as might be expected; in this case, the attacker can use a separate 802.15.4-compliant device, possibly modified to loose the adherence to the MAC protocol. Alternatively, an existing 802.15.4 device may be captured and subverted so as to be used for malicious purposes.

Attacks that follow the MAC protocol. A number of attacks may be conducted by an adversary which follows the 802.15.4 slotted CSMA-CA protocol to the letter, due to the openness of the wireless medium. A simple but not very efficient attack against network availability is to flood the network by simply transmitting a large number of packets. Packets should be large in size, perhaps the largest size allowed by the standard. In this manner, an adversary may degrade the network performance and drastically reduce throughput; the analysis in Chapters 3 and 4 indicate that the performance of an 802.15.4 network can be seriously affected by high packet arrival rates or by nodes operating in saturation regime.

An adversary may target different destination devices, possibly in a different PAN, with unnecessary packets, regardless of whether the destination PAN and/or device actually exist or not. This will waste the bandwidth and prevent legitimate devices from accessing the medium. If the goal of the attack is the depletion of the power source for a specific node (and the PAN coordinator), all injected packets may target that node. Since the downlink packets have to be explicitly requested from the coordinator, this will keep both the coordinator and the chosen destination device busy and eventually exhaust their respective power sources.

Note also that a node that succeeds in getting access to the medium will not increase its backoff exponent for the next transmission, while an unsuccessful one will increase it by one. Therefore, if the first attempt succeeded, the second one is even more likely to do so, which again clearly favors malicious nodes.

Attacks that use a modified MAC protocol. A number of additional attacks may be launched by simply modifying or disregarding certain features of the protocol. This can be accomplished either through dedicated hardware or by controlling an otherwise fully compliant 802.15.4 hardware device, as follows.

Smaller backoff exponents and shorter random backoff countdowns may be obtained by setting the *macBattLifeExt* variable to true but also by not incrementing the backoff exponent after an unsuccessful transmission attempt; however, the latter action would require changes in the firmware that implements the 802.15.4 protocol stack.

The random number generator can be modified to give preference to shorter backoff countdowns. Again, this allows the malicious node to capture the channel in a disproportionately high number of cases, and gives it an unfair advantage over regular nodes.

The number of required CCA attempts can be reduced to one instead of two, which would give the malicious node an unfair advantage over the regular nodes. It is worth noting that nodes that operate in unslotted CSMA-CA mode use a single CCA which is not aligned to the backoff period boundary.

The CCA check can be omitted altogether, in which case the node will start transmitting immediately after finishing the random backoff countdown. Even worse, the node can omit the random backoff countdown itself. In this manner, the malicious node can transmit its packets more often than a regular one. While not all of the messages will be sent successfully – there will be collisions in many cases – the malicious node probably doesn't even care, as long as the transmissions from regular nodes end up garbled and thus have to be repeated. Moreover, some of the attacker's transmissions may collide with acknowledgments. Again, this drains the power out of the affected device or devices, but also wastes the bandwidth of the entire network.

In case the acknowledgment is requested by the data frame or the beacon frame, a malicious node may simply refuse to send it. The PAN coordinator will retry transmission (up to a maximum of *aMaxFrameRetries*) and thus waste power and bandwidth.

Attacks that do not follow the MAC protocol. Finally, an adversary with appropriate resources might develop and use dedicated hardware which is compatible but not compliant with the 802.15.4 standard. In other words, the attacker would need a dedicated radio subsystem that is compatible with the PHY (and, to a certain extent, MAC) layer operation, yet not fully adherent to the rules of the 802.15.4 protocol. The availability of such hardware allows for a number of different attacks at varying level of sophistication, with or without the ability to impersonate legitimate nodes.

The addresses of the nodes that have pending downlink packets are announced within the beacon frame which all nodes receive. As this list is not encrypted, the adversary may learn which nodes have downlink packets, and it may send a data request packet posing as a legitimate node. With slight alteration of the MAC algorithm, as outlined in the previous subsection, the adversary may almost always succeed in sending the request before the legitimate destination node. The PAN coordinator, upon receiving the request, will proceed to send the downlink packet, which the adversary may even acknowledge; the coordinator will consider this transmission successful and delete the packet from its buffer. The legitimate destination device will not listen for the downlink packet before getting its request acknowledged by the coordinator, and it may miss the actual transmission while attempting in vain to send the data request packet. Therefore, the damage, in this case, includes not only wasted bandwidth and power, but also loss of information.

An adversary may try to destroy legitimate traffic by injecting packets of its own with the aim of garbling the legitimate packets and thus either destroy information or cause retransmission. Of course, the feasibility of such an attack depends on the importance of the network and information which the attacker is attempting to damage; it also depends on the amount of resources available to the attacker.

Since the packet headers are always transmitted in the clear – encryption protects the packet payload only – the attacker may figure out whether an acknowledgment is requested. If so, the jamming packet may be transmitted simultaneously with the regular acknowledgment packet. The absence of acknowledgment will cause the sender to repeat the transmission, which means that the original packet transmission is wasted; again, the long-term target of such attacks is the power supply of the regular nodes. We note that the efficiency of such attacks is better since the acknowledgment packets are shorter than the regular data packets, and the attacker can achieve the desired objective (i.e., cause retransmission) with smaller energy expenditure.

On the other hand, if the acknowledgment is not requested, the attacker must start its transmission immediately. In this case, the attacker needs a fast radio subsystem, since switching from reception to transmission takes time (we assume that the attacking device has only one such subsystem). Even when equipped with a fast radio, the success of this kind of attack depends on the length of the legitimate data packet. Shorter data packets may actually finish *before* the attacker can start a transmission of its own, and thus are more resilient to this kind of attack.

A particularly attractive target in beacon enabled PANs is the beacon frame itself, which is periodically sent by the PAN coordinator for synchronization and other purposes. A jamming packet sent at the precise time can collide with the beacon and thus disrupt the normal operation of the WPAN for prolonged periods of time.

Defending an 802.15.4 cluster. Even though the attacks listed above pose formidable risks to normal operation of an 802.15.4 WPAN, they are probably not very cost-effective to launch. Since individual 802.15.4 sensor nodes are small, low power, low cost devices, the development of dedicated compatible-but-not-compliant devices with modified behavior is likely to be prohibitively expensive – the potential attacker would probably find the use of simple devices for jamming at the PHY layer to be much more attractive. Let us consider just the 2450 MHz PHY option. From the specifications of the 802.15.4 standard (IEEE 2006), the processing gain is only around 8, and the Bit Error Rate is given with

$$BER = Q\left(\sqrt{\frac{E_b}{N_0}}\right)$$ (Garg et al. 1998), where $Q(u) \approx e^{-u^2/2}(\sqrt{2\pi}u)$, $u \gg 1$. Therefore, in

an interference-free environment, we should expect the BER values slightly below 10^{-4}; the probability that a given packet (data or acknowledgment) is much higher. In fact, Section 6.1.6 of the standard states that PER of 1% is expected on packets with length of 20 octets (IEEE 2006). However, the ISM band is also used by other wireless LAN/PAN standards such as 802.11b and 802.15.1 (Bluetooth), which means that higher BER values and, consequently, much higher PER values, may be expected; for example, at BER around 10^{-3}, the PER for a 20-octet packet increases to around 15%! Consequently, an attacker that chooses to jam the transmissions in an 802.15.4 sensor network can cause quite a lot of damage with modest energy expenditure.

Therefore, few remedies remain at the disposal of the 802.15.4 network designers. In terms of packet jamming, obviously shorter packets should be used whenever possible, as they are more difficult to jam. In terms of CCA checking, the standard allows CCA checks to be performed in one of three different modes: by energy only (mode 1), by carrier sense only (mode 2), or by energy and carrier sense (mode 3), as described in Section 2.1, p. 19. A legitimate node which uses mode 1 may experience CCA failures in a high interference environment, without an attacker specifically using 802.15.4 modulated signal. Obviously, CCA mode 3 should be used for highest resilience.

In terms of encryption described earlier in this chapter, the obvious weakness of the standard is that the encryption is applied to packet/frame payload only, but not to the information in packet headers; this holds for ordinary data packets but also for data request and beacon frames. In this manner, an attacker device may easily learn the node and PAN identifiers which can then be targeted in subsequent attacks.

Replay attacks can be identified by the coordinator as long as any form of security is used. While this would not prevent a malicious node from sending such packets, at least the coordinator could filter them out and avoid any further processing.

While the standard does not prescribe, or indeed even recommend, any particular key management scheme, the overall effectiveness of the security services provided by the standard depends very much on the choice of a suitable scheme (Stallings 2003). Key management will be discussed in the next chapter.

Intrusion detection techniques could help identify malicious nodes that might be trying to disrupt the normal operation of the PAN. By analyzing the traffic patterns, the PAN coordinator may become aware of the activity of such nodes, so that appropriate measures can be taken to minimize the disruption. Suspicious activities include amounts of traffic well above the average, traffic intensity that increases over time, possibly in an abrupt fashion, and sending packets to many destinations, possibly in different PANs. In more critical applications, devices with substantially higher computational capabilities (and operating on mains power, rather than battery power) could analyze the activities of individual nodes at the MAC level and identify potential intruder(s).

Quite a few applications, in particular those that deal with sensing and monitoring, involve operation with very low duty cycle of individual nodes; this makes intrusion detection comparatively easier to accomplish, and does not help the attackers which might be eager to achieve their objectives.

The standard does not provide for periodical checking of presence and/or integrity of individual devices. However, a sensing application might establish such checks on its own. Simple time-out counters, one per each associated device, would enable the coordinator to check for their continued presence; in addition, a simple challenge/response scheme could allow the coordinator to verify their integrity as well.

Of course, it is impossible to physically isolate an unwanted device in a wireless network; but at least the application should be made aware of the presence of such devices so that their impact on the normal operation of the network could be minimized.

12

The Cost of Secure and Reliable Sensing

Increased security means increased expenditure in terms of both computational and communication activities. In this chapter, we will extend our analytical model for the behavior of an 802.15.4 cluster with power managed nodes to the scenario in which keys are periodically updated. In this manner, we can evaluate the overhead of periodic key exchange in terms of medium behavior, total number of delivered packets, individual node utilization, and the impact on node and network lifetime.

Consider an 802.15.4 cluster with redundant nodes and activity management, similar to the one analyzed in Chapter 6. Let the cluster also employ a security scheme in which privacy and integrity of data communications are protected through the use of appropriate keys. In order to improve security, keys are updated after each n_k uplink communications which include alternating sleep and active phases. The key update procedure consists of n_s request-reply communications between the cluster coordinator and an individual node. (In case of the SKKE protocol described in Appendix A, $n_s = 2$.) Each request-reply communication includes a downlink and an uplink packet transfer, which must be performed according to the procedures described in Chapter 2. After n_s request-reply sub-phases, the key update includes the final acknowledgment sent by the coordinator to the node. This scenario is shown in Figure 12.1. All transmissions follow the slotted CSMA-CA protocol described in Section 2.3; furthermore, each downlink transmission requires the node to synchronize with the beacon, transmit the uplink request packet, and receive the downlink packet. Overall, the key update procedure involves $n_s + 1$ request packets, $n_s + 1$ downlink packets and n_s uplink packets, for a total of $S = 3n_s + 2$ packets exchanged using the slotted CSMA-CA procedure.

Wireless Personal Area Networks   Jelena Mišić and Vojislav B. Mišić
© 2008 John Wiley & Sons, Ltd

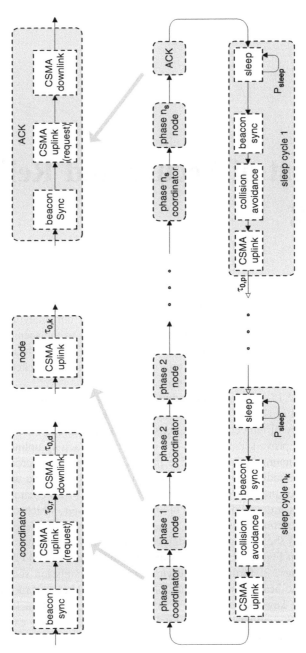

Figure 12.1 Markov chain for an individual node with key updates consisting of n_s request-reply phases, undertaken after each n_k transmission/sleep cycles. From J. Mišić and V. B. Mišić, 'The cost of secure and reliable sensing,' *Proc. Qshine 2007*, Vancouver, BC, Canada, © 2007 ACM. Reprinted with permission.

12.1 Analytical Model of a Generic Key Update Algorithm

Now, each of these CSMA-CA transmissions can be modeled with the Markov chain shown in Figure 12.2. As can be seen, the transmission may or may not start immediately after the current beacon, depending on the packet type; packets may be deferred to the next superframe if the remaining time in the current superframe is insufficient for the transaction, as described in Section 2.3. The probability that the packet will be transmitted immediately after the beacon is denoted with P_x where the index serves to distinguish

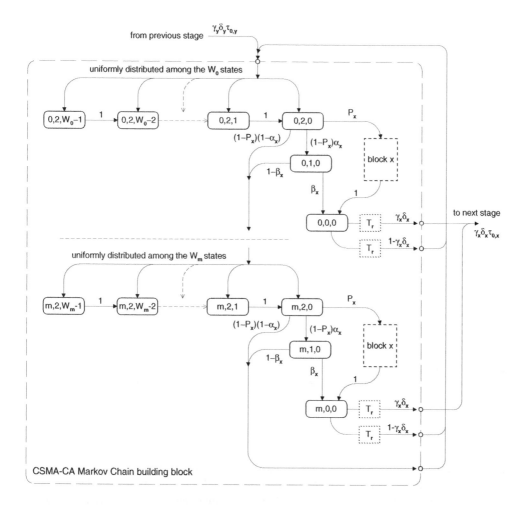

Figure 12.2 Markov sub-chain for a single CSMA-CA transmission from Figure 12.1. From J. Mišić and V. B. Mišić, 'The cost of secure and reliable sensing,' *Proc. Qshine 2007*, Vancouver, BC, Canada, © 2007 ACM. Reprinted with permission.

between data packets, $x = p$, request packets for key update, $x = r$, downlink packets with key information, $x = d$, and uplink packets with key information, $x = k$. The total packet transmission time is $\overline{D_x} = 2 + \overline{G_x} + 1 + \overline{G_a}$, including two CCAs, packet transmission time $\overline{G_x}$, waiting time for the acknowledgment and acknowledgment transmission time $\overline{G_a}$. (For simplicity, we assume that each packet type has a constant length.) The block labeled T_r denotes $\overline{D_x}$ linearly connected backoff periods needed for the actual transmission. The impact of noise and interference at the PHY layer is modeled via success probability $\delta_x = 1 - (1 - BER)^{\overline{G_x + G_a}}$, where BER denotes the bit error rate. Success probabilities at first CCA, second CCA, and the actual transmission will be denoted as α_x, β_x and γ_x, respectively. We also assume that acknowledgments are used in fully reliable mode, i.e., packet transmissions will be repeated until the packet is successfully acknowledged.

Within the transmission sub-chain, the process $\{i, c, k, d\}$ defines the state of the device at unit backoff period boundaries where

- $i \in (0 .. m)$ is the index of current backoff attempt, with m being a constant defined by the MAC with the default value of $macMaxCSMABackoffs = 4$;

- $c \in (0, 1, 2)$ is the index of the current CCA phase (two CCAs are used in the slotted CSMA-CA mode);

- $k \in (0 .. W_i - 1)$ is the value of the backoff counter, with W_i being the size of backoff window in i-th backoff attempt;

- $d \in (0 .. \overline{D_x} - 1)$ denotes the index of the state within the delay line in case packet transmission has to be deferred to the next superframe (in order to reduce notational complexity, it will be shown only within the delay line and omitted elsewhere).

Finally, the access probability of the general CSMA-CA block will be denoted as $\tau_{0,x}$.

Due to the differences among transmissions of packets with request, downlink key information, uplink key information and uplink sensing information, the following needs to be noted.

Data request packets. Upon hearing its address in the beacon frame, a node undertakes the transmission of a data request packet. This process begins with a random backoff countdown that starts in the first backoff period after the beacon. Since the size of first backoff window is between 0 and 7 backoff periods,

$$P_r = \frac{1}{8} \tag{12.1}$$

denotes the probability that the value of zero will be chosen, in which case the packet transmission will occur in the third unit backoff period after the beacon. (The medium is always idle in the first two unit backoff periods after the beacon frame, for reasons explained in Chapter 2.) For data request packets, the block denoted with x in Figure 12.2 is void, i.e., it directly connects the state $x_{0,2,0}^{(r)}$ to the state $x_{0,0,0}^{(r)}$. The duration of a data request packet is $\overline{G_r} = 2$ backoff periods, which gives a total transmission time of $\overline{D_r} = 5$ backoff periods.

From Figure 12.1, the probability that the first backoff phase in a transmission block will be successful is

$$x_{0,2,0}^{(r)} = \frac{\tau_{0,p}\gamma_p\delta_p + n_s\tau_{0,k}\gamma_k\delta_k}{n_s + 1} + \tau_{0,r}(1 - \gamma_r\delta_r) \tag{12.2}$$

where the first term stands for the input probability to the data request CSMA-CA block.

For brevity, let us introduce the auxiliary variables $C_{1,r}$, $C_{2,r}$, $C_{3,r}$ and $C_{4,r}$ using the equations

$$\begin{aligned}
x_{0,1,0}^{(r)} &= x_{0,2,0}^{(r)}(1 - P_r)\alpha_r = x_{0,2,0}^{(r)}C_{1,r} \\
x_{1,2,0}^{(r)} &= x_{0,2,0}^{(r)}(1 - P_r)(1 - \alpha_r\beta_r) = x_{0,2,0}^{(r)}C_{2,r} \\
x_{0,0,0}^{(r)} &= x_{0,2,0}^{(r)}((1 - P_r)\alpha_r\beta_r + P_r) = x_{0,2,0}^{(r)}C_{3,r} \\
C_{4,r} &= \frac{1 - C_{2,r}^{m+1}}{1 - C_{2,r}}
\end{aligned} \tag{12.3}$$

Using the variables $C_{i,x}$, the transmission sub-chain is described by the following:

$$\begin{aligned}
x_{i,0,0}^{(r)} &= x_{0,2,0}^{(r)}C_{3,r}C_{2,r}^i, & i &= 0..m \\
x_{i,2,k}^{(r)} &= x_{0,2,0}^{(r)}\frac{W_i - k}{W_i} \cdot C_{2,r}^i, & i &= 1..m; k = 0..W_i - 1 \\
x_{i,1,0}^{(r)} &= x_{0,2,0}^{(r)}C_{1,r}C_{2,r}^i, & i &= 0..m \\
x_{0,2,k}^{(r)} &= x_{0,2,0}^{(r)}\frac{W_0 - k}{W_0}, & k &= 1..W_0 - 1
\end{aligned} \tag{12.4}$$

Access probability for a single data request packet is

$$\tau_{0,r} = \sum_{i=0}^{m} x_{i,0,0}^{(r)} = C_{3,r}C_{4,r}x_{0,2,0}^{(r)} \tag{12.5}$$

By combining the last expression with Equation (12.2), we obtain

$$\tau_{0,r} = \frac{C_{3,r}C_{4,r}\dfrac{\tau_{0,p}\gamma_p\delta_p + n_s\tau_{0,k}\gamma_k\delta_k}{n_s + 1}}{1 - C_{3,r}C_{4,r}(1 - \gamma_r\delta_r)} \tag{12.6}$$

The sum of probabilities for request packet transmission sub-chain is

$$s_r = \sum_{i=0}^{m}\sum_{k=0}^{W_i-1} x_{i,2,k} + (\overline{D_r} - 2)\sum_{i=0}^{m} x_{i,0,0} + \sum_{i=0}^{m} x_{i,1,0} \tag{12.7}$$

which can be simplified to

$$s_r = x_{0,2,0}^{(r)}C_{4,r}\left(C_{3,r}(\overline{D_r} - 2) + C_{1,r}\right) + x_{0,2,0}^{(r)}\sum_{i=0}^{m}\frac{C_{2,r}^i(W_i + 1)}{2} \tag{12.8}$$

Downlink packets with key exchange information. Downlink packets are transmitted only upon successful acknowledgment of data request packets. The duration of a downlink packet will be denoted as $\overline{D_d}$. In this case, block x from Figure 12.2 contains the sub-chain for uniformly distributed random variable between 0 and $\overline{D_d}$ backoff periods and models the requirement from the standard that every transmission which can not be fully completed within the current superframe has to be delayed to the beginning of the next superframe. The probability that packet will be delayed is denoted as $P_d = \overline{D_d}/SD$, where SD denotes duration of active superframe part (in backoff periods). This delay is a uniformly distributed random variable taking values between 0 and $\overline{D_d} - 1$ backoff periods.

For downlink packet transmission, the probability of finishing the first backoff phase in the transmission block is

$$x_{0,2,0}^{(d)} = \tau_{0,r}\gamma_r\delta_r + \tau_{0,d}(1 - \gamma_d\delta_d) \tag{12.9}$$

As before, the first term in this equation denotes the input probability from the request block.

Similar to the case of data request packets, we obtain

$$\tau_{0,d} = \sum_{i=0}^{m} x_{i,0,0}^{(d)} = C_{3,d}C_{4,d}x_{0,2,0}^{(d)} = \frac{C_{3,d}C_{4,d}\gamma_r\delta_r\tau_{0,r}}{1 - C_{3,d}C_{4,d}(1 - \gamma_d\delta_d)} \tag{12.10}$$

The sum of probabilities for the downlink transmission sub-chain is

$$s_d = \sum_{i=0}^{m}\sum_{k=0}^{W_i-1} x_{i,2,k}^{(d)} + (\overline{D_d} - 2)\sum_{i=0}^{m} x_{i,0,0}^{(d)} + \sum_{i=0}^{m} x_{i,1,0}^{(d)} + \sum_{i=0}^{m}\sum_{l=0}^{\overline{D_d}-1} x_{i,2,0,l}^{(d)} \tag{12.11}$$

which can further be simplified to

$$\begin{aligned} s_d = \ & x_{0,2,0}^{(d)}C_{4,d}\left(C_{3,d}(\overline{D_d} - 2) + C_{1,d} + \frac{P_d(\overline{D_d} - 1)}{2}\right) \\ & + x_{0,2,0}^{(d)}\sum_{i=0}^{m} \frac{C_{2,d}^{i}(W_i + 1)}{2} \end{aligned} \tag{12.12}$$

Packets with uplink key information. Uplink packet with key information generally follow the downlink packets, as shown in Figure 12.1. Since these transmissions start at any time during the superframe, rather than immediately after the beacon frame, they may well occur in the second half of the superframe and, possibly, even get deferred to the next superframe. Therefore,

$$P_k = \frac{\overline{D_k}}{SD} \tag{12.13}$$

The input probability for the uplink transmission block is $\tau_{0,d}\gamma_d\delta_d$, and the probability of finishing the first backoff phase in transmission block for uplink key packet is

$$x_{0,2,0}^{(k)} = \tau_{0,d}\gamma_d\delta_d + \tau_{0,k}(1 - \gamma_k\delta_k). \tag{12.14}$$

and the access probability is

$$\tau_{0,k} = \sum_{i=0}^{m} x_{i,0,0}^{(k)} = C_{3,k}C_{4,k}x_{0,2,0}^{(k)} = \frac{C_{3,k}C_{4,k}\gamma_d\delta_d\tau_{0,d}}{1 - C_{3,k}C_{4,k}(1 - \gamma_k\delta_k)} \tag{12.15}$$

The sum of probabilities for the uplink key transmission sub-chain can be expressed as

$$\begin{aligned} s_k = & \ x_{0,2,0}^{(k)}C_{4,k}\left(C_{3,k}(\overline{D_k} - 2) + C_{1,k} + \frac{P_k(\overline{D_k} - 1)}{2}\right) \\ &+ x_{0,2,0}^{(k)}\sum_{i=0}^{m}\frac{C_{2,k}^i(W_i + 1)}{2} \end{aligned} \tag{12.16}$$

Data packets. When an ordinary node wakes up and finds that it has a data packet to transmit, it will first synchronize with the beacon, and then begin a short random backoff in the range of 0 to 7 unit backoff periods in order to avoid contention with data request packets, followed by the regular CSMA-CA random backoff countdown. Since the size of first backoff window is between 0 and 7 backoff periods, the probability that the transmission will occur in the third unit backoff period after the beacon frame (i.e., that zero will be chosen for the first backoff) is

$$P_p = \frac{1}{8}\cdot\frac{1}{8} = \frac{1}{64} \tag{12.17}$$

In this case, the block denoted with x in Figure 12.2 directly connects the state $x_{0,2,0}^{(p)}$ with the state $x_{0,0,0}^{(p)}$. Packets with sensing information have total length of $\overline{D_p}$ backoff periods.

The probability of finishing the first backoff phase in transmission block for an uplink data packet is equal to

$$x_{0,2,0}^{(p)} = \frac{\tau_{0,d}\gamma_d\delta_d + n_k\tau_{0,p}\gamma_p\delta_p}{n_k + 1} + \tau_{0,p}(1 - \gamma_p\delta_p) \tag{12.18}$$

where the first term is the input probability to the sensing block. The corresponding success probability is

$$\begin{aligned} \tau_{0,p} = & \ \sum_{i=0}^{m} x_{i,0,0}^{(p)} = C_{3,p}C_{4,p}x_{0,2,0}^{(p)} \\ = & \ \frac{C_{3,d}C_{4,d}\dfrac{\gamma_d\delta_d\tau_{0,d}}{n_k + 1}}{1 - C_{3,p}C_{4,p}\left(\dfrac{\gamma_p\delta_p n_k}{n_k + 1}(1 - \gamma_p\delta_p)\right)} \end{aligned} \tag{12.19}$$

The sum of probabilities for the data packet transmission sub-chain is

$$s_p = x_{0,2,0}^{(p)}C_{4,p}\left(C_{3,p}(\overline{D_p} - 2) + C_{1,p}\right) + x_{0,2,0}^{(p)}\sum_{i=0}^{m}\frac{C_{2,p}^i(W_i + 1)}{2} \tag{12.20}$$

Synchronization lines. We also need to include the time from the moment when the node wakes up till the next beacon frame. This synchronization time is uniformly distributed between 0 and $BI - 1$ backoff periods, and its PGF is

$$D_1(z) = \frac{1 - z^{BI}}{BI(1 - z)}. \tag{12.21}$$

The collision separation line is needed to reduce the probability of collision between transmissions from the nodes that wake up in the same superframe. It is uniformly distributed in the range from 0 to 7 backoff periods, with the PGF of

$$D_2(z) = \frac{1 - z^8}{8(1 - z)}. \tag{12.22}$$

Synchronization with the beacon is also needed to receive the final acknowledgment from the coordinator after successful key update; we assume that this acknowledgment is sent in a downlink packet.

Therefore, the sum of probabilities within the beacon synchronization line is

$$d_b = \left(\frac{n_k}{n_k + 1} \tau_{0,p} \gamma_p \delta_p + \frac{1}{n_k + 1} \tau_{0,d} \gamma_d \delta_d \right) \sum_{i=0}^{BI} \frac{i}{BI}$$

$$s_b = \left(\frac{1}{n_s + 1} \tau_{0,p} \gamma_p \delta_p + \frac{n_s}{n_s + 1} \tau_{0,k} \gamma_k \delta_k \right) \sum_{i=0}^{BI} \frac{i}{BI} \tag{12.23}$$

in the sleep cycle and the key cycle, respectively. The sum of probabilities for the collision avoidance line is equal to

$$d_c = 3.5 \frac{\tau_{0,d} \gamma_d \delta_d + n_k \tau_{0,p} \gamma_p \delta_p}{n_k + 1} \tag{12.24}$$

The sleep time for an ordinary node is geometrically distributed with the parameter $1/P_{sleep}$; then, the sum of probabilities of being in a single sleep cycle is

$$s_{s,1} = \frac{\tau_{0,d} \gamma_d \delta_d n_k \tau_{0,p} \gamma_p \delta_p}{(n_k + 1)(1 - P_{sleep})} \tag{12.25}$$

However, if the node wakes up to find its buffer empty, it will immediately begin a new sleep cycle. Let the probability of finding the buffer empty after a sleep as Q_c (the exact value will be derived later); the sum of probabilities of being in consecutive sleep then becomes

$$s_s = \frac{s_{s,1}}{1 - Q_c} \tag{12.26}$$

If the maximum number of packets sent using the same key for encryption and/or authentication is n_k, the normalization condition for the whole Markov chain becomes

$$n_k(s_s + d_b + d_c + s_p) + n_s(s_b + s_r + s_d + s_k) + s_r + s_k = 1 \tag{12.27}$$

The total access probabilities for packets of a particular type can be obtained by summing the access probabilities for each transaction:

$$
\begin{aligned}
\tau_p &= n_k \tau_{0,p} \\
\tau_r &= (n_s + 1)\tau_{0,r} \\
\tau_d &= (n_s + 1)\tau_{0,d} \\
\tau_k &= n_s \tau_{0,k}
\end{aligned}
\tag{12.28}
$$

12.2 Analysis of the Node Buffer

We may consider the MAC subsystem of a node as an $M/G/1/K$ queueing model with 1-limited service policy, in which case the node will never transmit more than a single packet before going to sleep again (Takagi 1991). In the discussion that follows, packets are arriving to each node following the Poisson process with the arrival rate of λ, and the node buffer can accommodate up to L packets. The PGF for packet service time will be denoted as $T_{t,x}(z)$, where x can take values p, r, d, or k, depending on the packet type; the exact value of $T_{t,x}(z)$ will be derived in Section 12.5.

In this model, relevant state variables are sleep time, synchronization time to the beacon, key exchange time, and packet service time. The state of the model is uniquely determined by the values of these variables at Markov points that correspond to certain events during the system operation. According to Figure 12.1, the following Markov points can be identified.

Packet departure time. The PGF for the number of packet arrivals to the node buffer during the packet service time is

$$
A(z) = \sum_{k=0}^{\infty} a_k z^k = T_{t,p}^*(\lambda - z\lambda)
\tag{12.29}
$$

and the probability of k packet arrivals to the node buffer during the packet service time is

$$
a_k = \frac{1}{k!} \frac{d^k A(z)}{dz^k} \bigg|_{z=0}
\tag{12.30}
$$

Also, let π_k, $k = 0..L-1$, denote the probability of having k packets in the buffer immediately after a packet departure.

End of sleep. After a packet transmission, and provided the number of packets transmitted using the current key is smaller than n_k, the node starts a sleep cycle in which the radio subsystem is turned off. This sleep time is geometrically distributed with the parameter $1/P_{sleep}$. A new sleep cycle will also begin if the node buffer is found to be empty upon returning from sleep. The PGF for one sleep period is

$$
V(z) = \sum_{k=1}^{\infty} (1 - P_{sleep}) P_{sleep}^{k-1} z^k = \frac{(1 - P_{sleep})z}{1 - z P_{sleep}}
\tag{12.31}
$$

According to Takagi (1991), the PGF for the number of packet arrivals to the node buffer during the sleep time is

$$F(z) = \sum_{k=0}^{\infty} f_k z^k = V^*(\lambda - z\lambda) \tag{12.32}$$

where $V^*()$ denotes the Laplace-Stieltjes Transform (LST) of the sleep time. Since the sleep time is a discrete random variable, the LST can be obtained by replacing the variable z with e^{-s} in the expression for $V(z)$. Then, the probability of k packet arrivals to the node buffer during the sleep time can be obtained as

$$f_k = \frac{1}{k!} \frac{d^k F(z)}{dz^k} \Big|_{z=0} \tag{12.33}$$

Finally, let q_k, where $k = 0..L$, denote the probability of having k packets in the node buffer after sleep time.

End of key update cycle. According to Figure 12.1, after every n_k packet transmissions the node enters the key update phase during which regular data packets are not transmitted. Therefore, the key update phase should be considered as a single vacation period for the server of the node data buffer, and its duration has the PGF of

$$E_{ku}(z) = (z^3 T_{t,r}(z))^{n_s+1} (D_1(z) T_{t,d}(z))^{n_s+1} (T_{t,k}(z))^{n_s} \tag{12.34}$$

Number of packet arrivals to the node's buffer during key exchange phase has the PGF of

$$E(z) = \sum_{k=0}^{\infty} e_k z^k = E_{ku}^*(\lambda - z\lambda) \tag{12.35}$$

and the probabilities of k packet arrivals to the node's buffer during key exchange, can be obtained as

$$e_k = \frac{1}{k!} \frac{d^k E(z)}{dz^k} \Big|_{z=0} \tag{12.36}$$

After the key update, regular transmission/sleep cycle will be resumed, and v_k, where $k = 0..L$, denotes the probability of having k packets in the node buffer at that time.

Beacon synchronization time. A node returning from sleep with a non-empty buffer has to synchronize with the next beacon frame; as explained earlier, the synchronization time is uniformly distributed between 0 and $BI - 1$ backoff periods and its PGF $D_1(z)$ is given by Equation (12.21). When the node finds the next beacon, another delay referred to as the collision separation time, is incurred before a regular random backoff countdown can begin. The PGF $D_2(z)$ for the collision separation time was given by Equation (12.22). The total idle time after the sleep has the PGF of

$$S_{sync}(z) = D_1(z) D_2(z) \tag{12.37}$$

and its LST will be denoted as $D^*(s)$. The PGF for the number of packet arrivals to the node buffer during the synchronization time is

$$D(z) = \sum_{k=0}^{\infty} s_k z^k = S_{ync}^*(\lambda - z\lambda) \tag{12.38}$$

and the probability of k packet arrivals to the buffer during the synchronization time, packet service time, and sleep time, can be obtained as

$$d_k = \frac{1}{k!} \frac{d^k S(z)}{dz^k} \bigg|_{z=0} \tag{12.39}$$

Also, let ω_k, $k = 0 .. L$ denote the probability of having k packets in the node buffer after the synchronization phase.

Buffer behavior in Markov points. By modelling the node buffer state in four types of Markov points, the steady state equations for state transitions can be obtained as

$$q_0 = (q_0 + p\pi_0 + v_0) f_0$$

$$q_k = (q_0 + \pi_0 + v_k) f_k + \sum_{j=1}^{k} p\pi_j f_{k-j} + \sum_{j=1}^{k} v_j f_{k-j}, \quad 1 \le k \le L-1$$

$$q_L = (q_0 + \pi_0 + v_0) \sum_{k=L}^{\infty} f_k + \sum_{j=1}^{L-1} p\pi_j \sum_{k=L-j}^{\infty} f_k$$

$$+ \sum_{j=1}^{L-1} v_j \sum_{k=L-j}^{\infty} f_k$$

$$\omega_k = \sum_{j=1}^{k} q_j d_{k-j}, \qquad\qquad 1 \le k \le L-1$$

$$\omega_L = \sum_{j=1}^{k} q_j \sum_{k=L-j}^{\infty} d_k \tag{12.40}$$

$$\pi_k = \sum_{j=1}^{k+1} w_j a_{k-j+1}, \qquad\qquad 0 \le k \le L-2$$

$$\pi_{L-1} = \sum_{j=1}^{L} w_j \sum_{k=L-j}^{\infty} a_{k-l}$$

$$v_k = \sum_{j=0}^{k} (1-p)\pi_j e_{k-j}, \qquad\qquad 1 \le k \le L-1$$

$$v_L = \sum_{j=0}^{k} (1-p)\pi_j \sum_{k=L-j}^{\infty} e_k$$

$$\sum_{k=0}^{L} (q_k + v_k) + \sum_{k=1}^{L} \omega_k + \sum_{k=0}^{L-1} \pi_k = 1$$

where $p = \frac{n_k}{n_k+1}$ and $1 - p = \frac{1}{n_k+1}$ denote the probability that a sleep and a key update, respectively, will follow a packet departure.

The probability distribution of the device queue length at Markov points can be found by solving the system of equations (12.40). In this manner, we obtain the probability that the Markov point corresponds to a return from the vacation and the queue is empty at that moment as

$$Q_c = \frac{q_0}{\displaystyle\sum_{i=0}^{L} q_i} \tag{12.41}$$

The average value of total inactive time of the node (including possible multiple consecutive sleeps) is

$$\overline{I} = \frac{1}{(1 - P_{sleep})(1 - Q_c)} \tag{12.42}$$

Given that there are n nodes in the cluster the total throughput is

$$\mathcal{R} = \frac{\gamma_p \delta_p \tau_p}{t_{boff}} \tag{12.43}$$

where t_{boff} denotes the duration of a unit backoff period, and the desired value of $\mathcal{R}$ is set by the sensing application.

Satisfying Equation (12.43) will result in minimal energy consumption. However, the periodic key update procedure will result in additional traffic, the packet rate of which is $(\tau_d \delta_d \gamma_d + \tau_r \delta_r \gamma_r + \tau_k \delta_k \gamma_k)/t_{boff}$.

12.3 Success Probabilities

We are now equipped to derive approximate probabilities of success at first and second CCA, as well as the probability of successful transmission (i.e., transmission without a collision). The exact values of success probabilities can be derived in the case where there is no activity management (i.e., the nodes are not sleeping), as we have done in Chapter 8, but the use of activity management under low traffic load allows substantial simplifications. Let us assume that the required throughput per isolated cluster is $\mathcal{R}$, the number of nodes in each cluster is n, and the threshold for key update is n_k. The probability that an individual node is active and transmitting a data packet is

$$p = \frac{\mathcal{R}}{n} \tag{12.44}$$

and the probability that a node is active and involved in the key update procedure is

$$q = \frac{\mathcal{R}}{n n_k} \tag{12.45}$$

By using the trinomial probability distribution, we obtain the probability that exactly r nodes are involved in data transmission while exactly w nodes are involved in key update as

$$P(n, r, w) = \frac{n!}{r! w! (n - r - w)!} p^r q^w (1 - p - q)^{n - r - w} \tag{12.46}$$

If $n \gg \mathcal{R}$, which is a reasonable assumption in applications where redundant nodes are used, and $n_k > \mathcal{R}$, the probability above is close to zero for nearly all combinations of r and w.

It has non-negligible value only in the case that several nodes are sending data packets while at most one node is active in the key update procedure, and even those values are rather small, below 0.05 or so. This provides the rationale for the inclusion of the random collision separation time after the beacon frame for data packets. Furthermore, we are able to model the arrival processes of medium access events for different packet types as Poisson processes. If the generic cluster consists of n nodes, we will focus on a target node and model the aggregate packet arrival rates of the remaining $n - 1$ nodes as background traffic. This approximation is justified when activity management is implemented and the cluster operates far below the saturation regime.

Packet arrival rates for background traffic. Due to the initial random backoff countdown and the inclusion of a random collision separation time, both of which are random variables in the range between 0 and 7 unit backoff periods, the transmission of a data packet can commence within the period of 16 unit backoff periods starting from the third backoff period after the beacon frame. The background packet arrival rate for sensing packets is thus equal to

$$\lambda_p = (n - 1)\tau_p \frac{SD}{16} \tag{12.47}$$

Transmission of data request packets follows just the random backoff countdown. Therefore, it can commence within the period of eight backoff periods starting from the third backoff period after the beacon frame. The background packet arrival rate of data request packets is thus equal to

$$\lambda_r = (n - 1)\tau_r \frac{SD}{8} \tag{12.48}$$

Transmission of downlink packets with key information can start after the request packet has been acknowledged. The corresponding background packet arrival rate is

$$\lambda_d = (n - 1)\tau_d \frac{SD}{SD - \overline{D_r} - 0.5(W_0 - 1)} \tag{12.49}$$

Transmission of uplink packets with key information follows the acknowledgment of the downlink packet; uplink packets can be transmitted at the end of the current superframe or at the beginning of the following superframe. The corresponding background packet arrival rate is

$$\lambda_k = (n - 1)\tau_k \tag{12.50}$$

Medium behavior. Success probabilities for data packets will be

$$\alpha_p = \frac{1}{16} \left(\sum_{i=0}^{7} e^{-i(\lambda_p + \lambda_r + \lambda_k)} + \sum_{i=0}^{16} e^{-i(\lambda_p + \lambda_k)} \right)$$

$$\beta_p = \frac{1}{2} \left(e^{-(\lambda_p + \lambda_r + \lambda_k)} + e^{-(\lambda_p + \lambda_k)} \right) \tag{12.51}$$

$$\gamma_p = \frac{1}{2} \left(e^{-(\lambda_p + \lambda_r + \lambda_k) \max(\overline{D_p}, \overline{D_k}, \overline{D_r})} + e^{-(\lambda_p + \lambda_k) \max(\overline{D_p}, \overline{D_k})} \right)$$

Success probabilities for request packets will be

$$\alpha_r = \frac{1}{8} \sum_{i=0}^{7} e^{-i(\lambda_p + \lambda_r + \lambda_k)}$$

$$\beta_r = e^{-(\lambda_p + \lambda_r + \lambda_k)} \tag{12.52}$$

$$\gamma_r = \beta_r^{\max(\overline{D_p}, \overline{D_k}, \overline{D_r})}$$

Success probabilities for downlink packets with key information will be

$$\alpha_d = \frac{1}{\theta_d} \left(\sum_{i=0}^{7} e^{-i(\lambda_p + \lambda_r + \lambda_k)} + \sum_{i=8}^{\theta_d - 1} e^{-i(\lambda_r + \lambda_k)} \right)$$

$$\beta_d = e^{-(\lambda_d + \lambda_k)} \tag{12.53}$$

$$\gamma_d = \beta_d^{\max(\overline{D_p}, \overline{D_k})}$$

where, for clarity, we have introduced the auxiliary variable $\theta_d = SD - \overline{D_r} - 0.5(W_0 - 1)$.

Finally, uplink packets with key information will experience the following success probabilities:

$$\alpha_k = \frac{1}{SD} \sum_{i=0}^{SD-1} e^{-i(\lambda_p + \lambda_r + \lambda_d + \lambda_k)}$$

$$\beta_k = e^{-i(\lambda_p + \lambda_r + \lambda_d + \lambda_k)} \tag{12.54}$$

$$\gamma_k = \beta_k^{\max(\overline{D_p}, \overline{D_k}, \overline{D_r}, \overline{D_d})}$$

12.4 Key Update in a Multi-Cluster Network

Let us now focus on the network consisting of multiple hierarchically connected clusters that use master-slave bridges; this network is essentially a form of the multi-cluster tree. Redundant nodes and activity management are employed to ensure that each cluster delivers a predefined throughput of $\mathcal{R}$ to the network sink. Consider the network shown in Figure 12.3, in which all three clusters operate in beacon enabled, slotted CSMA-CA mode under the control of their respective cluster coordinators. (Note that a similar network was used to illustrate the discussions in Chapter 8.) The coordinators of the middle and bottom clusters act as bridges to the top and middle clusters, respectively; they use CSMA-CA access, which means that the bridge has to compete for medium access with ordinary nodes in the upper cluster. We assume that the three clusters employ identical values for the beacon interval BI and superframe duration SD. As a result, the time between successive bridge visits to the 'upper' cluster is the same as the period between two beacons in its own, 'lower' cluster.

When security with periodic key updates is applied in this setup, data exchanges, which are shown in Figure 12.3(a), are interspersed with key update exchanges, which are shown in Figure 12.3(b). (The diagrams are separated for clarity.) The analysis is complicated by the presence of bridge traffic in middle and top clusters. Yet the general approach is valid nevertheless, and simple modifications to the results presented above will suffice. Let the

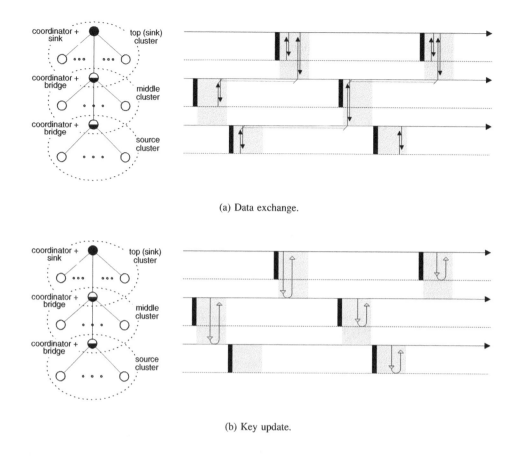

(a) Data exchange.

(b) Key update.

Figure 12.3 A multi-cluster network with periodic key updates, adapted from J. Mišić and V. B. Mišić, 'The cost of secure and reliable sensing,' *Proc. Qshine 2007*, Vancouver, BC, Canada, © 2007 ACM. Reprinted with permission.

clusters contain n_{bot}, n_{mid}, and n_{top} ordinary nodes, respectively, with the packet arrival rate of λ per node. (As usual, references to specific clusters will use the subscripts *bot*, *mid*, and *top*, respectively.) The top cluster coordinator acts as the network sink.

Bottom cluster. The analysis for the bottom cluster is simple, since there are virtually no differences from the single cluster case. In order to find access probabilities for each packet type in the bottom cluster we can use Equation (12.28) for $\tau_{x,bot}$ (where x indicates packet type). When access probabilities are known, we can calculate α_x, β_x, and γ_x.

Middle cluster. In this cluster, besides ordinary nodes we must account for the presence of the bridge, i.e., the coordinator from the bottom cluster. For an ordinary node, we apply the model from Section 12.1 to the environment of middle cluster and use Equation (12.28) for $\tau_{x,mid}$.

The bridge coming from the bottom cluster will have the same threshold for packet exchange. Its access probabilities can be modeled as

$$\tau_{x,bri,mid} = n_{bot}\tau_{x,bot} \tag{12.55}$$

The success probability for bridge transmissions depends on all the nodes in the middle cluster as

$$\gamma_{x,bri,mid} = (1 - \tau_{x,mid})^{\overline{D_x}n_{mid}} \tag{12.56}$$

The medium access event rate for an ordinary node in the middle cluster must also account for both the ordinary nodes and the bridge:

$$\lambda_{x,c,mid} = (n_{mid} - 1)\tau_{x,mid} + \tau_{x,bri,mid} \tag{12.57}$$

Parameters $\alpha_{x,mid}$, $\beta_{x,mid}$, and $\gamma_{x,mid}$ can, then, be calculated in a similar way as their bottom cluster counterparts.

Top (sink) cluster. Finally, success probabilities $\alpha_{x,top}$, $\beta_{x,top}$, and $\gamma_{x,top}$ for the sink (top) cluster can be found starting from

$$\tau_{x,bri,top} = (n_{bot}\tau_{x,bot} + n_{mid}\tau_{x,mid}) \tag{12.58}$$

12.5   Cluster Lifetime

While activity management achieves the extension of the lifetime separately for each cluster, individual cluster lifetimes will differ due to different traffic loads caused by bridge activity. In this case, the lifetime of the network is determined by the shortest cluster lifetime, and network lifetime will be maximized if individual cluster lifetimes are equalized. Equalization may be achieved by adjusting node populations in each cluster, using the iterative algorithm outlined in Chapter 9. We will now analyze the performance of activity management algorithm in the presence of periodic key updates.

First, the Laplace-Stieltjes Transform (LST) for the energy consumption during the j-th backoff time prior to transmission is

$$E_{B_j}^*(s) = \frac{e^{-s\omega_r W_i} - 1}{W_j(e^{-s\omega_r} - 1)} \tag{12.59}$$

Let the PGF of the data packet length be $G_x(z) = z^{k_x}$, where k_x is a constant representing the length of the packet of type x. Let $G_a(z) = z$ stand for the PGF of the duration of an ACK packet. Finally, let the PGF of the time interval between the data packet and subsequent ACK packet be $t_{ack}(z) = z^2$, and the PGF for the packet transmission time and receipt of acknowledgment be

$$T_x(z) = G_x(z)t_{ack}(z)G_a(z) \tag{12.60}$$

Then, the PGF for the time needed for one complete transmission attempt of packet type x, including backoffs, becomes

$$A_x(z) = \frac{\sum_{i=0}^{m} \left(\prod_{j=0}^{i} B_j(z) \right) (1 - \alpha_x \beta_x)^i z^{2(i+1)} \alpha_x \beta_x T_x(z)}{\alpha_x \beta_x \sum_{i=0}^{m} (1 - \alpha_x \beta_x)^i} \qquad (12.61)$$

The LST for the energy consumption during pure packet transmission time is $e^{-sk\omega_t}$, while the LST for the energy consumption during the two CCAs is $e^{-s2\omega_r}$; the LST for the energy consumption during waiting for and receiving the acknowledgment packet is $e^{-s3\omega_r}$.

Since the cluster coordinators must send the information about the number of live nodes and requested per-cluster throughput within the beacon frame, the duration of the beacon frame must exceed the necessary minimum of two unit backoff periods. If the duration of the beacon frame is assumed to be three backoff periods, as explained in Chapter 9, the LST for the energy consumption while receiving it is $e^{-s3\omega_r}$, which is the same as the LST for the energy consumption during the reception of a data packet with the same duration.

Then, the LST for the energy consumption during a transaction consisting of a data packet (which is three unit backoff periods long) and subsequent acknowledgment, will be denoted with

$$T_x^*(s) = e^{-sk_x \omega_t} e^{-s3\omega_r} \qquad (12.62)$$

The LST for energy consumption for one transmission attempt then becomes

$$\mathcal{E}_{A_x}^*(s) = \frac{\sum_{i=0}^{m} \left(\prod_{j=0}^{i} E_{B_j}^*(z) \right) (1 - \alpha_x \beta_x)^i e^{-s2\omega_r (i+1)} \alpha_x \beta_x T_x^*(s)}{\alpha_x \beta_x \sum_{i=0}^{m} (1 - \alpha_x \beta_x)^i} \qquad (12.63)$$

By taking packet collisions into account, the PGF of the probability distribution of the total packet service time is

$$\begin{aligned} T_{t,x}(z) &= \sum_{k=0}^{\infty} (A_x(z)(1 - \gamma_x))^k A_x(z) \gamma_x \\ &= \frac{\gamma_x A_x(z)}{1 - A_x(z) + \gamma_x A_x(z)} \end{aligned} \qquad (12.64)$$

Note that this is actually PGF of a geometric distribution where the argument is the PGF of one transmission attempt.

Then, the LST for the energy spent on a packet service time is

$$E_{T_x}^*(s) = \frac{\gamma_x \mathcal{E}_{A_x}^*(s)}{1 - \mathcal{E}_{A_x}^*(s) + \gamma_x \mathcal{E}_{A_x}^*(s)}. \qquad (12.65)$$

In the bottom cluster, the LST for the energy spent in packet service is obtained by substituting those values in Equation (12.65). The average energy consumed for the service of a single packet is

$$\overline{E_{T_x,bot}} = -\frac{d}{ds}E^*_{T_x,bot}(s)\Big|_{s=0} \tag{12.66}$$

The total energy consumption during a single key exchange cycle in the bottom cluster is

$$\begin{aligned}
\overline{C_{bot}} = \ & n_k\left(\overline{D_1}\omega_r + \overline{D_2}\omega_r + 3\omega_r + \overline{I_{bot}}\omega_s + \overline{E_{T_p,bot}}\right)\\
& + (n_s + 1)(\overline{D_2}\omega_r + 3\omega_r + \overline{E_{T_r,bot}} + \overline{E_{T_d,bot}}) + n_s\overline{E_{T_k,bot}}
\end{aligned} \tag{12.67}$$

The PGF for the duration of the node cycle time, including both key exchange and transmission of data, is

$$\begin{aligned}
D_{bot}(z) = \ & (D_1(z)D_2(z)z^3 I_{bot}(z)T_{t,p,bot}(z))^{n_k}\cdot\\
& \cdot(D_2(z)z^3 T_{t,r,bot}(z)T_{t,d,bot}(z))^{n_s+1}(T_{t,k,bot}(z))^{n_s}
\end{aligned} \tag{12.68}$$

and the average duration of this key exchange cycle is

$$\begin{aligned}
\overline{D_{bot}} = \ & n_k(\overline{D_1} + \overline{D_2} + 3 + \overline{I_{bot}} + \overline{T_{p,bot}})\\
& + (n_s + 1)(\overline{D_1} + 3 + \overline{T_{r,bot}} + \overline{T_{d,bot}}) + n_s\overline{T_{k,bot}}
\end{aligned} \tag{12.69}$$

The average battery energy consumption per backoff period can be found as

$$u_{bot} = \frac{\overline{C_{bot}}}{\overline{D_{bot}}} \tag{12.70}$$

Given the battery budget of b, the average number of key exchange cycles in bottom cluster can be found as

$$n_{c,bot} = \left\lceil \frac{b}{\overline{C_{bot}}} \right\rceil \tag{12.71}$$

Given the law of large numbers, the PGF for total lifetime of the node in bottom cluster becomes

$$L_{bot}(z) = (D_{bot}(z))^{n_{c,bot}} \tag{12.72}$$

Then, we can move on to the middle cluster. By using appropriate values of $\alpha_{x,mid}$, $\beta_{x,mid}$, and $\gamma_{x,mid}$, the PGFs for a single transmission attempt and for the overall packet transmission time can be calculated as $\mathcal{A}_{x,mid}(z)$ and $T_{x,mid}(z)$, respectively, both of which depend on the number of nodes n_{mid}.

The average battery energy consumption per backoff period in the middle cluster is calculated as

$$u_{mid} = \frac{\overline{C_{mid}}}{\overline{D_{mid}}} \tag{12.73}$$

In order to equalize the lifetime for two adjacent clusters, the average energy consumed by a node per backoff period must be the same in both clusters, $u_{mid} = u_{bot}$. From this condition we can obtain the initial population of the middle cluster n_{mid}. Given the battery budget of b, the average number of transmission/sleep cycles in bottom cluster can then be found. Finally, the PGF for total lifetime of the node in bottom cluster can be found using Equation (12.72).

The procedure is then repeated for the top cluster, This algorithm is scalable since overall model can be broken in individual cluster models with input from all clusters at lower level. The condition for the correctness for this approximation is that all clusters are not operating in the saturation condition.

12.6 Evaluation of Lifetimes and Populations

Finally, let us present the numerical results obtained by solving the system of equations presented in Sections 12.1 and 12.5. The solution consists of system parameters $\tau_{0,x}$, τ_x, P_{sleep}, α_x, β_x, γ_x and Q_c. We assumed that each node is powered by two AA batteries with a supply voltage between 2.1 and 3.6 V and total energy of $b = 10260J$. We assume that the network employs the 2450 MHz PHY option, and thus operates in the ISM band with a raw data rate of 250 kbps and bit error rate of $BER = 10^{-4}$. The superframe size was controlled with $SO = 0$ and $BO = 1$. The packet size has been fixed at $\overline{G_p} = 12$ unit backoff periods for data packets, $\overline{G_r} = 2$ unit backoff periods for request packets, $\overline{G_d} = 12$ unit backoff periods for downlink packets with key information, and $\overline{G_k} = 12$ backoff periods for uplink packets with key information; the device buffer had a fixed size of $L = 2$ packets. The packet sizes indicated above include the Message Integrity Code and all PHY and MAC layer headers. Other parameters at the MAC layer were kept at the default values prescribed by the standard (IEEE 2006).

The impact of the key update protocol. In order to investigate the impact of the parameters of the key update protocol on cluster lifetimes, we have set the population to 100 nodes in each of the clusters, while the required throughput of each cluster was set to $\mathcal{R} = 10$ packets per second. Then, we have varied the length of the key update $S = 3n_s + 2$ between 8 and 20 packet transmissions. In addition, the threshold to start the key update between the cluster coordinator and its nodes (including bridges) was varied between 20 and 100 packets. The diagrams in Figure 12.4 show the resulting cluster lifetimes. As can be seen, the cluster lifetime generally decreases with increased complexity and decreased threshold of the key update. The lifetime also decreases toward the sink because the amount of inter-cluster traffic which has to be securely transmitted (and, henceforth, supported with updated keys) gradually increases from the bottom cluster to the sink cluster.

The next two diagrams, Figure 12.5, present node populations in middle and sink clusters, respectively, which lead to equal cluster lifetimes when the bottom cluster has 100 nodes. As can be seen, significant increases in cluster populations are needed to equalize the cluster lifetimes when key updates are done frequently and/or require complex interactions.

Impact of throughput. We have also investigated impact of the required throughput $\mathcal{R}$ on the cluster lifetimes. For this experiment, the number of packet transmissions for the key update protocol was fixed to $S = 8$. In the non-equalized case, shown in the first three diagrams of Figure 12.6, we see that the cluster lifetime decreases hyperbolically with $\mathcal{R}$, which is to be expected since the average period between packet transmissions is equal to $1/\mathcal{R}$.

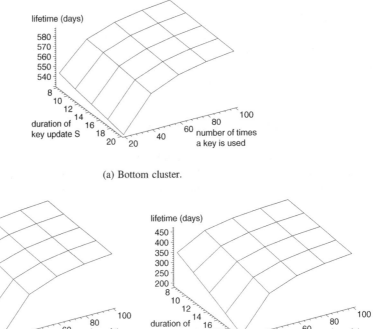

(a) Bottom cluster.

(b) Middle cluster.

(c) Sink cluster.

Figure 12.4 Cluster lifetimes as functions of frequency and duration of key exchanges, equal cluster populations, adapted from J. Mišić and V. B. Mišić, 'The cost of secure and reliable sensing,' *Proc. Qshine 2007*, Vancouver, BC, Canada, © 2007 ACM. Reprinted with permission.

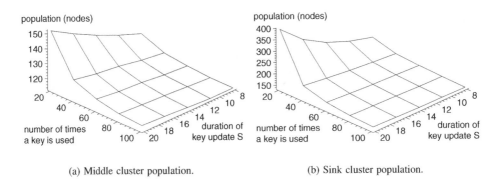

(a) Middle cluster population.

(b) Sink cluster population.

Figure 12.5 Cluster populations that lead to equalized lifetimes, as functions of frequency and duration of key exchanges. From J. Mišić and V. B. Mišić, 'The cost of secure and reliable sensing,' *Proc. Qshine 2007*, Vancouver, BC, Canada, © 2007 ACM. Reprinted with permission.

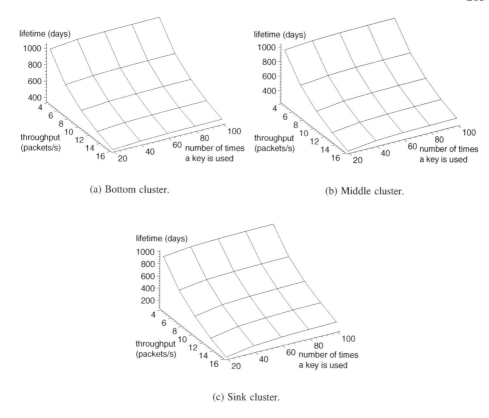

(a) Bottom cluster.

(b) Middle cluster.

(c) Sink cluster.

Figure 12.6 Cluster lifetimes as functions of required per-cluster throughput $\mathcal{R}$, equal cluster populations. From J. Mišić and V. B. Mišić, 'The cost of secure and reliable sensing,' *Proc. Qshine 2007*, Vancouver, BC, Canada, © 2007 ACM. Reprinted with permission.

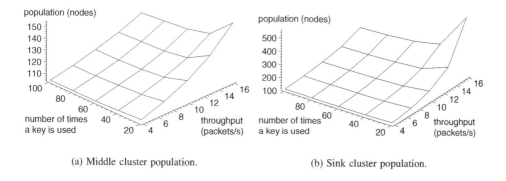

(a) Middle cluster population.

(b) Sink cluster population.

Figure 12.7 Cluster populations that equalize cluster lifetimes, as functions of desired per-cluster throughput $\mathcal{R}$. From J. Mišić and V. B. Mišić, 'The cost of secure and reliable sensing,' *Proc. Qshine 2007*, Vancouver, BC, Canada, © 2007 ACM. Reprinted with permission.

Finally, the diagrams in Figure 12.7 show the populations in middle and sink clusters needed to equalize cluster lifetimes when the population of the bottom cluster is fixed to 100 nodes. As can be seen, higher values of the required throughput $R \geq 8$ necessitate a larger number of nodes in both middle and sink clusters; the increase is particularly noticeable in the sink cluster, which has to carry the traffic coming from all three clusters.

Part IV Summary and Further Reading

In this Part we have presented the main security provisions of the 802.15.4 standard (IEEE 2006) and discussed a general analytical model for performance evaluation of periodic key updates (or exchanges) in a CSMA-CA-based network. Our discussion of threats in the 802.15.4 environment is mostly based on the analysis by Mišić et al. (2005e), where the impact of different attacks is also considered. The impact of periodic key exchanges, evaluated through the queueing theoretic analysis, is based on (Mišić and Mišić 2007); similar results, but obtained through simulation, were presented in (Mišić et al. (to appear)).

A different view, albeit with similar conclusions, is provided by Sastry and Wagner (2004), where several security risks in the initial version of the 802.15.4 specification (IEEE 2003b) were highlighted. Some remedies for these problems could be alleviated by small changes in the MAC protocol, as proposed by Heo and Hong (2005). A discussion of secure MAC protocols with emphasis on intrusion detection and possible countermeasures can be found in (Ren and Liang 2004). Intrusion detection is also discussed by Zhang and Lee (2000), Brutch and Ko (2003), Doumit and Agrawal (2003), (Huang and Lee 2003), Agah et al. (2004), Deng et al. (2004), Kachirski and Guha (2003), da Silva et al. (2005), Stamouli et al. (2005), Subhadrabandhu et al. (2005), and Xiao et al. (2005), among others. A simple scheme for intrusion detection was proposed by Onat and Miri (2005); a variation of that scheme, suitable for resource-limited devices such as the ones used in 802.15.4 networks, was described in (Mišić and Begum 2007).

As mentioned in Chapter 11, more general considerations of security in sensor networks can be found in recent survey papers such as those by Chan and Perrig (2003), Hu and Sharma (2005), and Shi and Perrig (2004). Still, all of these are small contributions to an increasingly important and complex area which has yet to receive the full attention it deserves.

Wireless Personal Area Networks   Jelena Mišić and Vojislav B. Mišić
© 2008 John Wiley & Sons, Ltd

Part IV Summary and Further
Reading

Appendices

Appendices

Appendix A

An Overview of ZigBee

The ZigBee Alliance is an industrial consortium that aims to promote and develop wireless networks for industrial monitoring and control purposes, but also for home networking, medical sensor applications, gaming, and other application areas where flexible, interoperable, low cost and low power network connectivity and interoperability are needed. To that end, the ZigBee Alliance has developed (ZigBee Alliance 2004) and subsequently revised the ZigBee specification (ZigBee Alliance 2006). This specification is an open standard that uses the MAC and PHY layer functionality of the 802.15.4 standard to build the required functionality at the network (NWK) layer, as well as the foundation framework for the application (APL) layer.

In general, a ZigBee network consists of a number of devices interconnected using one of the three topologies described in Section A.3. These devices use the 802.15.4 MAC and PHY.

An outline of the ZigBee protocol architecture is shown in Figure A.1.

A.1 ZigBee Functionality

The ZigBee NWK layer provides functionality which corresponds to the network layer of the ISO/OSI seven-layer protocol stack (OSI 1984). To that end, it includes mechanisms for:

- starting a new network

- joining and leaving the network, and assigning addresses to newly associated devices

- discovery of one-hop neighbors and storage of pertinent information about them

- discovery and maintenance of routes between devices

- routing of frames to their intended destinations.

Note that the 802.15.4 specification provides only partial support for the first two of these, and completely ignores the others.

Wireless Personal Area Networks   Jelena Mišić and Vojislav B. Mišić
© 2008 John Wiley & Sons, Ltd

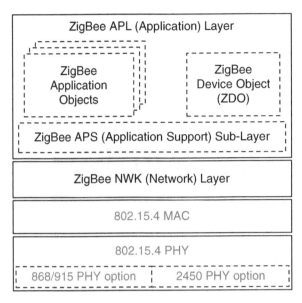

Figure A.1 The ZigBee protocol stack.

The ZigBee APL layer consists of the application support sub-layer (APS), the application framework (AF), the ZigBee device object (ZDO) and manufacturer-defined application objects.

The APS sub-layer allows the creation of a link between two devices based on their needs and services they provide, and provides facilities for subsequent operation and maintenance of that link.

The ZDO allows the device to define its role within the ZigBee network, which in turn allows it to initiate binding requests to, and respond to such requests received from, other devices in the ZigBee network. To that end, it makes use of the facilities provided by the NWK layer with respect to device and service discovery. The ZDO also allows the device to establish a secure relationship with other devices, in the manner that will be explained below.

Application objects are beyond the scope of the ZigBee specification, which allows up to 240 such objects to be defined, each with a uniquely numbered endpoint. Endpoint 0 denotes the data interface to the ZDO itself, whereas endpoint 255 is used for broadcasts that target all application objects; the remaining 14 endpoints are reserved for future use.

A.2 Device Roles

Each Zigbee device must be capable of joining a ZigBee network as a member, and leaving the ZigBee network to which it currently belongs. The request to join the network may originate from the lower layer of the network protocol stack (i.e., the MAC layer of the underlying 802.15.4 firmware); alternatively, the application executing on the ZigBee device may explicitly request that the device joins a specific network. The request to leave the network may originate from the ZigBee coordinator, or from the application.

In addition, some ZigBee devices have the capability to permit other devices to join an existing network and to permit a device to leave the network to which it currently belongs. Such devices can function as ZigBee coordinators, as explained below, or as ZigBee routers. (Devices that have no such capabilities are often referred to as end devices.) Devices of either type can take part in the process of assigning logical addresses to other network devices, and are capable of maintaining a list of neighboring devices. The list of neighboring devices is useful in the process of network discovery, where the information about potential routers is needed to identify candidate parents (see below). It is also useful during regular operation, when it stores the information needed to perform routing; this information may be updated after each received frame.

ZigBee coordinators have the additional capability to establish a new network, provided they are not already associated with an existing network. The process of establishing a new network begins with an energy scan, performed by the MAC layer, in order to learn about used and free channels. The scan can be executed over all channels, or a specified subset thereof. Once a suitable channel is found, the device undertakes an active scan. The channel with no detected networks, or the one with the lowest number of existing networks, should be used for the new network. The device then proceeds to assign a network address (i.e., PAN identifier) for the new network and informs the higher layers of the network protocol stack. Thereafter, the ZigBee coordinator may permit other devices to join the network. As can be seen, this procedure closely parallels the one for 802.15.4 networks, described in Section 2.8.

However, a ZigBee device which is a member of an existing network can allow a new device to join; this is allowed only if the device is a ZigBee coordinator or a ZigBee router. The new device can associate with the network by itself, in case it is not a member of any network; alternatively, a new device can be invited by an existing device. In the latter case, the new device is referred to as the child, and the device that has allowed the child to join the network is referred to as its parent.

A device may request that it joins the network as a router, provided it has the necessary capabilities; however, this request may be rejected if a specified number of routers already exists. In this case, the device may join the network as an end device.

A.3   Network Topologies and Routing

In terms of network topologies, ZigBee supports three different topologies, referred to as star, tree, and mesh networks.

Star network.   In a star network, one device or node functions as the ZigBee coordinator, and its duties include various tasks related to the creation and maintenance of the network. All communications must be routed through the ZigBee coordinator. The star topology corresponds to the single cluster with star topology of the 802.15.4 standard. Star networks operate in beacon enabled, slotted CSMA-CA access mode, and the responsibilities of the ZigBee coordinator closely correspond to those of the PAN coordinator in the 802.15.4 standard.

Tree network. In a tree network, there is again one ZigBee coordinator which is responsible for the entire network. There are also a number of routers which transfer data and control messages and, thus, extend the network. (Note that the role of a router requires that the device is capable of acting as a ZigBee router.) Since the ZigBee tree network operates in beacon enabled, slotted CSMA-CA access mode, it closely corresponds to the multi-cluster tree defined in Section 2.8. In this topology, individual clusters are essentially sub-networks, while the routers are master-slave bridges (Chapter 8) that double as coordinators for those clusters.

The routers repeat the beacon frames received from the ZigBee coordinator, after a suitable delay, and relay the data and command frames between the sub-networks. The beacon frame contains information about the device/sub-network depth, router capacity (i.e., whether the router is capable of accepting join requests from router-capable devices or not), end-device capacity (i.e., whether the router is capable of accepting join requests from router-capable devices or not), and the time difference between the current beacon and the beacon transmission of the parent. For compatibility reasons, the beacon frame also includes the information about the version of the ZigBee protocol supported by the router device.

Typically, the beacon interval in a tree network will be much longer than the superframe duration, so as to allow a number of sub-networks to co-exist without interfering with each other.

Furthermore, device addresses are unique within the network, and each parent device is given a distinct subset of available addresses (i.e., an address sub-block) for its children. Some parents may exhaust their address sub-blocks before the others, and a new device may have to find a parent that still has unallocated addresses before it can join the network.

A notable characteristic of a ZigBee tree network is that the maximum values for the number of children a device may have, the depth of a tree, and the number of routers that a parent may have as children, may be prescribed and subsequently enforced.

Mesh network. A ZigBee mesh network operates in a peer-to-peer topology, using non-beacon enabled, unslotted CSMA-CA access mode. In this setup, the network operates in a full peer-to-peer mode, and virtually any device can function as a router. Note that the absence of beacon frames in a mesh network (or, indeed, any 802.15.4 peer-to-peer topology network) means that there are no superframes, and no active or inactive periods. Since incoming data may occur at any time, the devices cannot go to sleep for prolonged periods of time. As a result, energy efficiency cannot be improved through the use of redundant nodes and/or activity management.

The ZigBee specification provides detailed procedures for joining, leaving, and maintaining a network; the interested reader should consult ZigBee Alliance (2006) for further details.

Routing. Data and commands can be exchanged between a parent and its children nodes following the standard procedures outlined in Chapter 2. However, ZigBee networks provide support for routing within tree and mesh network topologies. Route discovery and maintenance are performed on the basis of the list of neighboring devices and the link/path cost metric. Namely, for each path between two devices, the cost of the path is calculated as the sum of the costs of individual links between communicating nodes. The cost $C(L)$

of a link L is calculated as

$$C(L) = \begin{cases} 7 \\ \min\left(7, \text{int}\left(\frac{1}{p_L^4}\right)\right) \end{cases} \tag{A.1}$$

where p_L is the probability of packet delivery over the link L which should be re-evaluated after every attempt to use the link, be it successful or otherwise. The value of the Link Quality Indicator, LQI, may be used in the calculation.

It is worth noting that the choice of the exact formula to use is left to device and firmware manufacturers, so as to allow them to find the best tradeoff between functionality and complexity (and the associated cost) of the device.

A ZigBee router or ZigBee coordinator may maintain a routing table, where for each destination address, the status and type of the route, the address of the next hop, and other relevant information is recorded. A separate route discovery table to store ephemeral data needed during a single route discovery operation.

In general, routing is much simpler in a tree network, on account of its regular structure and ease of addressing, than in a mesh network.

A.4 Security

Security functions and security management are implemented in both NWK and APL layers; they use some of the security primitives from the MAC layer which were described in more detail in Chapter 11. Depending on the security level chosen, the entire frame at the APL layer may be protected before passing it on to the NWK layer; furthermore, the entire frame at the NWK layer may be protected before passing it on to the MAC layer for transmission. In both cases, protection is performed using the CCM* algorithm provided by the 802.15.4 standard.

The general approach. A notable characteristic of the security approach adopted by the ZigBee standard is that it recognizes the limitations imposed by the low cost and low complexity of practical devices. In particular, different applications and protocol layers that contribute to the implementation of security procedures and services are not independent of one another, and there is no cryptographic task separation within a single device. This fact has a number of immediate and less-than-immediate consequences:

- First, an open trust model must be adopted in which different applications and protocol layers have to trust each other.

- Keying material can be reused among different protocol layers.

- End-to-end security must be implemented on a device-to-device basis, rather than between pairs of layers or applications on two communicating devices.

A particularly important concept related to the security management is the distinction between link and network keys. Link keys are shared between two devices that communicate, and they may be used to protect any communication between them. Network keys are shared by all devices on the network, and they are used to protect broadcast communications. It is assumed that the destination device is always aware of the security arrangement used.

Link keys may be acquired by key transport (i.e., through communication with other nodes, possibly protected), key establishment (or update), or pre-installation by the manufacturer or operator. Network keys may be acquired by key transport or pre-installation. Key establishment procedure used to update link keys is based on the availability of a master key; again, this key may be acquired by key transport or pre-installation.

Trust and trust center role.   A device with sufficient capabilities may act as the trust center. The main task of the trust center is to store, safeguard, and distribute keys to be used by other devices in the network. Each secure network has exactly one trust center, and each trust center serves exactly one network. Typically, the role will be played by the ZigBee coordinator itself, but another device may be assigned that role as well, depending on the requirements of the application.

The operation of the trust center and the manner in which it serves the members of its network depend on the type of security adopted for the particular network. In commercial, high security applications, devices may be equipped with the initial master key and the hardwired address of the trust center device; in this manner, subsequent communications and key exchanges can be adequately protected from the very start. The trust center, on the other hand, maintains a list of devices and keys of all types (link keys, network keys, and master keys) needed to control admission to the network and the process of updating the network keys. Sufficient computational and memory resources must be available to support these activities.

In residential, low security applications, devices communicate with the trust center under the protection provided by the network key; this key can be provided via pre-installation, or sent through an ordinary data communication, if a momentary vulnerability can be tolerated. In this setup, the trust center will still maintain the network key and control admission to the network; resources permitting, it may also maintain the other keys and the list of devices.

Either way, a device should not accept an initial master or network key unless it comes from the designated trust center device.

Joining a secure network.   The procedure through which a device may join a secure network involves some additional features that are not present in unsecure networks. It begins with an active or passive scan that allows the device to learn about the networks in its vicinity. Among the networks that allow association, the device chooses one that it want to join; if the device already has a network key for this network, it should use it to secure the association request command. Upon receiving an association request, the router or coordinator decides whether to provisionally admit the device or not. If admitted, the device is declared 'joined but unauthenticated,' and an authentication routine immediately follows. Authentication is performed by contacting the trust center (if the router or coordinator is not the trust center), which makes the final decision about the admission. The decision depends on whether the trust center currently allows new devices to join; whether the network operates in commercial or residential mode; whether the trust center has a master key that corresponds to the new device (in commercial mode) or whether the device is joining unsecured or secured (in residential mode); and on other factors that may be relevant in the current context.

Upon deciding to admit the device, the trust center of a network operating in residential mode sends the device the current network key. (Note that the device may already possess that key.) If the device is not pre-configured with a network key, the trust center will send the network key. This transmission is unsecured, i.e., without encryption, at low power, and targets both the router and the new device. Since this is done only once, the security risk is deemed minimal. If the new device does not receive the key within a predefined time period, it should restart the joining procedure.

In the network operating in commercial mode, authentication depends on whether the new device possesses the trust center master key or not. In the latter case, the trust center sends the master key to the new device; this is again done without encryption, at low power. Once the new device shares the master key with the trust center, the trust center will begin the establishment of the link and network keys in a secure manner, using the SKKE algorithm described below. The SKKE algorithm is also used when the keys need to be updated.

Once the authentication procedure is successfully completed, the new device is declared 'joined and authenticated,' and it may take part in regular operation of the network.

Key update and the SKKE algorithm. In networks operating in residential, low security mode, the network key is updated by the trust center and broadcast to the network. In networks operating in commercial mode, the trust center will first send the new network key to each device on its list of network devices, and then ask each device to switch to the new key. Upon receiving the appropriate command, individual devices will check to see if it comes from the known trust center and, if so, update the network key and reset the frame counter(s).

If the value of the network key is somehow lost, the affected device may send an explicit request to recover it from the trust center. The trust center will send back the network key, followed by the command to switch to the new key.

The link and master keys are established and subsequently updated using a procedure that involves an initiator device, a responder device, and the trust center. First, the initiator device requests the link key from the trust center, indicating the responder device with which it will share this key. The trust center may reply with a link key or a master key from which the link keys will be subsequently derived, depending on its current configuration. The trust center sends the same information to the responder device as well.

Now, if the trust center has provided the link key, the initiator and responder devices may immediately use it to protect their communications. If the trust center has provided the master key, the initiator device begins the procedure to generate the link key using the Symmetric-Key Key Establishment (SKKE) algorithm.

The SKKE algorithm is based on the cryptographic primitives defined by ANSI (2001), and it involves three distinct steps: first, the devices involved generated a shared secret; second, the devices make sure that they indeed share the same secret; third, the shared secret is used to generate the new key. Two types of hash functions are used: the block-cipher cryptographic hash, denoted with $E(v)$, using the Matyas-Meyer-Oseas hash function (Menezes et al. 1997), and the keyed hash, denoted with $H_K(v)$, using the HMAC specified in (FIPS 2002), where v denotes the value to be processed and K denotes the appropriate key.

Let u_1 and u_2 denote the unique identifiers of the initiator node (typically, the cluster coordinator) and the responder node (typically, an ordinary node), respectively; also, let the master key shared among the nodes be denoted with Mk. Using this notation, the actual key establishment protocol may be described as follows.

First, the initiator device generates a challenge QEU_1, which is a bit string of length len. The challenge is generated as a random bit string from the challenge domain D, defined as the set of all strings of a length between predefined limits of min(len) and max(len). In the current version of the Zigbee specification, min(len) = max(len) = 128 bits (ZigBee Alliance 2006).

The challenge is sent to the responder, which validates it by checking that min(len) $\leq$ max(len) and that the length of the challenge is within the predefined limits of min(len) and max(len). If the validation is successful, the responder generates a challenge of its own, QEU_2, and sends it back to the initiator for validation.

Both devices then generate a shared secret using their unique identifiers, the master key, and the challenges they have generated and received. This is accomplished by first concatenating the node identifiers with the challenges:

$$v = u_1||u_2||QEU_1||QEU_2 \tag{A.2}$$

from which the keyed hash is calculated, using Mk as the key:

$$Z = H_{Mk}(v) \tag{A.3}$$

The value Z is the shared secret from which the link key will be derived, but both devices need to make sure that they indeed share the same secret. To this end, both nodes generate two cryptographic hashes of the shared secret as

$$\begin{aligned} h_1 &= E(Z||01_{16}) \\ h_2 &= E(Z||02_{16}) \end{aligned} \tag{A.4}$$

Then, the initiator node generates the strings

$$\begin{aligned} MD_1 &= 02_{16}||u_1||u_2||QEU_1||QEU_2 \\ MD_2 &= 03_{16}||u_1||u_2||QEU_1||QEU_2 \end{aligned} \tag{A.5}$$

and sends the tag

$$t_2 = H_{h_1}(MD_2) \tag{A.6}$$

to the responder. (The string MD_2 may optionally be followed by a text string.)

The responder, on the other hand, generates those same strings MD_1 and MD_2. Upon receiving the tag t_2, it verifies that it is indeed obtained from the string MD_2, and than replies with the tag

$$t_1 = H_{h_1}(MD_1) \tag{A.7}$$

The initiator validates this tag in the same manner.

The message exchange is schematically shown in Figure A.2, where each of the four SKKE frames, SKKE-1 to SKKE-4, is labeled with the key information it contains.

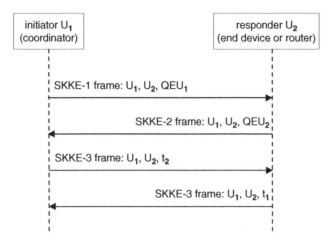

Figure A.2 Message exchange in the SKKE algorithm.

The hash value h_2 will be the link key shared by two devices. Note that the procedure enables the devices to generate the shared secret and make sure that they have both generated the same secret, but without actually exchanging it.

Once both devices generate the link key, they can proceed with secured data communication. As mentioned in Section 11.3, the 802.15.4 standard limits the number of frames that can be secured with a single link key to 2^{61}; in practice, however, the value of the link key will probably be updated much more frequently.

Appendix B

Probability Generating Functions and Laplace Transforms

In this section we will briefly introduce the definitions and notation related to probability generating functions and their Laplace-Stieltjes transforms. For a more detailed introduction, the reader should consult one of the numerous texts on probability and queueing theory (Grimmett and Stirzaker 1992; Kleinrock 1972; Takagi 1991; Wilf 1994).

Random variables can be classified into discrete and continuous types. Discrete random variables take values from a countable set, while continuous random variables take values from a continuous range.

The probability distribution of a discrete random variable is determined by the probabilities p_k that the variable will take value k. Probabilities p_k are sometimes called mass probabilities.

The probability distribution function (PDF) of a random variable C is the function $C(x)$, defined as probability that the value of the random variable is less than some number x:

$$C(x) \overset{\text{def}}{=} P[C \leq x] \tag{B.1}$$

The PDF of a discrete variable is a jump function.

The probability density function (pdf) for a continuous random variable C is defined as

$$c(x) \overset{\text{def}}{=} \frac{dC(x)}{dx} \tag{B.2}$$

We also note that

$$C(x) = \int_{-\infty}^{x} c(y)dy \tag{B.3}$$

and

$$\int_{-\infty}^{\infty} c(x)dx = 1 \tag{B.4}$$

Wireless Personal Area Networks   Jelena Mišić and Vojislav B. Mišić
© 2008 John Wiley & Sons, Ltd

The probability generating function (PGF) of a discrete random variable G_p is defined as

$$G_p(z) \stackrel{\text{def}}{=} E[z^C] = \sum_{k=0}^{\infty} p_k z^k \qquad (B.5)$$

i.e., it is a z-transform of the sequence of mass probabilities p_k. The PGF has the property that $G_p(1) = 1$. First and second moments of the probability distribution are given with

$$\overline{G_p} = G'_p(1) \qquad (B.6)$$

$$\overline{G_p^2} = G''_p(1) + G'_p(1) \qquad (B.7)$$

The Laplace-Stieltjes transform (LST) of the probability density function (pdf) of a random variable C is defined as

$$\begin{aligned} C^*(s) &\stackrel{\text{def}}{=} E[e^{-sC}] \\ &= \int_0^{\infty} e^{-sx} c(x) dx \end{aligned} \qquad (B.8)$$

where s is a complex variable.

Moments of the random variable C can be further obtained as:

$$\begin{aligned} \overline{C^i} &= \int_0^{\infty} x^i c(x) dx \\ &= (-1)^i \frac{d^i C^*(s)}{ds^i} \bigg|_{s=0}, \quad i = 1, 2, \ldots \end{aligned} \qquad (B.9)$$

Bibliography

Achir M and Ouvry L 2004 Power consumption prediction in wireless sensor networks. *ITC Specialist Seminar on Performance Evaluation of Wireless and Mobile Systems*, Antwerp, Belgium.

Agah A, Das SK, Basu K and Asadi M 2004 Intrusion detection in sensor networks: a non-cooperative game approach. In *Third IEEE International Symposium on Network Computing and Applications*, pp. 343–346, Boston, MA.

Akan OB and Akyildiz IF 2005 ESRT: event-to-sink reliable transport in wireless sensor networks. *ACM/IEEE Transactions on Networking* **13**(5), 1003–1016.

Akyildiz IF, Su W, Sankarasubramaniam Y and Cayirci E 2002 Wireless sensor networks: A survey. *Computer Networks* **38**, 393–422.

ANSI 2001 ANSI X9.63-2001, Public Key Cryptography for the Financial Services Industry – Key Agreement and Key Transport Using Elliptic Curve Cryptography American Bankers Association.

ANSI/IEEE 1999 Standard for part 11: Wireless LAN medium access control (MAC) and physical layer (PHY) specifications. IEEE Std 802.11, IEEE, New York, NY.

Antonopoulos C, Chondros P, Athanasopoulos A and Koubias S 2006 Comparative performance evaluation of 802.15.4 and 802.11b for usage in ad-hoc wireless networks *Proc. 2006 IEEE International Workshop on Factory Communication Systems*, pp. 223–226.

Athanasopoulos A, Topalis E, Antonopoulos C and Koubias S 2007 802.15.4: The effect of different back-off schemes on power and QOS characteristics. In *Proc. Third International Conference on Wireless and Mobile Communications (ICWMC'07)*, p. 68, Gosier, Guadeloupe.

Bellardo J and Savage S 2003 802.11 denial-of-service attacks: Real vulnerabilities and practical solutions. In *Proc. 12th USENIX Security Symposium*, Washington, DC.

Bertsekas DP and Gallager R 1991 *Data Networks* 2nd edn. Prentice-Hall, Englewood Cliffs, NJ.

Bianchi G 2000 Performance analysis of the IEEE 802.11 distributed coordination function. *IEEE Journal on Special Areas in Communications – Wireless Series* **18**(3), 535–547.

Bluetooth SIG 2003 *Specification of the Bluetooth System*, Version 1.2.

Bluetooth SIG 2004 *Core Specification of the Bluetooth System*, Version 2.0 + EDR.

Bougard B, Catthoor F, Daly DC, Chandrakasan A and Dehaene W 2005 Energy efficiency of the IEEE 802.15.4 standard in dense wireless microsensor networks: Modeling and improvement perspectives. In *DATE '05: Proceedings of the Conference on Design, Automation and Test in Europe*, vol. 1, pp. 196–201, Munich, Germany.

Brutch P and Ko C 2003 Challenges in intrusion detection for wireless ad-hoc networks. In *SAINT: Symposium on Applications and the Internet*, pp. 368–373.

Callaway, Jr. EH 2004 *Wireless Sensor Networks, Architecture and Protocols*. Auerbach Publications, Boca Raton, FL.

Chan H and Perrig A 2003 Security and privacy in sensor networks. *IEEE Computer* **36**(10), 103–105.

Cheng L and Bourgeois AG 2007 Efficient channel reservation for multicasting GTS allocation and pending addresses in IEEE 802.15.4. In *Proc. Third International Conference on Wireless and Mobile Communications (ICWMC'07)*, p. 46, Gosier, Guadeloupe.

Cho DH, Song JH and Han KJ 2006 An adaptive energy saving mechanism for the IEEE 802.15.4 LR-WPAN. In *Proc. First International Conference on Wireless Algorithms, Systems, and Applications (WASA2006)*, pp. 38–46, Xi'an, China.

da Silva APR, Martins MHT, Rocha BPS, Loureiro AAF, Ruiz LB and Wong HC 2005 Decentralized intrusion detection in wireless sensor networks. *Proceedings of the 1st ACM International Workshop on Quality of Service & Security in Wireless and Mobile Networks*, pp. 16–23.

Deng J, Han R and Mishra S 2004 Intrusion tolerance and anti-traffic analysis strategies for wireless sensor networks. *International Conference on Dependable Systems and Networks*, pp. 637–646.

Ding G, Sahinoglu Z, Orlik P, Zhang J and Bhargava B 2006 Tree-based data broadcast in IEEE 802.15.4 and ZigBee networks. *IEEE Transactions on Mobile Computing* **5**(11), 1561–1574.

Doumit SS and Agrawal DP 2003 Self-organized criticality and stochastic learning based intrusion detection system for wireless sensor networks. In *MILCOM: IEEE Military Communications Conference*, pp. 609–614.

FIPS 2001 Pub 197: *Advanced Encryption Standard (AES)*. Technical report, US Department of Commerce/NIST, Springfield, VA.

FIPS 2002 Pub 198: *The Keyed-hash Message Authentication Code (HMAC)*. Technical report, US Department of Commerce/NIST, Springfield, VA.

Garg VK, Smolik K and Wilkes JE 1998 *Applications of CDMA in Wirless/Personal Communications*. Prentice Hall, Upper Saddle River, NJ.

Golmie N, Cypher D and Rebala O 2005 Performance analysis of low rate wireless technologies for medical applications. *Computer Communications* **28**(10), 1266–1275.

Grimmett GR and Stirzaker DR 1992 *Probability and Random Processes* 2nd edn. Oxford University Press, Oxford.

Gupta V, Krishnamurthy S and Faloutsos M 2002 Denial of service attacks at the MAC layer in wireless ad hoc networks. In *Proc. Military Communications Conference MILCOM 2002*, Anaheim, CA.

Gutiérrez JA, Callaway, Jr. EH and Barrett, Jr. RL 2004 *Low-Rate Wireless Personal Area Networks*. IEEE Press, New York, NY.

Ha J, Kwon WH, Kim JJ, Kim YH and Shin YH 2005 Feasibility analysis and implementation of the IEEE 802.15.4 multi-hop beacon enabled network. In *Proc. 15th Joint Conference on Communications & Info*.

Heo J and Hong CS 2005 An efficient and secured media access mechanism using the intelligent coordinator in low-rate WPAN environment. In *Proc. 9th International Conference on Knowledge-based Intelligent Information and Engineering Systems (KES 2005)*, pp. 470–476, Melbourne, Australia.

Howitt I and Gutiérrez JA 2003 IEEE 802.15.4 low rate-wireless personal area network coexistence issues *Wireless Communications and Networking, 2003. WCNC 2003. 2003 IEEE*, vol. 3, pp. 1481–1486.

Howitt I, Neto R, Wang J and Conrad JM 2005 Extended energy model for the low rate WPAN. In *Proc. IEEE Int. Conf. on Mobile Ad-hoc and Sensor Systems MASS2005*, pp. 315–322, Washington, DC.

Hu F and Sharma NK 2005 Security considerations in ad hoc sensor networks. *Ad Hoc Networks* **3**(1), 69–89.

Hu YC and Perrig A 2004 A survey of secure wireless ad hoc routing. *IEEE Security & Privacy Magazine* **2**(3), 28–39.

Huang Y and Lee W 2003 A cooperative intrusion detection system for ad hoc networks. In *SASN: Proceedings of the 1st ACM Workshop on Security of Ad Hoc and Sensor Networks*, pp. 135–147.

IEEE 2002 Wireless medium access control (MAC) and physical layer (PHY) specifications for wireless personal area networks (WPAN). IEEE Std 802.15.1-2002, IEEE, New York, NY.

IEEE 2003a Wireless MAC and PHY specifications for high rate WPAN. IEEE Std 802.15.3, IEEE, New York, NY.

IEEE 2003b Wireless MAC and PHY specifications for low rate WPAN. IEEE Std 802.15.4-2003, IEEE, New York, NY.

IEEE 2005 Wireless medium access control (MAC) and physical layer (PHY) specifications for wireless personal area networks (WPAN). IEEE Std 802.15.1 (Revision of IEEE Std 802.15.1-2002), IEEE, New York, NY.

IEEE 2006 Wireless MAC and PHY specifications for low rate WPAN. IEEE Std 802.15.4-2006 (Revision of IEEE Std 802.15.4-2003), IEEE, New York, NY.

Intanagonwiwat C, Govindan R, Estrin D, Heidemann J and Silva F 2003 Directed Diffusion for Wireless Sensor Networking. *ACM/IEEE Transactions on Networking* **11**(1), 2–16.

Johansson N, Körner U and Tassiulas L 2001 A distributed scheduling algorithm for a Bluetooth scatternet. In *Proceedings of the International Teletraffic Congress – ITC-17*, pp. 61–72, Salvador de Bahia, Brazil.

Jones CE, Sivalingam KM, Agrawal P and Chen JC 2001 A survey of energy efficient network protocols for wireless networks. *Wireless Networks* **7**(4), 343–358.

Jung ES and Vaidya NH 2005 A power control MAC protocol for ad hoc networks. *Wireless Networks* **11**(1–2), 55–66.

Kachirski O and Guha RK 2003 Effective intrusion detection using multiple sensors in wireless ad hoc networks. In *Proceedings of the 36th Annual Hawaii International Conference on System Sciences*, pp. 57–65.

Karlof C, Sastry N and Wagner D 2004 TinySec: A link layer security architecture for wireless sensor networks. In *Proc. Second ACM Conference on Embedded Networked Sensor Systems SenSys 2004*, pp. 162–175, Baltimore, MD.

Kim M 2006 Performance analysis of service differentiation for IEEE 802.15.4 slotted CSMA/CA. In *Proc. Asia-Pacific Network Operations and Management Symposium (APNOMS 2006)*, pp. 11–22, Busan, Korea.

Kim TH and Choi S 2006 Priority-based delay mitigation for event-monitoring IEEE 802.15.4 LR-WPANs. *IEEE Communication Letters* **10**(3), 213–215.

Kim TH, Ha JY, Choi S and Kwon WH 2006a Virtual channel management for densely deployed IEEE 802.15.4 LR-WPANs. In *Proc. IEEE PerCom 2006*, Pisa, Italy.

Kim TO, Kim H, Lee J, Park JS and Choi BD 2006b Performance analysis of IEEE 802.15.4 with non-beacon-enabled CSMA/CA in non-saturated condition. In *Proc. 2006 IFIP International Conference on Embedded And Ubiquitous Computing (EUC'2006)*, pp. 884–893, Seoul, Korea.

Kiri Y, Sugano M and Murata M 2006 Performance Evaluation of Intercluster Multi-hop Communication Large-Scale Sensor Networks *The Sixth IEEE International Conference on Computer and Information Technology (CIT'06)*, Seoul, Korea.

Kleinrock LJ 1972 *Queuing Systems* vol. I: *Theory*. John Wiley and Sons, New York.

Kohvakka M, Kuorilehto M, Hännikäinen M and Hämäläinen TD 2006 Performance analysis of IEEE 802.15.4 and ZigBee for large-scale wireless sensor network applications. In *Proceedings of the 3rd ACM International Workshop on Performance Evaluation of Wireless Ad Hoc, Sensor and Ubiquitous Networks (PE-WASUN'06)*, pp. 48–57.

Koubaa A, Alves M and Tovar E 2006a i-GAME: an implicit GTS allocation mechanism in IEEE 802.15.4 for time-sensitive wireless sensor networks. *18th Euromicro Conference on Real-Time Systems*, vol. CD-ROM.

Koubaa A, Alves M and Tovar E 2006b Modeling and worst-case dimensioning of cluster-tree wireless sensor networks. *27th IEEE International Real-Time Systems Symposium (RTSS'06)*, pp. 412–421, Rio de Janeiro, Brasil.

Koubaa A, Alves M and Tovar E 2007a Energy and delay trade-off of the GTS allocation mechanism in IEEE 802.15.4 for wireless sensor networks. *International Journal of Communication Systems* **20**(7), 791–808.

Koubaa A, Cunha A and Alves M 2007b A time division beacon scheduling mechanism for IEEE 802.15.4/ZigBee cluster-tree wireless sensor networks. In *19th Euromicro Conference on Real-Time Systems (ECRTS'07)*, pp. 125–135, Pisa, Italy.

Kwon Y and Chae Y 2006 Traffic adaptive IEEE 802.15.4 MAC for wireless sensor networks. In *Proc. 2006 IFIP International Conference on Embedded And Ubiquitous Computing (EUC'2006)*, pp. 864–873, Seoul, Korea.

Latré B, De Mil P, Moerman I, Van Dierdonck N, Dhoedt B and Demeester P 2005 Maximum throughput and minimum delay in IEEE 802.15.4. In *Proc. International Conference on Mobile Ad-hoc and Sensor Networks (MSN'05)*, pp. 866–876, Wuhan, China.

Latré B, De Mil P, Moerman I, Van Dierdonck N, Dhoedt B and Demeester P 2006 Throughput and delay analysis of unslotted IEEE 802.15.4. *Journal of Networks* **1**(1), 20–28.

Lee JS 2005 An experiment on performance study of IEEE 802.15.4 wireless networks. In *10th IEEE Conference on Emerging Technologies and Factory Automation ETFA 2005*, vol. 2, pp. 451–458, Catania, Italy.

Lu G, Krishnamachari B and Raghavendra C 2004 Performance evaluation of the IEEE 802.15.4 MAC for low-rate low-power wireless networks. *Proc. Workshop on Energy-Efficient Wireless Communications and Networks EWCN'04*, Phoenix, AZ.

Menezes A, van Oorschot P and Vanstone S 1997 *Handbook of Applied Cryptography*. CRC Press, Boca Raton, FL.

Mirza D, Owrang M and Schurgers C 2005 Energy-efficient wakeup scheduling for maximizing lifetime of IEEE 802.15.4 networks. In *First International Conference on Wireless Internet*, pp. 130–137.

Mišić J 2007 Algorithm for equalization of cluster lifetimes in a multi-level beacon enabled 802.15.4 sensor network. *Computer Networks* **51**, 3252–3264.

Mišić J and Fung CJ 2007 The impact of master-slave bridge access mode on the performance of multi-cluster 802.15.4 network. *Computer Networks* **51**, 2411–2449.

Mišić J and Mišić VB 2005 *Performance Modeling and Analysis of Bluetooth Networks: Network Formation, Polling, Scheduling, and Traffic Control*. CRC Press, Boca Raton, FL.

Mišić J and Mišić VB 2007 The cost of secure and reliable sensing. *Proc. QShine 2007*, vol. CD-ROM, Vancouver, BC.

Mišić J and Udayshankar R (to appear) Analysis of cluster interconnection schemes in 802.15.4 beacon enabled networks. *International Journal of Sensor Networks (IJSNet)*.

Mišić J, Amini F and Khan M (to appear) Performance implications of periodic key exchanges and packet integrity overhead in an 802.15.4 beacon enabled cluster. *International Journal of Sensor Networks (IJSNet)*.

Mišić J, Chan KL and Mišić VB 2004a Admission control in Bluetooth piconets. *IEEE Transactions on Vehicular Technology* **53**(3), 890–911.

Mišić J, Fung J and Mišić VB 2005a Interconnecting 802.15.4 clusters in master-slave mode: queueing theoretic analysis. In *Proc. I-SPAN 2005*, Las Vegas, NV.

Mišić J, Fung CJ and Mišić VB 2006a On bridge residence times in master-slave connected 802.15.4 clusters. In *Proc. IEEE 20th conference on Advanced Information Networking (AINA2006)*, Vienna, Austria.

Mišić J, Mišić VB and Shafi S 2004b Performance of IEEE 802.15.4 beacon enabled PAN with uplink transmissions in non-saturation mode – access delay for finite buffers. In *Proc. BroadNets 2004*, pp. 416–425, San Jose, CA.

Mišić J, Shafi S and Mišić VB 2005b Avoiding the bottlenecks in the MAC layer in 802.15.4 low rate WPAN. In *Proc. HWISE2005*, vol. 2, pp. 363–367, Fukuoka, Japan.

Mišić J, Shafi S and Mišić VB 2005c The impact of MAC parameters on the performance of 802.15.4 PAN. *Ad hoc Networks* **3**(5), 509–528.

Mišić J, Shafi S and Mišić VB 2005d Maintaining reliability through activity management in 802.15.4 sensor networks. In *Proc. Qshine 2005*, Orlando, FL.

Mišić J, Shafi S and Mišić VB 2006b Admission control in beacon enabled 802.15.4 clusters. In *Proc. ACM IWCMC 2006*, vol. CD-ROM, Vancouver, Canada.

Mišić J, Shafi S and Mišić VB 2006c Cross-layer activity management in a 802.15.4 sensor network. *IEEE Communications Magazine* **44**(1), 131–136.

Mišić J, Shafi S and Mišić VB 2006d Maintaining reliability through activity management in an 802.15.4 sensor cluster. *IEEE Transactions on Vehicular Technology* **55**(3), 779–788.

Mišić J, Shafi S and Mišić VB 2006e Performance of beacon enabled IEEE 802.15.4 cluster with downlink and uplink traffic. *IEEE Transactions on Parallel and Distributed Systems* **17**(4), 361–376.

Mišić J, Shafi S and Mišić VB 2006f Performance limitations of the MAC layer in 802.15.4 low rate WPAN. *Computer Communications* **29**(13-14), 2534–2541.

Mišić J, Shafi S and Mišić VB 2006g Real-time admission control in 802.15.4 sensor clusters. *International Journal of Sensor Networks (IJSNet)* **1**(1), 34–40.

Mišić VB and Begum J 2007 Evaluating the feasibility of traffic-based intrusion detection in an 802.15.4 sensor cluster *Proc. 21st IEEE International Conference on Advanced Information Networking and Applications (AINA-07)*, Niagara Falls, ON.

Mišić VB, Fung J and Mišić J 2005e MAC layer security of 802.15.4-compliant networks. In *Proc. WSNS'05*, Washington, DC.

Moteiv Corporation 2006 tmote_sky low power wireless sensor module.

Myoung KJ, Shin SY, Park HS and Kwon WH 2007 IEEE 802.11b performance analysis in the presence of IEEE 802.15.4 interference. *IEICE Transactions on Communications* **E90-B**(1), 176–179.

Neugebauer M, Plönnigs J and Kabitzsch K 2005 A new beacon order adaptation algorithm for IEEE 802.15.4 networks. In *Proceeedings of the Second European Workshop on Wireless Sensor Networks*, pp. 302–311, Istanbul, Turkey.

Newsome J, Shi E, Song D and Perrig A 2004 The Sybil attack in sensor networks: Analysis and defenses. In *Proceedings of IEEE International Conference on Information Processing in Sensor Networks (IPSN 2004)*, pp. 259–268, Berkeley, CA.

O'Hara B and Petrick A 1999 *IEEE 802.11 Handbook: A Designer's Companion*. IEEE Press, New York, NY.

Onat I and Miri A 2005 An intrusion detection system for wireless sensor networks. In *IEEE International Conference on Wireless and Mobile Computing, Networking and Communications*, pp. 253–259.

OSI 1984 ISO 7498. Basic Reference Model for Open Systems Interconnection, International Standard.

Park TR, Kim TH, Choi JY, Choi S and Kwon WH 2005 Throughput and energy consumption analysis of IEEE 802.15.4 slotted CSMA/CA. *Electronics Letters* pp. 1017–1019.

Pebbles, Jr. PZ 1993 *Probability, Random Variables, and Random Signal Principles*. McGraw-Hill, Inc., New York, NY.

Perkins CE (ed.) 2001 *Ad Hoc Networking*. Addison-Wesley, Boston, MA.

Petrova M, Riihijärvi J, Mähönen P and Labella S 2006 Performance study of IEEE 802.15.4 using measurements and simulations. In *Proc. Wireless Communications and Networking Conference WCNC 2006*, pp. 487–492.

Pollin S, Ergen M, Ergen SC, Bougard B, Van Der Perre L, Catthoor F, Moerman I, Bahai A and Varaiya P 2006 Performance analysis of slotted carrier sense IEEE 802.15.4 medium access layer *IEEE GLOBECOM'06*, pp. 1–6, San Francisco, CA.

Ram Murthy C and Manoj B 2004 *Ad Hoc Wireless Networks, Architecture and Protocols*. Prentice Hall, Englewood Cliffs, NJ.

Ramachandran I, Das AK and Roy S 2007 Analysis of the contention access period of IEEE 802.15.4 MAC. *ACM Transactions on Sensor Networks* **3**(1), 4.

Ren Q and Liang Q 2004 Secure media access control (MAC) in wireless sensor networks: Intrusion detections and countermeasures. In *Proc. 15th IEEE International Symposium on Personal, Indoor and Mobile Radio Communications PIMRC 2004*, vol. 4, pp. 3025–3029, Barcelona, Spain.

RSoft Design, Inc. 2003 *Artifex v.4.4.2*. RSoft Design Group, Inc., San Jose, CA.

Sankarasubramaniam Y, Akan ÖB and Akyildiz IF 2003 ESRT: event-to-sink reliable transport in wireless sensor networks. In *Proc. 4th ACM MobiHoc*, pp. 177–188, Annapolis, MD.

Sastry N and Wagner D 2004 Security considerations for IEEE 802.15.4 networks. In *WiSe '04: Proceedings of the 2004 ACM workshop on Wireless security*, pp. 32–42, Philadelphia, PA.

Saxena N, Tsudik G and Yi JH 2003 Admission control in peer-to-peer: design and performance evaluation. In *Proc. SASN'03*, pp. 104–113, Fairfax, VA.

Saxena N, Tsudik G and Yi JH 2004 Access control in ad hoc groups. In *International Workshop on Hot Topics in Peer-to-Peer Systems*, pp. 2–7, Volendam, The Netherlands.

Shah SH, Chen K and Nahrstedt K 2005 Dynamic bandwidth management in single-hop ad hoc wireless networks. *Mobile Networks and Applications* **10**, 199–217.

Shi E and Perrig A 2004 Designing secure sensor network. *IEEE Wireless Communications* **11**(6), 38–43.

Shin SY, Park HS, Choi S and Kwon WH 2005 Packet error rate analysis of IEEE 802.15.4 under IEEE 802.11b interference. In *Proc. Third International Conference on Wired/Wireless Internet Communications WWIC 2005*, pp. 279–288, Xanthi, Greece.

Sohrabi K, Gao J, Ailawadhi V, and Pottie GJ 2000 Protocols for self-organization of a wireless sensor network. *IEEE Personal Communications* **7**(5), 16–27.

Stallings W 2002 *Wireless Communications and Networks*. Prentice Hall, Upper Saddle River, NJ.

Stallings W 2003 *Cryptography and Network Security: Principles and Practice* 3rd. edn. Prentice Hall, Upper Saddle River, NJ.

Stamouli I, Argyroudis PG and Tewari H 2005 Real-time intrusion detection for ad hoc networks *Sixth IEEE International Symposium World of Wireless Mobile and Multimedia Networks*, pp. 374–380.

Subhadrabandhu D, Sarkar S and Anjum F 2005 RIDA: Robust intrusion detection in ad hoc networks. In *Proceedings of The 4th International IFIP-TC6 Networking Conference*, pp. 1069–1082.

Takagi H 1991 *Queueing Analysis* vol. 1: *Vacation and Priority Systems*. North-Holland, Amsterdam, The Netherlands.

Takagi H 1993 *Queueing Analysis* vol. 2: *Finite Systems*. North-Holland, Amsterdam, The Netherlands.

Timmons NF and Scanlon WG 2004 Analysis of the performance of IEEE 802.15.4 for medical sensor body area networking. In *Proc. IEEE Conference on Sensor and Ad Hoc Communications and Networks (SECON'04)*, pp. 16–24, Santa Clara, CA.

Toh CK 2002 *Ad Hoc Mobile Wireless Networks: Protocols and Systems*. Prentice-Hall PTR, Upper Saddle River, NJ.

van Dam T and Langendoen K 2003 An adaptive energy-efficient MAC protocol for wireless sensor networks. In *Proc. ACM SenSys03*, pp. 171–180, Los Angeles, CA.

Waterloo Maple, Inc. 2005 *Maple 10*, Waterloo, ON, Canada.

Wilf HS 1994 *Generatingfunctionology* 2nd edn. Academic Press, New York.

Wood AD and Stankovic JA 2002 Denial of service in sensor networks. *IEEE Computer* **35**(10), 54–62.

Xiao D, Chen C and Chen G 2005 Intrusion detection based security architecture for wireless sensor networks *IEEE International Symposium on Communications and Information Technology*, vol. 2, pp. 1412–1415.

Xing Z, Wang H, Zeng P and Liang W 2006 IC-MAC: A dynamic scheduling supported MAC protocol optimized for intra-cluster communication *IEEE International Conference on Information Acquisition*, pp. 144–148.

Yang Y and Kravets R 2005 Contention-aware admission control for ad hoc networks. *IEEE Transactions on Mobile Computing* **4**(4), 363–377.

Ye W, Heidemann J and Estrin D 2004 Medium access control with coordinated adaptive sleeping for wireless sensor networks. *ACM/IEEE Transactions on Networking* **12**(3), 493–506.

Zhang Y and Lee W 2000 Intrusion detection in wireless ad-hoc networks. In *MobiCom '00: Proceedings of the 6th Annual International Conference on Mobile Computing and Networking*, pp. 275–283.

Zheng J and Lee MJ 2006 A comprehensive performance study of IEEE 802.15.4. In *Sensor Network Operations* (ed. Phoha S, La Porta TF and Griffin C), pp. 218–236. IEEE Press.

Zhou Y, Wu S and Nettles SM 2004 Analyzing and preventing MAC-layer denial of service attacks for stock 802.11 systems. In *Proc. BroadWISE 2004: Workshop on Broadband Wireless Services and Applications*, pp. 22–88, San Jose, CA.

ZigBee Alliance 2004 ZigBee specification. ZigBee Document 053474r06, Version 1.0, ZigBee Alliance, San Ramon, CA.

ZigBee Alliance 2006 ZigBee-2006 specification. ZigBee document 053474r13, ZigBee Alliance, San Ramon, CA.

Index

Printed and bound in the UK by
CPI Antony Rowe, Eastbourne

Printed and bound by CPI Group (UK) Ltd, Croydon, CR0 4YY

16/04/2025

14658554-0003